CRYOPRESERVATION
Applications
in Pharmaceuticals
and Biotechnology

**Drug Manufacturing
Technology Series
Volume 5**

Kenneth E. Avis
Carmen M. Wagner
Editors

informa
healthcare

New York London

In Memoriam

This book is dedicated to Kenneth E. Avis, D.Sc.,
a distinguished leader and educator,
and an inspiration to many pharmaceutical
professionals worldwide.

CONTENTS

7. The Design and Use of Thermal Transport Containers   435

Thomas C. Pringle

FOREWORD

The *Drug Manufacturing Technology Series,* of which this book is a part, is the by-product of Kenneth E. Avis's vision and creativity. Four volumes have already been published, and we are sad that Dr. Avis did not have a chance to share in the excitement of completing the fifth book of the series. As his co-editor and friend, I think of this book as a gift to him and his family in gratitude for all that he did for my development and that of many other professionals in the health field.

The series was conceived from many conversations between Amy Davis, former Editorial Director of Interpharm Press, and Dr. Avis about the need for furthering the manufacturing knowledge in the pharmaceutical and biotechnology industries. The volumes' content was strongly influenced by Dr. Avis's knowledge, experience, and enthusiasm. For Dr. Avis, a recognized educator and leader in the field of sterile product manufacturing and control, the field of biotechnology manufacturing held a special fascination because of all that was yet to be learned and conquered. My participation came as the series was starting, and I was honored to be invited by Dr. Avis to be one of his co-editors. What an excitement to work with such a scholar and such an inspiring mentor!

Three of the five books in the series are related to biotechnology. This fifth volume is dedicated to cryopreservation. The seven chapters in this book provide new and often unique information in the biological application of the principles of cryopreservation for the preservation and storage of viable cells and cell materials used in biopharmaceutical applications. Cryopreservation technology is essential for continued progress in the application of biological materials to the development of new and innovative therapeutics.

Dr. Avis and I enjoyed editing this book and bringing these topics together. In his words, "the field of biotechnology has some unique challenges in producing biopharmaceutical products," and he sincerely hoped that this volume would be helpful to employees trying to carry out their responsibilities in biopharmaceutical processing and quality control. Even though this is Dr. Avis's last book, his vision for the series will endure as we continue to find ways to improve these processes.

I hope you will enjoy reading it but, as Dr. Avis often said, "your pleasure and enjoyment will come to fruition only if you find it beneficial in the furtherance of your work." I sincerely hope this book and the whole series will do just that.

Carmen M. Wagner, Ph.D.
Co-editor
May 1999

Other Volumes in **Drug Manufacturing Technology Series**

Volume 1:   *Sterile Pharmaceutical Products:*
              Process Engineering Applications
              Edited by Kenneth E. Avis

Volume 2:   *Biotechnology and Biopharmaceutical Manufacturing*
              Processing and Preservation
              Edited by Kenneth E. Avis and Vincent L. Wu

Volume 3:   *Pharmaceutical Unit Operations: Coating*
              Edited by Kenneth E. Avis, Atul J. Shukla,
              and Rong-Kun Chang

Volume 4:   *Biotechnology: Quality Assurance and Validation*
              Edited by Kenneth E. Avis, Carmen M. Wagner,
              and Vincent L. Wu

AUTHOR BIOGRAPHIES

KENNETH E. AVIS

Dr. Kenneth Avis was a distinguished leader in pharmaceutical science. He was president of the PDA and served the U.S. Pharmacopeia and U.S. FDA in numerous advisory capacities. Dr. Avis consulted with more than 50 pharmaceutical companies, universities, hopitals, and governmental agencies; published over 30 peer-reviewed research papers; and served as coordinating Editor for the *Drug Manufacturing Technology Series* for Interpharm Press while maintaining his affiliation with the University of Tennessee, Memphis, as Emeritus Professor in Pharmaceutics.

STEVEN S. LEE

Steven S. Lee, Ph.D., currently works for Merck & Co., Inc., as Director of Sterile Process Technology of Merck Manufacturing Division, Vaccine Technology and Engineering Group, West Point, Pennsylvania. Dr. Lee received his B.S. degree in Agricultural Chemistry from National Taiwan University, Taipei, Taiwan, and subsequently an M.S. degree in Chemical Engineering from the University of Michigan, Ann Arbor, Michigan. After some brief industrial experience, he returned to Massachusetts Institute of Technology, Cambridge, Massachusetts, and obtained a Ph.D. in Biochemical Engineering. In 1990, Dr. Lee joined Merck as a member of Bioprocess Research and Development, Merck Research Laboratories. His broad experience with biological processes encompasses small and large scale microbial fermentation to product recovery and purification, and to formulation, filling, and lyophilization in both laboratory and cGMP manufacturing environment. Besides his strong interests in integrated bioprocesses and systems, Dr. Lee currently focuses his technical efforts on lyophilization of live virus vaccines.

THOMAS C. PRINGLE

Mr. Pringle is President and CEO of ISC, Inc., based in Phoenix, Arizona. Prior to joining ISC in 1989, he served as International Director for a U.S. healthcare division of Boehringer Mannheim and prior to that in marketing and manufacturing roles at Beckman Instruments. Mr. Pringle holds a degree in Biology from Williamette University in Salem, Oregon, and has spent over 30 years in the healthcare field. He chairs a packaging industry committee related to transport of temperature sensitive pharmaceuticals and has published papers and lectured on the principles of temperature controlled packaging for the past several years.

WILLIAM H. SIEGEL

William H. Siegel is a technical service representative for BioWhittaker, Inc., Walkersville, Maryland, a leading global supplier of cell culture and media products to the academic, biotechnology, and pharmaceutical industries. For the previous 18 years he was a member of the Cell Culture Department at the American Type Culture Collection, where he acquired extensive and diverse experience in propagating and cryopreserving numerous cell lines of mammalian, amphibian, avian, and insect origin. He received a B.S. in Chemistry from the University of Minnesota Institute of Technology.

CARMEN M. WAGNER

Dr. Carmen M. Wagner is currently the Director of Quality for Wyeth-Lederle Vaccines and Pediatrics in Sanford, North Carolina, a manufacturer of vaccines. Dr. Wagner has held a variety of quality-related positions in the health industry and has worked in the area of biological manufacturing for more than 10 years. She started her professional career as an Assistant Medical Research Professor at Duke University Medical Center in Durham, North Carolina. From academia, Dr. Wagner moved to small, and then large, pharmaceutical companies, including tenures at E.I. DuPont and Johnson & Johnson. Her work experience includes positions in research,

research and development, manufacturing technical support, quality assurance, regulatory compliance, and validation. For the last 10 years, Dr. Wagner has worked on creating and improving quality systems that can help companies become more efficient in the transfer of new products/processes from development to operations. Dr. Wagner is an active member of the PDA and the ISPE.

RICHARD WISNIEWSKI

Richard Wisniewski is a research scientist at the NASA Ames Research Center in Mountain View, California. He has a background in mechanical and chemical engineering and in biotechnology and protein chemistry. Previously he was with Genentech, Inc., So. San Francisco, California, and with Wyeth Ayerst, Philadelphia, Pennsylvania. His interests include cryobiology, fermentation and cell culture, and protein purification and final processing and preservation (formulation and lyophilization), as well as technology development and equipment and system design for these fields. He is a scientific advisor to Integrated Biosystems, Benicia, California.

FANGDONG YIN

Fangdong Yin, Ph.D., currently works for Merck & Co., Inc., as Senior Engineer of Sterile Process Technology of Merck Manufacturing Division, Vaccine Technology and Engineering Group. Dr. Yin received a B.S. Degree in Environmental Chemistry from Peking University, Beijing, China, and later a Ph.D. in Chemical Engineering from California Institute of Technology, Pasadena, California. In 1990, Dr. Yin joined Merck Manufacturing Division, to develop novel calorimetric methods for evaluating hazardous reactions and was recognized as an expert in loss prevention and process safety. In 1996, he joined the Sterile Process Technology group at West Point, Pennsylvania, and has since been involved in various aspects of manufacturing biological products, especially formulation, filling, and lyophilization processes. Currently, Dr. Yin is working closely with Dr. Steven Lee on lyophilization of live virus vaccines.

1

INTRODUCTION

Kenneth E. Avis

The University of Tennessee

Carmen M. Wagner

Wyeth-Lederle Vaccines and Pediatrics

Biotechnology is the applied biological science that utilizes living mammalian and microbial cells and their products. The elements required for these cells to grow and multiply are provided under controlled conditions so that healthy cells reproduce themselves in the same way each and every time, and products of their metabolism or specific cell components are produced reliably time after time in the culturing process. Concurrently, this process must be protected from the intrusion of contaminating microorganisms, viruses, and other pathogenic or otherwise undesirable living entities. In other words, pure, healthy cultures of a single cell entity are the goal of each repetition of the process.

As biotechnology has developed beyond the laboratory and its small-scale operations, the need for preserving living cells and their by-products has become more apparent and critical. In addition, the preparation and preservation of larger quantities of cell cultures and their by-products have become necessary to further their application to and development into therapeutically active pharmaceutical dosage forms. The most effective means developed thus far for preserving these materials is cryopreservation.

Cryopreservation is the process whereby chemical and biological entities are preserved from degradation at temperatures well below freezing. It has long been known that chemical reactions can be slowed significantly when the temperature is lowered. Similarly, biological reactions can be slowed by lowering the ambient temperature of the reaction. However, when living cells are involved, preserving the integrity of cell membranes, other cell structures, and cell functions is usually critical and much more complex. Further, it must be recognized that water is the primary component of the living cell and its growth environment; therefore, a major factor in preserving the integrity of viable cells is the control of ice crystals formed during freezing.

It is relatively simple to select a low temperature at which biologically derived materials can be preserved—for example, in liquid nitrogen at −196° C. The extremely difficult aspects of the process are lowering the temperature of the materials to that level and then, at the time of use, raising the temperature of the materials to room or body temperatures. In other words, the control of the conditions during the freezing and thawing processes is extremely critical for preserving the function and/or the viability and integrity of living cells and other biological materials. In some instances, the addition of ingredients functioning as chemical or physical stabilizers —cryoprotectants—will aid in protecting cells against the stresses encountered.

Storage of biological materials while maintaining their inherent characteristics for extended periods of time becomes essential for several reasons:

- Cell stock should be capable of storage for long periods of time without a negative impact on recovery of its original essential characteristics and with minimal risk of contamination.

- Long-term storage is needed to allow time for evaluating cell properties, including purity, and for searching for new applications for the cells and cell products.

- Cell fragments, including proteins and peptides in solution, should be storable and recoverable with the maintenance of their integrity.

- Cell concentrates and other biological materials should be shippable to other locations for processing.

- Production materials need to be storable until the appropriate time for introduction into campaign processing schedules.
- A period of time is needed to accomplish the processing, quality control, and distribution of pharmaceutical products produced from these materials.

The cryopreservation of materials for low-temperature storage over extended periods of time is fraught with the problems associated with the maintenance of such low temperatures. Mechanical refrigeration systems can reach low temperatures of –50 to –60° C with reliability, but relatively large and complex machinery is required. Such machinery is subject to failure or electrical outage; therefore, backup systems are usually needed to assure maintenance of the process. These systems are not only costly but also require rigid operational maintenance. On the other hand, refrigerant systems such as those using liquid nitrogen or solid carbon dioxide (dry ice) to achieve and maintain low temperatures are not subject to failure as long as an adequate supply of the refrigerant is available. Further, lower temperatures can be achieved and maintained (–196° C with liquid nitrogen; –60 to –70° C with dry ice). However, personal safety awareness, relative to the extremely low temperatures, must be understood and practiced. Although these refrigerant systems are not as complex and maintenance intensive as mechanical refrigeration systems, adequate equipment must be provided to operate any system.

Another approach to cryopreservation is freeze-drying or lyophilization. Freeze-drying, as the name suggests, involves freezing the material and then removing its water (drying) while in the frozen state by the process of sublimation. An advantage of using this process is that the finished product has been deprived of its moisture and usually can be safely stored at room temperature once it has been sealed in suitable individual containers. Therefore, no special requirements exist for maintaining stability during extended storage.

Essentially the same problems exist in the freezing of the product as those mentioned above. However, there are some additional requirements for the product formulation to maintain its stability in the dry state. Here, also, cryoprotectants are often incorporated in the formulation to help protect the cells from the stresses encountered during the process. In some instances, stabilizers that enhance stability in the reconstituted liquid state are added to the aqueous liquid used to solubilize the product prior to use. However, not all biological materials can withstand the specific stresses associated with freeze-drying.

The preservation of single, pure cell lines that are free from contamination, particularly from other biological entities, is highly important, not only for research purposes but also for production of biopharmaceutical products. For a biopharmaceutical product, a commercial pharmaceutical entity will have been developed and described in a regulatory license application for a specific therapeutic effect in human or animal patients. Therefore, the product must be capable of being reproduced consistently for every production batch, necessitating the availability of well-controlled starting cells. Thus, the preservation of pure cell lines is absolutely essential. To accomplish this purpose, starter cells are usually banked in private collections and in a backup location such as the American Type Culture Collection (ATCC), which is a nationally centralized collection. Banked cells should be well characterized, physically secured to ensure their long-term availability, and adequately documented relative to their identity and historical traceability.

Though this book does not focus on methods for the identification and characterization of cells and the contaminates that are frequently encountered, it should be noted that the development and validation of such methods are highly critical and necessary. Methods for identifying and eliminating contaminants such as pathogenic viruses are essential for the quality assurance of safe biopharmaceutical products. Failure to identify and eliminate toxic contaminants during the early development of some biological pharmaceuticals has led to tragic consequences in some patients. Therefore, it is imperative that testing methodologies be developed early in the history of a product and be continued concurrently with the development of the dosage form of the product.

The preceding brief summary highlights the importance of the cryopreservation of viable mammalian and microbiological cells and their products. This book is intended to provide practical information for those involved in the development and processing of biological products for therapeutic and diagnostic human and veterinary use. As technology has improved, much knowledge has been gained, making possible the approval of more and more finished products for human and veterinary administration. This book will provide the reader with an understanding of the current perspectives on this significant technology and its application to biopharmaceutical processing.

THE CHAPTER CONTENTS

Six chapters follow this introductory chapter. Of these, Chapter 2 requires special comment because of its content and special significance. Although this book generally follows the concept of the series of books of which it is a part (that is, applied treatments of the included topics), Chapter 2 is an extensive review of "The Principles and Scale-Up of Cryopreservation for Cells, Microorganisms, Protein Solutions, and Biological Products." Because an understanding of the principles involved in cryopreservation of biological materials is so critical for effective biological processing, it was decided to include a chapter written by Dr. Richard Wisniewski, an eminently qualified scientist, which presents in detail the basic principles applicable to cryopreservation of mammalian and microbial cells.

Dr. Wisniewski reviews first the basic physiologic effects of the freezing and thawing process on viable cells. This provides a foundation for understanding how the cryopreservation process must be performed to successfully preserve viable cells. Similarly, the author discusses the stressful effects of freezing and thawing on protein molecules in aqueous solutions. Subsequently, he reviews the physicochemical phenomena inherent in the process and follows this with insights into methods for accomplishing the desired product results. He provides information on the addition of cryoprotectants to aid in preventing degradation during the freezing and thawing processes. Considerations in the initial scale-up from laboratory-sized batches to production-sized lots are also presented. The amount of information contained in this chapter is very extensive and impressive and is further extended by a voluminous list of 386 literature references from around the world. This is a foundational, must-read chapter.

The third chapter, also written by Dr. Wisniewski, is a very practical discussion entitled "Large-Scale Cryopreservation: Freezing and Thawing of Large Volumes of Cell Suspensions, Protein Solutions, and Biological Products." In small-scale operations, the thermal distribution and heat transfer throughout a small volume of cell suspension or other biological preparation can be controlled so that all parts of the liquid volume are subjected to essentially the same temperature changes during freezing and thawing. This level of control is increasingly difficult to maintain as the volume increases, but it must be maintained if the viability of the cells is to be preserved. On the other hand, large batch volumes are essential as product processing moves up in scale from laboratory to production levels of operation.

Dr. Wisniewski reviews the advantages and disadvantages of the various processes available for low-temperature processing, pointing out particularly their limitations in the freezing of large volumes of fragile, viable mammalian and microbial cells and other biological products. He then carefully develops the unique design of the equipment and the process for freezing large volumes of aqueous suspensions and solutions of viable cells and protein preparations. Using three case studies, he presents the practical details of accomplishing required processing objectives.

The first case study focuses on the use of liquefied refrigerant gases with the appropriate equipment. The second case study uses equipment designed with novel heat transfer surfaces (fins) to successfully achieve a rapid and uniform heat transfer in bulk freezing of the biological materials. The third case study utilizes biological liquid products in bags in intimate contact with plates with extended heat transfer surfaces (fins) that greatly increase the efficiency of heat transfer and the control of the freezing process. He concludes the chapter with a discussion of the thawing process, the freezing process in reverse, utilizing the equipment discussed previously.

The author provides practical engineering and processing information in great detail. He also provides an extensive list of 114 references from the worldwide literature.

The fourth chapter is entitled "Method Selection Considerations for Long-Term Preservation of Mammalian Cells and Microorganisms" and was written by Dr. Carmen Wagner. Dr. Wagner discusses the necessity of preserving viable mammalian and microbial cells over a long period of time and under carefully controlled conditions to assure the identity, integrity, and historical documentation of the cell entities. She discusses the development of the Master Seed Bank and its management to provide an inventory of pure, stable cell stock. She discusses in detail the selection of the method of preservation (i.e., low-temperature freezing or freeze-drying). The latter method has the advantage of needing no specialized storage conditions or equipment, but not all cells maintain their viability when subjected to the added stress of drying.

Dr. Wagner emphasizes that the performance of the specimen, following thawing and recovery, must be assessed and is a measure of the effectiveness of the preservation procedure. This assessment is done by testing the recovered cells to determine their viability (growth and reproduction capability), their functional performance (the ability to produce a given by-product), and their genetic stability (particularly for mammalian cells). The author gives detailed procedures for the preservation of mammalian cells, bacteria, bacterial vaccines, and viral-infected cells. She also provides a list of eleven guidance documents available from the Center for Biologics Evaluation and Research. This practical chapter provides very useful information concerning the necessary controls and procedures required for the long-term preservation of mammalian and microbial cells for use in research and biopharmaceutical production.

The fifth chapter is written by Dr. William Siegel and is entitled "Cryopreservation of Mammalian Cell Cultures." The author states at the outset that the primary reason for freezing mammalian cells is risk management. The process reduces the risk of contamination or accidental death from the use of defective growth medium during continuous culture or due to equipment failure, but conversely, provides insurance that reliable cell stock can be recovered when needed.

Dr. Siegel presents practical matters such as safety issues for operators and preparing the cells for freezing, including the use of additives to enhance preservation and recovery of the cells. He discusses the selection and use of glass ampuls for packaging dispensing units. The author describes the cooling process and its control and then, at the time of recovery, the thawing process. He gives insight into the relative reliability and usefulness of mechanical refrigeration versus the use of the refrigerant liquid nitrogen. He concludes by presenting issues relative to storage and the reliability required. The final section of the chapter is a list of 55 references from the relevant worldwide literature. This is an excellent, practical, and readable resource.

Drs. Hamilton, Yin, and Lee have written a very informative and interesting chapter entitled "Freeze-Drying of Live Virus Vaccines." They begin their presentation in Chapter 6 with a brief review of the physical and chemical characteristics of viruses, summarizing them in a convenient table. The instability of live virus vaccines (LVVs) to such external stimuli as chemical agents, heat, pH, and ultrasonic disruption is well known, but there are some non-enveloped viruses that exhibit quite good stability. Components of formulations can contribute significantly to the stability of LVVs, and some of these effects are conveniently summarized in a table of data.

The authors discuss the freeze-drying process (lyophilization) in detail and identify variables that need to be controlled to enhance the reproducibility of the process and its outcome. One of the process parameters that draw the attention of the authors was the effect of radiation heating on nonuniformity of drying of LVVs in vials at various locations on the shelves of the freeze-dryer. As a result, they developed a unique shielding arrangement to improve the uniformity of radiation heating, resulting in improved uniformity and stability of dried LVVs in vials.

The authors also discuss the parameters of validation of the lyophilization process and its importance in assuring reproducibility from batch to batch. They suggest, further, that once the fundamental heat and mass transfer processes taking place during the freeze-drying cycle of a given product are understood, a simulation model can be developed leading to computerization of the process. Once established, computerized control should be able to assure enhanced reproduction of lyophilized product and markedly reduce the variability of past production processes. The chapter concludes with a list of 56 pertinent references from the literature.

Because of the importance of the need to transport frozen biological materials without the risk of thawing and concurrent loss of stability and activity, Tom Pringle has provided an appropriate, valuable, and unique final chapter entitled "The Design and Use of Thermal Transport Containers." He states, "Selecting a transport method to maintain the 'cold chain' in transporting temperature-sensitive materials requires a fundamental understanding not only of the material's acceptable temperature range in transit but also of the transit environment and the capabilities and capacities of the insulated transport options that are available."

In the section on transport system selection criteria, he discusses the temperature range requirement, refrigerant and transport method options, shipping methods, and internal (company) and external (carrier and regulatory) handling issues, including the quantity and form of the material that will constitute a single shipment. He then moves into a discussion that defines the insulated container/transporter and refrigerant options. Practical considerations of commonly used insulating materials and refrigerants—such as liquid and vapor (dry) nitrogen, dry ice, and gel packs—are reviewed in detail. Also reviewed are the multiple aspects of shipping dynamics and the qualifications of insulated container/transporters.

Throughout the chapter, the author utilizes sketches to illustrate designs and principles and provides data in graphs and tables. This is particularly so in the section containing examples of the thermal performance range of commercially available transport systems. He gives examples of: (1) dry ice in an expanded polystyrene (EPS) molded container, (2) dry ice in a polyurethane molded container, (3) a eutectic solution in a polyurethane molded container, (4) a low phasing gel ice in a polyurethane molded container, and (5) liquid nitrogen in a "dry" nitrogen shipping container. In the next section, he provides information concerning test methods for evaluating package performance, followed by a consideration of user responsibilities. A useful section follows with the presentation of information concerning temperature monitors, including electronic and chemical monitors, and their use in frozen shipments. The electronic monitors are the most reliable and most widely used for accurate determinations.

In the final two sections, the author identifies common problems in the cold chain, followed by a list of sources for products and testing services for shipment of frozen materials. An appendix is also provided with a list of regulations, guides, and other information useful for those responsible for shipping frozen materials.

This book provides new and often unique information in the biological application of the principles of cryopreservation to the preservation of viable cells and cell materials being developed for biopharmaceutical applications. This rapidly developing technology is critically needed for continued progress in the application of biologically innovative materials to the medical armamentarium of current-day therapeutics. It is anticipated that this book will be an effective, contributory tool in these developments.

2

PRINCIPLES OF LARGE-SCALE CRYOPRESERVATION OF CELLS, MICROORGANISMS, PROTEIN SOLUTIONS, AND BIOLOGICAL PRODUCTS

Richard Wisniewski

NASA Ames Research Center, Mountain View, CA

PRINCIPLES IN THE FREEZING AND THAWING OF CELLS

Large-scale cryopreservation (freezing and thawing operations) of biological materials is becoming increasingly important as current biotechnology industries (including biopharmaceuticals and biologicals) introduce the concept of a multiproduct facility operating on a campaign basis. The preservation by freezing step may be applied at various stages of the manufacturing process. The product generated in a short campaign may be stored for 1 to 2 years. The frozen product then can be thawed and filled and finished according to market demands. As a result, the manufacturing processes can become decoupled and the overall product life extended.

There are two major areas where the freezing and thawing operations can be applied. The first area is the cryopreservation of whole cells, microorganisms, and cell fragments or components. The process involves a suspension of viable or active biological material at various possible concentrations (from dilute fermentation broths to concentrated products of filtration or centrifugation). The freezing and thawing step may also be applied in certain processes as an aid to product release from cells. The second area is the cryopreservation of solutions of biomolecules (such as proteins and peptides). The process may involve raw cell-free supernatants, partially purified products (storage in frozen state as an intermediate step), finally purified products, or formulated bulk preparations ready for filling and finishing. The cryopreservation step also can be applied to fermentation and cell culture media or media concentrates. In all these cases, there might be a time- and temperature-dependent product degradation (for example, when proteases are present). Rapid and well-controlled freezing of large volumes of product also may become a solution to this problem.

The freezing and thawing of cell suspensions in large volumes are more difficult than freezing and thawing of large volumes of proteins or peptides. In protein freezing, the cryoconcentration, cooling/freezing rate, and molecular interactions such as protein-protein, protein-ice, protein-solute, and solute-ice are the phenomena that may affect the product activity (Wisniewski and Wu 1996; Wisniewski, 1998a). The protein may be affected by cold unfolding, pH shifts in the liquid phase prior to solidification, increase in concentration of solutes, and protein-ice surface interactions. The freezing and thawing of cells involve these issues as well as additional factors of cell physiology, cell cultivation conditions, cell membrane behavior during freezing and thawing, osmotic transport and cell water content in the environment of freezing and thawing, intracellular ice formation, action of penetrating versus nonpenetrating cryoprotectants, cell capture or rejection by the freezing front, ice recrystallization, and cell settling (Wisniewski, 1998b). The focus of this chapter will be mostly on the freezing and thawing of cells and microorganisms and, to a lesser degree, on the freezing and thawing of solutions of biomolecules, which were already introduced in the earlier publications (Wisniewski and Wu, 1992; 1996; Wisniewski, 1998a). Many of the described principles of cell and microorganism cryopreservation may also be applied to cryopreservation of biomolecular solutions.

The thrust of biopharmaceutical cryobiology (including large-scale operations) has been centered on the behavior of cells and microorganisms at low temperatures and during freezing and thawing as well as required conditions of low temperature storage and preservation. Typical recommended final freezing and storage temperatures have been −196° C (liquid nitrogen boiling point temperature), −140° C (typically obtained using liquid nitrogen vapor), and for mechanical refrigeration-based systems, −70° to −86° C or lower. For large-scale, walk-in freezers, higher temperatures are also used, for example, in the range of −20° to −30° C. Essential information on low-temperature preservation technology and available equipment can be found in the publications by the International Institute of Refrigeration (Recommendations 1986). In addition, the food technology literature may provide information that can be utilized in system design for freezing and thawing of tissues and cells.

Basic physicochemical phenomena in low-temperature biology were outlined by Taylor (1987). He discussed properties of water and ice, chemical and biochemical effects in solutions at low temperatures, pH effects, and preservation without freezing. These phenomena can have pronounced effects on the freezing and thawing of protein solutions, microorganisms, cells, and tissues. Heat transfer phenomena occurring during freezing of biological materials were reviewed by Rubinsky and Eto (1990). Examples of general works and reviews on the effects of low temperature on cells and organisms, including the phenomena of freezing and the mechanisms of freezing injury to cells, are those by Calcott (1985); Daw et al. (1973); Diettrich et al. (1987); Diller and Raymond (1990); Franks (1985); Greaves and Davies (1965); Mazur (1984); Meryman (1974); and Pegg (1976).

A phenomenon of solute concentration in the liquid phase may occur during the freezing process. As a result, the solubility of buffer-forming salts may change and some may precipitate. This can also lead to a rapid change in the pH of the solution. Sensitive biological materials, such as serum components for cell culture applications, may be adversely affected by the physicochemical phenomena and chemical reactions occuring during freezing and thawing. Special freezing procedures are undertaken to minimize these effects (HyClone 1992).

Koerber (1988) presented a thorough review of his own investigations and literature data on the phenomena at the advancing ice-liquid interface, including behavior of solutes, particles, and biological cells. He reported on: freezing of salt solutions with phenomena of solute redistribution, behavior of dissolved gases (including gas bubble formation), ice-liquid interface interaction with particles and biological cells, and the effects of ice formation on biological cells (including osmotic transport and intracellular ice formation).

Certain cellular characteristics should be known/investigated before designing the large-scale process for freezing and thawing. If cell viability is the major concern, factors affecting cell cryoinjury are important. If the material within the cells is of interest (for example, inclusion bodies), then the cell viability and integrity may not be important since a cell rupture step typically follows the freeze-thaw operations. In such a case, typically, cell inactivation (for example, heat killing) may precede the freezing step.

Grout and Morris (1987) presented a review of the subject of cellular organization and freezing phenomena. They discussed the subjects of freezing cells in suspension, freezing injury and viability of cells, effects of cooling rate, events occurring in the frozen state, thawing stresses, sites of freezing injuries, and cryoprotection. Sharp (1984) reviewed methods of preservation of genetically unstable microorganisms and cryopreservation of fermentation seed cultures. Malik (1987) published a thorough review on preservation of microorganisms in culture collections. Graumann and Marahiel (1996) reviewed the response of microorganisms to cold shock. The cold shock generates induction of large numbers of proteins as a response to lowering the temperature. For example, in *E.coli*, growth stops after lowering the temperature from 37° to 10° C, and the synthesis of a majority of proteins ceases, but after a few hours the synthesis of most proteins is resumed. Established cryopreservation protocols pertain to small samples containing cells in aqueous solutions, usually including cryoprotectants. The freezing involves transport of water and solutes across the cellular membrane. Efforts have been made to estimate such transport for various freezing protocols (see, for example, Walcerz 1995). Cell dehydration and supercooling are the major factors considered in the models of freezing. No intracellular ice formation may be predicted by the model, although the conditions of intracellular ice formation may exist.

Most of the phenomena encountered during freezing and thawing of cell suspensions and biological materials have one factor in common: the presence of ice crystals. The molecular structure of the surface of ice crystals may play an important role in ice-solutes interactions (Wisniewski and Wu 1996; Wisniewski 1998a,b). The usual crystalline form of ice is a hexagonal structure (ice Ih). Very rapid freezing of small droplets can produce cubic ice structure (ice Ic). A review of the ice surface structure was published by Devlin and Buch (1995). This review includes results of spectroscopic and computer modeling studies. The ice surface may be less regular than the crystal interior (Delzeit et al. 1996). Schaff and Roberts (1996) reported on differences between surface properties of amorphous and crystalline ice. The surface chemical properties of ice are determined by the presence of OH groups on the ice surface and formation of hydrogen bonds with solutes at the ice-liquid interface. In the bulk ice, the water molecule can form hydrogen bonds with 4 other water molecules. On the ice surface (in vacuo) the water molecule can form only 3 hydrogen bonds with other water molecules. At ice surface temperatures above 240 K, a quasi-liquid layer surface structure could be detected, but not at temperatures below 240 K. The level of solute incorporation into the ice lattice is typically low or close to zero. The interpretations of mechanisms of solute incorporation into ice include hydrogen bond formation and matching the hexagonal pattern of ice (Ih) structure. Recent discoveries of the behavior of water molecules at the ice surface (Materer et al. 1995) shed light on the explanation of molecular interactions between ice and solutes or cells (via cell surface molecules). The oxygen atoms located at the ice Ih surface were found to be strongly vibrating. These vibrations may significantly affect formation of intermolecular bonding between the ice surface and solutes/molecules contained in solution. Recent work by Materer et al. (1997) included the molecular dynamics simulations, which showed large amplitude of vibration of the uppermost (surface) water molecules with stable bulk ice structure. Further investigation of these effects and their importance for freezing and thawing of solutions and suspensions of biological materials is needed.

The important phenomena during freezing and thawing of cells are the response of the cell membrane, behavior of solutes, formation and growth of ice crystals, and the mutual interactions between the crystals. Membrane fluidity and permeability are the important factors at changing temperatures. When the fluidity decreases (membrane becomes rigid) with the decrease in temperature, the mechanical stress caused by growing ice crystals may have deteriorating effects on membrane integrity. Water structure at the membrane lipid layer is perturbed in comparison to the bulk water (Cascales et al. 1996). The freezing pattern of such water is probably different from the freezing of the bulk water. Proteins are a large part of cytoplasmic bacterial membranes. The *E.coli* membrane consists of phospholipids with a large amount of proteins (up to 70 percent of membrane mass is represented by about 200 different kinds of proteins (Neidhardt et al. 1990)). The membrane proteins are proteins with hydrophobic parts locked within the membrane and with hydrophilic parts protruding into the bulk external solution. These external parts may interact with ice crystals during the freezing process—most likely being ejected from the ice structure (Wisniewski and Wu 1996). Complex polysaccharides are anchored in the cell wall membranes of bacteria and may extend into the bulk solution; for example, the lipopolysaccharide chains may extend to about 30 nanometers (nm) (Hammond et al. 1984).

Bacterial surface appendages (flagella and fimbriae) may first interact with the forming extracellular ice crystals. The major part of a flagellum is a filament, which is approximately 20 nm in diameter and 10–20 micrometers long. The filament consists of a single protein, flagellin. The cell surface proteins may be affected by an increased concentration of solutes between the ice crystals prior to complete solidification and by interacting with ice. Binding of proteins to ice crystals occurs within a group of antifreeze proteins and requires specific structural characteristics (Davies and Hew 1990). The ice-binding mechanisms are attributed to regularly spaced polar residues matching the ice lattice for linear alpha-helical protein surfaces and to binding sites that can form hydrogen bonds with the ice surface oxygen atoms (Jia et al. 1996). In general, one should not anticipate binding of the cell surface proteins to ice. The filament structure suggests that it might not be incorporated into the ice lattice, but the flagellum can be trapped between ice crystals and severed during stress buildup or ice recrystallization. In addition to flagella, many bacteria possess other surface appendages such as fimbriae. The fimbriae can be 3–14 nm thick and 0.2–20 micrometers long. The fimbriae are composed of protein and are relatively fragile (can be shed by mechanical agitation). Fimbriae have hydrophobic properties and are involved in bacterial adhesion. During freezing, they may contribute to bacterial aggregation. The freezing and thawing process may inflict damage to the fimbriae due to mechanical stress and ice recrystallization.

Dehydrated prokaryotic cells may reach water content low enough to minimize damage by intracellular ice crystals. Extracellular ice crystals, however, may cause cell deformation and damage. The extracellular ice crystals may also damage cell flagella or fimbriae. The freezing step is also very important in freeze-drying microorganisms (Souzu 1992). The microbial viability in reconditioned sample and product stability during storage strongly depend on the freezing protocol.

Escherichia coli

One of the *organisms* on which a large amount of work has been done in the biopharmaceutical industry is *Escherichia coli*—see a thorough review work on this bacterium edited by Neidhardt (1987). In that work, effects of temperature, pH, water activity, and pressure on the cell (Ingraham 1987); outer membrane (Nikaido and Vaara 1987); osmotic shock-sensitive transport systems (Furlong 1987) and cytoplasmic membrane (Cronan et al. 1987) are reviewed. Related topics of interest that can be found in that work are the heat-shock response, chemical composition of *Escherichia coli*, its periplasm, and protein secretion.

The structure of an outer membrane seems to play a role during the freezing and thawing steps. Gram-negative bacteria (such as *E. coli*) can have complicated polysaccharide structures attached to their outer membranes (Whitfield and Valvano 1993). The lipopolysaccharides are the main component of the outer membrane in contact with the outside medium. They consist of a hydrophobic lipid part located within the membrane and two saccharide parts: a hydrophilic polysaccharide protruding into the outside medium and a core oligosaccharide in between (Nikaido and Vaara 1987). These exposed saccharides interact with the water molecules from the bulk and may interact with the ice surface. The lipopolysaccharide (LPS) ends with an O-antigen domain. It is a repeating (frequently a tetrasaccharide subunit) oligosaccharide attached to a glucose residue in the core of the LPS. The number of subunits may be as high as 40. The O-antigen structures are variable, and some LPS molecules may lack them. In the *E. coli* K-12, the O-antigen is missing (Raetz 1990). These large carbohydrate structures would most likely be rejected from the ice lattice. The lipopolysaccharide is a unique component found in gram-negative bacteria with a resulting rejection of hydrophobic compounds by the membrane (not common among other bacteria types). This rejection should be considered during cryoconcentration of solutes in freezing.

Calcott et al. published several papers on the subject of freezing and thawing of *E. coli* cells (Calcott and MacLeod 1974; Calcott and MacLeod 1975a; Calcott and MacLeod 1975b; Calcott and Calcott 1984; Calcott et al. 1984). The results showed that during freezing and thawing cycles, damage can be done to the outer membranes of *E. coli* and gram-negative bacteria in general. They postulated that possibly two effects may be responsible for the freeze and thaw damage to the outer membrane. These effects were an alteration of the structure of membrane lipopolysaccharides and an alteration of protein components of the membrane, especially the hydrophilic porins. MacLeod and Calcott (1976) presented results of their early works in a review on the subject of cold shock and freezing damage to microorganisms. They reported that the cell survival was dependent on cooling and warming rates. They cited the results on release of periplasmic enzymes from *E. coli* cells after a freeze-thaw cycle as evidence of freezing damage and an increase in the penetrability of the outer membrane of the cell. Calcott (1985) presented a review on cryopreservation of microorganisms that included discussion of the effects of freeze-thaw cycles with a focus on the cell wall structure. He pointed out a sensitivity of the outer membrane of gram-negative microorganisms to a freeze-thaw cycle and its permeability change to certain chemicals (such as surfactant sodium dodecyl sulfate [SDS]), which can cause cell lysis. Frequently such damage to the outer membrane can be repaired during an incubation in medium containing phosphate and Mg^{++} ions. The areas in the outer cell wall membrane sensitive to a freeze-thaw cycle were the exposed lipid regions, which are not protected by the lipopolysaccharide (LPS) chains. The concept of ice penetration through membrane pores was supported by results of freeze-thaw experiments with the standard *E. coli* and porin-deficient strains. The porin-deficient strains showed much less damage to the outer membrane than the standard strains. These effects of a freeze-thaw cycle on the gram-negative bacteria might be utilized in a partial cell lysis process. The outer cell membrane may be damaged in a selective way by a well-controlled cell freeze-thaw cycle.

Saccharomyces cerevisiae

Yeast preservation by lyophilization may not give very good results; therefore, cryogenic methods of preservation were introduced

(Wellman and Stewart 1973). The yeast cells (and particularly *Saccharomyces cerevisiae*) have been well characterized regarding behavior during freeze-thaw cycles (Mazur 1984). Yeast cell walls are mostly composed of two classes of polysaccharides: mannoproteins (polymers of mannose covalently linked to peptides) and glucans [polymers of glucose]. In mannoproteins, an inner core of mannose molecules is linked to the peptide. The inner core contains 15–17 mannose units. An outer chain of 100–200 mannose residues is attached to the core (Cabib et al. 1982). Investigations of freeze-thaw resistance of yeasts pointed out its dependence on plasma membrane lipid composition (Calcott and Rose 1982) and, in particular, composition of fatty acids in plasma membranes (Sajbidor et al. 1989). The yeasts produce extracellular polysaccharides and glycoproteins that may provide a certain degree of cryoprotection (Breierova and Kockova-Kratochvilova 1992). Isolated extracellular glycoproteins and polysaccharides were used as nonpenetrating cryoprotectants. A combination of extracellular yeast glycoprotein with the penetrating agent DMSO showed noticeable cryoprotective properties. The secreted glycoproteins possess N-glycosidically linked carbohydrate, which consists of an oligosaccharide core with 8–13 mannose residues to which an outer chain of about 50 mannose units is attached.

Matsutani et al. (1990) described physical and biochemical properties of selected freeze-tolerant mutants of *S. cerevisiae*. The freeze-tolerant mutant cells were of more rigid structure than the wild-type yeast. Freeze tolerance was partially induced by an increase in the rigidity of cell surface. After six freeze-thaw cycles, the viability of wild form cells declined to around 25 percent and to 75 percent for freeze-tolerant mutants. Schu and Reith (1995) evaluated preparation parameters and protocols for the production and cryopreservation of seed cultures with recombinant *S. cerevisiae*. Freeze-induced dehydration and the concentration of solutes were considered major causes of insufficient yeast recovery after freeze storage as well as intracellular ice formation during fast freezing. Glycerol concentrations of 2–5 percent gave the best recovery, while concentrations of 20 percent and 50 percent drastically reduced viability. Ultrafast freezing led to very low recovery levels. The second freezing protocol; cooling at –20° C for 2 hours (given a cooling rate of 0.5° C/min) followed by rapid cooling in liquid nitrogen (cooling rate of 200° C/min), also gave very low recovery. The third protocol, cooling at −70° C for 2 hours (given a cooling rate of 2.4° C/min) with subsequent rapid cooling in liquid nitrogen, gave the highest recovery rates (about 78 percent). The fourth protocol, using freezing and storage in liquid nitrogen vapor phase (cooling rate of 12° C/min), resulted in viability rates of 50–60 percent. A rapid thawing process (at a warming rate of approximately 100° C/min) provided the highest viability.

Scale-Up

Most of the work on freezing and thawing of cells has been conducted on a small scale where conditions of freezing and cooling have not changed significantly within the samples. Reported scale-up has not involved any single volumes within the tenth- hundredth-liters range. A typical approach to scaling up was to go from the vial to the bag. For example, Kasai and Mito (1993) reported a large-scale cryopreservation of isolated dog liver cells (hepatocytes) with the scale-up from 3-mL vials to 100-mL bags. Cooling rates were within the range of 1–15° C/min. Cell viability was comparable for small and large sample size and was higher for the lower cooling rates (1–10° C/min.) than for the higher rates (above 10° C/min) by approximately 12–18 percent.

In a process of scaling up, any increase in the dimensions of the specimen would cause substantial differentiation in cell environment across the sample during freezing and thawing. Meryman (1966) pointed out the importance of location of thermocouples during freezing and thawing tests. Location near the surface of the sample will show almost no temperature plateau, whereas location near its center will show an extended period of temperature plateau (Figure 2.1).

Freezing of aqueous solutions in large volumes can involve complex phenomena such as dendritic growth of ice crystals, cryo-concentration effects, and natural convection in the liquid phase. Addition of cells may add to the complexity of freezing and thawing due to their discrete nature, sedimentation effects, osmotic phenomena, effects of ice recrystallization, and extra- and intracellular freezing. At low cell concentrations, suspensions can freeze in a manner similar to that of aqueous solutions. However, when cell concentration increases, the freezing process can become similar to the freezing of pasty foodstuffs or biological tissue.

Figure 2.1. Freezing histories of the sample surface and center (t = time; T = temperature). The sample surface temperature follows closely the cooling environment temperature, while the center temperature remains at the plateau close to 0° C until the solid-liquid interfaces reach the central zone.

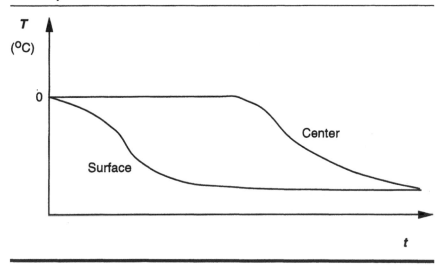

Whereas in the freezing of protein solutions the molecular interactions are of primary importance (Wisniewski and Wu 1996), in the freezing and thawing of cell suspensions, the physicochemical, biochemical, and macroscopic effects may become equally important phenomena, involving cell membrane characteristics and overall physiology and molecular biology of the cell.

Meryman (1966) formulated the concept of an optimum cooling rate, with the slow cooling injuries based on the time of exposure to highly concentrated solutes as well as cell dehydration and the fast cooling injuries based on intracellular ice crystallization. He cited the results of studies with erythrocytes using the electron microscope. Intracellular ice was present in the cells after rapid cooling and absent after slow cooling. Meryman also summarized available information on the influence of warming rates on cell survival. In certain cases, the warming rate had no influence if the cooling rates were slow, but could have a significant influence at rapid cooling rates, with lower survival levels at slower warming rates than at rapid ones. Meryman also indicated a possibility of growth of ice crystals in cells during slow warming. In addition to the freezing and thawing effects associated with cell components and organelles, the cooling, freezing, and thawing may affect cell and solution molecules such as proteins. Cold lability may be anticipated if the protein is multimeric. The dissociation of these complexes may be related to weakening of the hydrophobic interactions among the subunits with temperature decrease. (For review of protein freezing, see Franks and Hatley 1992, and Wisniewski and Wu 1996.) The cell damage during freezing and thawing depends on type of cells, cooling-freezing and warming-thawing protocols, and the presence of solutes (small and large molecules, cryoprotectants). The size and shape of the sample may affect localized damage to the cells.

Intracellular Ice Formation

It has been widely acknowledged that the major damage done to cells during the freezing process can be intracellular ice formation (Karlsson et al. 1993; Muldrew and McGann 1990; Pitt 1992). The intracellular ice formation may occur in cells within the range of $-5°$ to $-20°$ C in the presence of extracellular ice. There are several hypotheses regarding the mechanisms of intracellular ice formation (Chandrasekaran and Pitt 1992; Karlsson et al. 1993; Muldrew and McGann 1990); such as intracellular freezing as a result of critical undercooling, nucleation of intracellular ice crystals through aqueous pores in the membrane as a result of electrical transients at the ice-liquid interface, and damage to the plasma membrane due to the gradient in osmotic pressure. Karlsson et al. (1993) summarized the results of intracellular ice formation, including dependency on cooling rate and on temperature when the intracellular ice formation occurs. The following hypotheses of intracellular ice formation may be considered:

- Critical level of undercooling with ice nucleation by membrane/internal cell components;
- Ice nucleation through the aqueous pores in the cell membrane;
- Ice nucleation due to membrane damage caused by electrical transients at the ice surface;
- Ice nucleation due to membrane damage by a gradient in osmotic pressure across the membrane;

- Ice nucleation due to membrane damage caused by mechanical stress imposed by extracellular ice crystals.

Muldrew and McGann (1990) suggested that intracellular freezing may be caused by damage to the plasma membrane due to the critical gradient of osmotic pressure across the membrane. The damaged plasma membrane loses its ability to act as a barrier to extracellular ice. Toner et al. (1991) investigated intracellular ice formation during freezing of oocytes without cryoprotectants. The presence of extracellular ice was indicated as a triggering event for intracellular ice formation. The second mechanism of intracellular ice formation was the ice nucleation by internal cell structures. Toner (1993) published a summary of current knowledge of nucleation of ice crystals inside biological cells in a thorough review article. Detailed information on intracellular ice formation was provided for mouse oocytes, one-cell mouse embryos, *Drosophila melanogaster* embryos, isolated protoplasts from *Secale cereale*, and rat hepatocytes. A design of multistep freezing protocols was discussed. A heterogeneous nucleation model was proposed. It was hypothesized that intracellular ice formation is catalyzed by local sites on the interior surface of the plasma membrane in the presence of external ice or by internal molecular structures, depending on the freezing conditions.

Although bacterial cells themselves may serve as ice nuclei (Wolber 1993), this mechanism has not been of interest during freezing of microorganisms. However, the bacterial cell wall proteins could in certain cases trigger the ice nucleation in undercooled solution. The surface proteins, which can bind water in an arrangement resembling a small ice crystal, might become the nucleation sites.

Cell Membrane, Its Permeability and Freeze-Thaw Phenomena

The behavior of cell membranes under varying physicochemical conditions can be one of the most important factors in studies on cell damage during the freeze-thaw cycles. Transport of water and solutes through a membrane during the presence of osmotic imbalance, transport of penetrating cryoprotectants, and membrane properties (such as fluidity, flexibility, strength, and permeability) are of great importance in the freeze-thaw processes. Membranes should maintain sufficient permeability, flexibility, and fluidity, as well as resistance to concentrated solutes and strength (resistance to mechanical damage from forces caused by growing ice crystals). Studies of membrane properties at low temperatures and studies on membrane behavior under real freezing and thawing conditions are important for optimization of cell freezing and thawing processes.

The background information on microbial membranes and outer wall characteristics can be found in works by Hammond et al. (1984) or Neidhardt et al. (1990). The physical chemistry of biological membranes was summarized by Silver (1985). Water transport across membranes occurs through both the lipid bilayer and specific water transport proteins. The transport rate through the lipid bilayer depends on lipid structure and the presence of sterols. Water flow is passive and directed by osmosis (Haines 1994). Water transport depends on ion pumps, ion channels, and ion exchange proteins. The diffusion through lipid bilayers occurs only above the melting (fluidity level) temperature of the bilayer. The lipid composition of membranes in living organisms may frequently keep this temperature approximately 10° C below ambient. Permeation of molecules across the lipid bilayer is thought to occur in 3 steps: partitioning of the substance from the aqueous phase on one side into the bilayer, diffusion across the bilayer, and partitioning from the bilayer into the aqueous phase on the other side (Lande et al. 1995). During the cooling and freezing processes, the fluidity limit of the membrane may be reached, and the water diffusion through the membrane may cease. The water transport may subsequently continue via the membrane transport proteins.

The bacterial cell walls carry a strong negative charge, and this charge may play an important role in interactions with ions, charged molecules, and surfaces. The hydrophobic effect is also thought to play an important role in bacterial interactions with the outside environment (Doyle and Rosenberg 1990). The existence of a hydrophobic effect, although indicated for various bacterial species, may not be the most important effect in bacteria-ice interaction. During freezing-out of water and cryoconcentration of cell suspensions and solutes in the remaining liquid phase, the hydrophobic interactions may cause cell aggregation. (For more about freezing of cell aggregates, see the later section on cell concentration effects during freezing.) Cell aggregation during freezing may also occur if there is a pH change leading to a decrease in electrokinetic potential (under ambient conditions near neutral pH, the cell surfaces are typically negatively charged; thus, there is a repulsive cell-cell electrostatic interaction). Many yeast strains have a tendency to flocculate—for example, yeast used in fermentation processes—and this is attributed to hydrophobicity (Mozes et al. 1988; Mozes and Rouxhet 1990).

Pringle and Chapman (1981) reviewed biomembrane structures and the temperature effects on membranes. Schubert (1987) reviewed the biophysical approach to studies of biological membranes, including the lipid structure in membranes, lipid phase transitions, and protein-lipid interactions. A general review on the structure and functions of the outer membrane of *E. coli* and *S. typhimurium* organisms was given by Nikaido and Vaara (1987), including diffusion of hydrophilic compounds through the membrane. They reported the damaging effects to outer membranes caused by the presence of divalent cations (as Ca^{++}) at high concentrations and at low temperatures. The presence of Ca^{++} ions can solidify the acidic phospholipids by increasing their thermal transition temperature. This effect may already occur at temperatures near 0° C, even before the freezing process begins. The authors suggested that Ca^{++} ions may also affect the LPS monolayer in the same way—by solidifying it and causing membrane cracks. Since such cations are rejected from the ice during freezing, the cryo-concentration phenomena occur in the liquid phase, and cells may be exposed to a high level of ionic species.

General information on the cell membrane of yeast (*Saccharomyces*) can be found in the works by Ballou (1983), Cabib et al. (1982), Henry (1983), or Schekman and Novick (1983). Cell structure of *S. cerevisiae* was reviewed by Cid et al. (1995). Kruuv et al. (1978) reported on the effect of fluidity of membrane lipids and their composition on freeze-thaw survival of yeast. The yeast with gamma-linolenic acid in their membranes survived better than cells containing stearic acid. The phospholipid composition can vary, depending on yeast culture conditions (Carman and Zeimetz 1996). Quinn (1985) discussed a model of low-temperature damage to biological membranes by analyzing the stability of lipids in membrane structure. He suggested that phase separation of lipids occurs due to cooling below the lamellar gel temperature to a liquid-crystalline phase transition temperature. Some lipids can remain in a gel phase even at very low temperatures. During warming, a reverse process may not restore exactly the same lipid bilayers, and change in the permeability of the membrane may occur. Zlotnik et al. (1984) investigated wall porosity of *S. cerevisiae*. The authors concluded that the porosity is determined by an external cell layer formed by mannoproteins. These proteins are linked by disulfide and hydrophobic interaction bonds in the vicinity of outer cell surface. An article by de Nobel et al. (1990a,b) reported on the cell wall porosity of yeast. They suggested that the wall porosity depends upon glucanase-soluble mannoproteins and upon the number of disulfide bridges.

During osmotic water removal from cells, driven by the external cryoconcentration of solutes in liquid phase, the transport phenomena across membranes determine kinetics of this equilibration process. Principles of the membrane transport phenomena can be found in general texts on biophysics (Schultz 1980). A more specific approach can be found in works dedicated to membrane transport in microorganisms and cells. Nikaido (1979) reviewed the general subject of nonspecific transport phenomena through the bacterial outer membrane. He provided summaries on the subjects of transmembrane diffusion of hydrophobic and hydrophilic molecules. McGrath (1988) reviewed membrane transport related to the freezing and thawing of cells. Hancock (1984) reviewed outer membrane permeability and ways of its possible alterations. He proposed a hypothesis explaining the mechanism of those alterations at sites where divalent cations noncovalently crossbridge the adjacent lipopolysaccharide (LPS) molecules. Mazur and Rigopoulos (1983) showed an influence of warming rate on membrane damage. The slowly thawed cells survived freezing of a higher fraction of extracellular water than did the rapidly thawed cells. Cell size and cell solute concentration were found to be independent of the fraction of unfrozen extracellular water. However, cell survival was strongly dependent on that fraction.

Porins and Cell Wall Permeability

The outer membrane of gram-negative bacteria possesses unique properties that make it different from other biological membranes. The outer membrane of *E. coli* bacteria has an exceptionally low permeability to hydrophobic solutes (Vaara et al. 1990) despite the general feature of lipid bilayer type membranes being highly permeable to small hydrophobic molecules. The authors suggest that this low permeability of *E. coli* membranes depends on the presence of lipopolysaccharide (LPS) monolayer in the outer membrane. The outer cell wall works as a barrier that allows diffusion of hydrophilic molecules smaller than a certain critical size. The major route of such substances is through the pore-forming proteins called porins. Typical diffusion porins form trimers within the membrane, and their monomer molecular weight is within the range of 30–48 kDa. These porins are associated with the lipopolysaccharide and the peptidoglycan layer in a noncovalent fashion.

Benz (1988) reviewed structure and function of porins in gram-negative bacteria. The author described the structure and composition of bacterial outer membranes and the role of porins. He reported results of investigations of porins *in vivo* and purified and reconstituted porins. The hydrophobic substances had reduced permeability, and small hydrophilic solutes were highly permeable due to the hydrophilic character of porins. Nikaido and Vaara (1985) reviewed the permeability of outer bacterial membranes. Since the porins are nonspecific diffusion channels, they may facilitate penetration of ice through the membrane. The size of these pores can be approximated from testing their diffusional permeability. It was found that the porins do not pass solutes of molecular weight larger than 700 daltons and that a significant permeability begins below 500 daltons. In the *E. coli* K-12, the pore size has been found to be 1.2 nm for OmpF and 1.1 nm for OmpC porins. These two porins are weakly cationic in selectivity. The pore size for the *Pseudomonas aeruginosa* PAO1 porin is around 2 nm. Such a range of pore size requires substantial undercooling of water to allow an ice front to penetrate through the pore.

Nakae (1986) presented a thorough review on the subject of outer membrane permeability in bacteria. The author discussed the structure of the outer membrane, transport through the outer membrane in gram-negative bacteria, and the types of pores. The physicochemical structure of porins was described, and the porin sizes were listed for several organisms. Reported porin diameters were within the range of 0.7– 2.2 nm, depending on the organism and type of porin. The permeability of pores by hydrophilic substances was a basis for the conclusion that the porins are filled with water. Hancock (1987) presented a short review on the role of porins in outer membrane permeability. The major function of porins was a sieving action for solutes in aqueous phase. Benz and Bauer (1988) presented a review on the permeation of hydrophilic molecules through outer membranes of gram-negative bacteria with emphasis on bacterial porins. They pointed out that most porins in outer membranes are diffusion porins (nonspecific) and are filled with water. The authors listed the size of porin channels for gram-negative bacteria. The reported range of porin diameter was 1.1–2.0 nm, depending on the organism. Nikaido et al. (1991) reported on the porin characteristics in *P. aeruginosa*.

Despite the presence of large porins (wider than for *E. coli*), the permeability of the outer membrane was low. The authors proposed an explanation of that apparent anomaly: The *P. aeruginosa* porins have a uniform diameter along the channel, whereas the *E. coli* porins have only a very short narrow part on the periplasmic side and a wide mouth on the extracellular side; therefore, less drag resistance along the porin channels should be anticipated for *E. coli* than for *P. aeruginosa*. Details of the molecular structure of one of the *E. coli* porins—PhoE, with similar molecular structure of a narrow channel and a wide mouth—were described by Jap (1989). Jap and Walian (1990) thoroughly reviewed the biophysics of the structure and function of porins. Results of pore size estimates based on exclusion limits, permeability rates, and channel conductances were summarized.

Sen et al. (1988) reported on the effects of the Donnan potential across outer membranes on porin channels in *E. coli*. The permeability of porin channels was measured for the intact cells suspended in solutions with varied ionic strengths. Since porin channels allow for a rapid flux of small ions, the only potential that can exist across the outer membrane is the Donnan potential. The Donnan potential was controlled by changing the ionic strength of the suspension medium. Change in the Donnan potential from 5 mV to 100 mV did not cause any change in the porins' (OmpF and OmpC) permeability. The authors concluded that the majority of porin channels remain wide open during changes in the Donnan potential. This finding may suggest an easy diffusional transport through the porins within a wide range of ionic strength changes and therefore, a relatively rapid cell accommodation to the changes in environment. The authors also reported on the ion effects on dissipation of the Donnan potential. They concluded that divalent ions (such as Mg^{++}) can dissipate the Donnan potential more effectively than monovalent ions (such as Na^+). This is because of the presence of negatively charged groups in the porins. Buechner et al. (1990) reported that only a fraction of porins remain open, although the

Figure 2.2. The *E. coli* OmpF porin molecule with an internal channel.

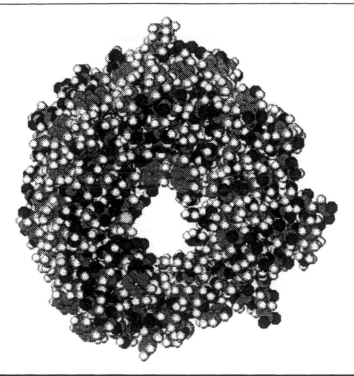

porins may open under stimulation by voltage, by osmotic pressure, or by specific solutes. Porins may also be affected by interactions with lipopolysaccharides, peptidoglycans, or periplasmic proteins. Figure 2.2 shows a view of the *E. coli* OmpF porin molecule along the axis to show the porin channel (displayed using HyperChem v. 5 molecular modeling software). Figure 2.3 shows the close-up view of the internal channel (from the same point of view) with the internal residues labeled (displayed using Hyper-Chem v. 5 molecular modeling software).

Karshikoff et al. (1994) proposed an electrostatic model for the two *E. coli* porins, OmpF and PhoE. They used a finite difference technique to calculate the pKa values of titratable groups and electrostatic field in the porin channel. The electrostatic potential for the OmpF porin near the restrictive end was close to zero, whereas the potential for the PhoE porin was positive. This difference in

Figure 2.3. The details of the *E. coli* OmpF internal channel with the residues labeled.

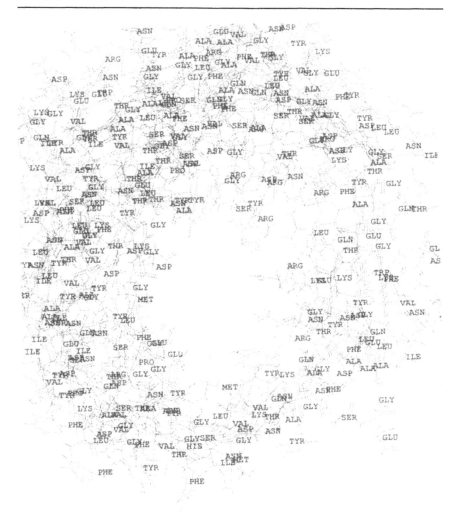

electrostatic field may contribute to the observed difference in ion selectivity. Additional information on the electrostatic phenomena in biological membranes can be found in a comprehensive review by Cevc (1990).

The porin characteristics in gram-negative bacteria may change under various pHs of the environment. Todt et al. (1992)

reported on the pH effects on function of *E. coli* porins OmpF, OmpC, and PhoE. The small channels stabilized at low pH, and the larger channels were detected at high pH. For example, the OmpC average channel size was about 0.9 Angstrom (Å) at pH 5.4, was about 1.9 angstroms at pH 6.5 to 7.5, and increased to 2.4 angstroms at pH 9.4. Since the pH in a liquid phase during freezing may be changing, the porin's permeability of microorganisms involved may also change. Because there is a cryoconcentration effect during freezing and the ionic strength outside the cell increases, these findings may aid in interpretation of cell molecular transport through the porins. Recently, Jap and Walian (1996) published an updated review on the structure and functional mechanism of porins. The *E. coli* porins are trimers (except the OmpA, which is a monomer); they have an approximate channel length of 35 angstroms; the constriction zone length is about 9 Angstroms; and its cross section is about 7 × 11 Angstroms. The major structural block of the porins is a beta-sheet cylinder. Trimer porins consist of three such cylinders. The external part of the trimer (about 25 angstroms thick), which is embedded into the wall, has nonpolar residues exposed. Along the internal channel, the charged amino acid residues are located on the sides of the channel forming an electric field. The arrangement of charged residues determines the transport characteristics of the porin. The charged residues are densely located on the extracellular side of the channel. The potential gradient may vary from 1 mV/Angstrom at the channel inlet to about 20 mV/Angstrom at the constriction zone. The pore size may be affected by the presence of charged solutes. The channel electrostatic properties are also affected by pH. The electric field may affect the orientation of water clusters within the channel. The arrangement of water molecules within porins has an effect on transport of solutes. The hydration shells around solutes interact with water within the channels. The hydrophobic solutes transport may be retarded through porins due to the interactions among the water structures associated with the channel and those formed around the hydrophobic residues of the solute.

When water is undercooled below approximately –40° C, freezing will occur even in the absence of nucleation centers (Franks 1985). At temperatures close to this level, ice can easily penetrate even through very narrow capillaries similar to porin channels. McGrath (1981) proposed a computer model for cell membrane damage due to freezing. He used human erythrocytes as model cells. The model was based on thermodynamic principles associated with biological membranes. Diller and Lynch (1984) analyzed cell freezing with membrane-permeable additives. They applied an irreversible thermodynamic approach. The model was proposed to analyze changes in the cell volume and changes in the concentrations of additives inside and outside the cells. Cell membranes can be affected by other than ionic solutes in the surrounding medium. For example, the presence of ethanol may increase the fluidity and permeability of the yeast plasma membrane (Jones and Greenfield 1987). Membrane fluidity increases with an increase in ethanol concentration. Genetic transformations of microorganisms can affect cell wall porosity and permeability (Brzobohaty and Kovac 1986).

Cryoconcentration Effects during Freezing and Chemical Reactions in the Liquid Phase

Exclusion of solutes from ice during freezing causes a cryoconcentration effect (freeze concentration) of those solutes in the remaining liquid phase. Such changes in concentration may lead to precipitation of solutes, shifts in pH, or chemical reactions among solutes in the liquid phase. Typically, chemical reactions in the liquid phase are slowed at lower temperatures. Lowering the temperature may lead to solidification of the liquid phase as a eutectic state or in the form of a very viscous, rubbery, or glassy, amorphous state. In the eutectic state, liquid can coexist with two crystal phases at a specific concentration (C_e) and temperature (T_e). The solidified eutectic may produce a lamellar form of alternating layers of two crystalline phases. The scale of these layers is usually very small and may be comparable with the cell size. In the solid phase, the chemical reactions may be inhibited due to limited mobility of molecules. The concentrated liquids solidifying in a very viscous state may also slow the chemical reactions to an insignificant level.

In the partly frozen aqueous solution, some chemical reactions may accelerate. Such an acceleration may depend on multiple effects, such as freeze concentration effect, a possible catalytic effect of ice crystals, higher proton mobility in ice than in water, a substrate-catalyst orientation as a result of freezing, or a higher dielectric constant for water than for ice (Takenaka et al. 1996). An inequality of distribution of solute anions and cations between ice and solution may create an electric potential between ice and solution. Ions accumulate near the interface due to electrostatic forces, the potential is neutralized by mobile OH^- or H_3O^+ ions, and the pH of the solution can change. Acceleration of chemical reactions may occur due to the freeze concentration and hydrolysis caused by the OH^- and H_3O^+ ions. The pH shift may typically be caused by the presence of freezing potential, but sudden pH changes are possible if buffer components precipitate. The freezing potential difference and its sign may depend on the solutes, their concentration, and the freezing rate. For example, in the case of NaCl solution, the level of Cl^- ions in ice is higher than that of Na^+, and therefore, the ice is negatively charged in relation to the solution (a similar situation exists in the case of KCl). In the case of NH_4Cl solution, the ice potential versus the liquid phase becomes positive. The chloride ion partition between the liquid and solid phases may be considered as an example of the science of ion solvation leading to possible understanding of how the ions could be incorporated in the ice structure. In general, levels of ion entrapment in ice lattice are low. The traditional approach to ion solvation (hydration) has been a use of the presence of hydration shells of water molecules around ions and effects associated with those ion-affected layers of water (Marcus 1985). In his later works Marcus has modeled hydration of ions (Marcus 1987) and calculated partial molar volumes of hydrated ions (Marcus 1993). Marcus has given the number of entirely immobilized water molecules at Na^+ and at K^+ ions as 3, at Mg^{++} ion as 10 and at Ca^{++} ion as 7. The SO_4^{--} ion has had 3 water molecules and PO_4^{---} 4 water molecules, while Cl^- 2 water molecules. The ionic radii were given for Na^+ as 0.102 nm, for K^+ 0.138 nm, for Mg^{++} 0.072 nm, for Ca^{++} 0.100 nm and for NH_4^+ 0.168 nm. The ionic radius of Cl^- was given as 0.181 nm, for SO_4^{--} as 0.230 nm and for PO_4^{---} as 0.238 nm. Recent approaches intensively use computer simulation of interactions among the charged ions and associated clusters of water molecules (Rips 1992). The modeling of water with clusters held by hydrogen bonds has been an intensively studied subject by itself (for example, see papers by Wales and Ohmine [1993] and by Saito and Ohmine [1994]), and it may be directly related to the ion hydration models involving water clusters. In small water clusters (number of water molecules about or less than 30), the hydrogen bonds network strongly controls the cluster dynamics.

The example of Cl⁻ ion hydration could be used to present current approaches to ion solvation. The polarizable and nonpolarizable potential models have been used for interactions between water clusters and chloride ion, and differences in approach have been pointed out (Jorgenson and Severence 1993; Perera and Berkowitz 1993; Stuart and Berne 1996). It has been suggested that for small water clusters the chloride ion Cl⁻, may be asymmetrically solvated by the water molecules and be located at the surface of water cluster. This point has been disputed, considering that polarization may only apply to ions with strong electric fields. Recently, Stuart and Berne (1996) attempted to resolve the dispute by modeling $Cl^-(H_2O)_n$ clusters with n up to 255, using polarizable and nonpolarizable potential models. The results showed that by using polarizable models, even for the large water clusters, the Cl⁻ ion is preferentially hydrated near the surface of the cluster. The nonpolarizable models have not shown this surface effect. The suggested interactions have been as follows: The determination of whether the ion is hydrated inside or on the surface of the cluster depends on balances of water-water and ion-water interactions. If the water-water forces dominate, the ion moves toward the cluster surface (the ion is unable to form hydrogen bonds). If the ion-water bonds are strong enough to compensate for the structure-breaking effects on the surrounding water, ions may remain inside the cluster. The ions may stay close to the surface for the small water clusters and move inside for the large clusters, although using a polarizable model resulted in the presence of Cl⁻ ion close to the cluster surface even for large water clusters (larger than 100–255 molecules). The polarizable model concept has been recommended for studying ion hydration. The gas phase electric dipole moment of a water molecule is 1.85 debye, whereas for ice the dipole is 2.6 D (debye unit) (Bader and Berne 1996). In liquid water, the molecule dipole moment is thought to be near 2.4 D, higher than in the gas phase due to the influence of surrounding dipole molecules. Cannon et al. (1994) studied the sulfate anion (SO_4^{--}) in water using the molecular mechanics modeling. The first hydration shell was found to contain 13 water molecules (the diffraction data were showed numbers between 6 and 14.3). Approximately 3 water molecules were coordinated to each sulfate oxygen. The sulfate anion bound tightly to 13 waters in the first hydration shell—the waters were oriented with one hydrogen pointing toward a sulfate oxygen and other hydrogen pointing toward surrounding water oxygens.

Further studies with various ions and water cluster sizes may determine thresholds of cluster sizes to accomplish intercluster hydration. Such studies would also shed light on the phenomena of ion incorporation into ice and ionic partition between the ice and liquid phases, cryoconcentration, ice nucleation in the presence of solutes, and electrical freezing potential.

Low temperature (below 0° C) aqueous liquid phases may promote covalent linking of oligonucleotides (Gryaznov and Letsinger 1993) or oligomerization of amino acids that at room temperature may not occur (Liu and Orgel 1997).

Oxidation reactions also may be affected by an increase in dissolved oxygen concentration in the liquid phase during freezing (concentration of oxygen in ice is lower than in the aqueous solution, and occurrence of gas bubbles in the liquid phase has been observed during freezing). Depending on the composition of solutes, the foregoing phenomena may not lead to any significant chemical reactions or precipitation of substances, but may still affect the cells present in solution due to the changing ionic strength, changing ion composition, changes in pH, and electrostatic effects.

Osmotic Effects during Freezing and Thawing of Cells

During the freezing process, solutes present in aqueous solution may be excluded from ice and concentrated in the remaining liquid phase. As a result, the cells present in the liquid phase will be exposed to concentrated solutions. Water from inside the cells will migrate via osmotic transport through the cell membrane (Csonka and Hanson 1991) into bulk liquid and may be subsequently frozen there.

The osmotic behavior of cells during freezing and thawing was reviewed by Levin (1988). Koch (1984) reported on the shrinkage of *E. coli* cells due to osmotic challenge. By an increase in concentration of external osmotic agents, the living cells could shrink to approximately 50 percent of their original volume. Morris et al. (1986) reported on the effects of osmotic stress on the ultrastructure and viability of the yeast *S. cerevisiae*. Cell shrinkage led to a loss in cell viability after being exposed to NaCl solutions of different strengths. The viability declined with decrease in cell volume and dropped to around 50 percent when cell volume decreased by half. The authors also noted significant changes in the cell wall after osmotic shrinkage. Most of these changes were reversible after returning to an isotonic solution. However, the rate of enzymatic digestion of the cell wall changed, indicating changes in the wall structure.

The concentration of solutes in the liquid phase during freezing may be evaluated if the phase diagrams are available (Cocks and Brower 1974; Shepard et al. 1976). Typically available phase diagrams may involve water and salt. Binary salt solution eutectic temperature may be lowered by the addition of cryoprotective agents, such as glycerol, and the formation of ternary systems. For example, the eutectic temperature of $NaCl/H_2O$ may be lowered from 252 K to about 193 K by the addition of glycerol. Many ternary phase diagrams for solute and cryoprotective agents are available in cryobiology literature (Daldrup and Schoenert 1988; Shalaev and Kanev 1994; Shalaev et al. 1996; Suzuki and Franks 1993). Such phase diagrams may be, however, unavailable for most of the complex biological solutions. Studholme and McGann (1996) described a prediction method for an estimate of the phase diagrams for simple compositions of solutes used in cryobiology. Results for aqueous solutions of DMSO, ethylene and propylene glycols, ethanol, methanol, glycerol, D-glucose, sucrose, 1- and 2-propanol, and acetone were reported. A set of equations for the phase diagram estimate was derived from the Gibbs' free energy equation, using changes in enthalpy and entropy during freezing. The required parameters were the molecular weights, molecular volumes, van der Waal's gas constant applied to liquids, and a solubility parameter δ. The predictions closely fitted the experimental data.

Cooling and Warming Rates during Freezing and Thawing

The temperature change rate plays an important role in the freezing and subsequent thawing of microorganisms. Malik (1987) reviewed protocols used for freezing and thawing microorganisms in the culture collections. For bacteria, yeast, and fungi, typical procedures recommend cooling rates from 1° to 10° C/min and a very rapid thawing. However, all of these procedures were applied in a small scale (vials, ampoules, straws, capillaries, etc.) and, therefore, were of limited use in large scale operations.

Varying rates of temperature change during freezing can cause ice crystals of different sizes to form within the cells. The behavior of water near the freezing point and phenomena related to water undercooling and nucleation and growth of ice crystals are of great importance in studies of cell freezing and thawing. The basic reviews on these topics are by Luyet (1966), Franks (1982), and Angell (1982). Mazur has suggested that the cell membrane can act as a barrier to ice nucleation and growth of ice crystals until the temperature reaches approximately –10° C. Below this temperature, intracellular ice would form (Mazur 1963). He analyzed the ice permeability of biological membranes based on a pore size and the corresponding data from freezing of supercooled water in capillaries and as droplets in the micron size range. He also investigated the kinetics of water loss from cells and intracellular freezing and proposed a mathematical approach for water loss calculation. Fahy (1981) simplified Mazur's equation for cell water content as a function of temperature. He compared the results to those obtained using Mazur's equation and found that the differences were negligible.

In a thorough review, Mazur (1966) discussed the freezing and thawing phenomena applied to *E. coli* and *S. cerevisiae* cells. The *E. coli* cells have had a good survival rate at a wide range of cooling rates. Only at very slow and very rapid cooling rates did the survival levels decline. The warming rate during thawing had a significant impact on survival of rapidly frozen *E. coli* cells. Slow warming caused a significant decline in survival rate. Mazur summarized the earlier work on freezing living cells in a review article (Mazur 1984). He reported results on cell survival for different types of cells, depending on the cooling and warming rates. At slow cooling rates, up to 90 percent of intracellular water can be removed by osmotic transport through the membrane. The remaining 10 percent may not freeze even after significant supercooling. Mazur pointed out the importance of ice recrystallization during slow warming and its damaging effects on cells. Levin et al. (1979) investigated the water permeability of yeast cells at sub-zero temperatures. They proposed a model for water flux from the yeast cells during freezing of extracellular water. The experimental results determining water remaining in the cells at different cooling rates were higher than from the predictions according to Mazur's approach. Experimental work was conducted in very small samples.

Schwartz and Diller (1983a, 1983b) investigated the osmotic response of cells during freezing. The organisms were yeasts that were exposed to varying cooling rates (from 9° to 82° C/min) and supercooling (from 1° to 36° C). Even at a low cooling rate of 9° C/min, the cells did not follow an equilibrium curve for normalized cell water volume versus temperature change. At higher cooling rates, the cells were found to be farther and farther away from an equilibrium; and at rates above 35° C/min, they seemed to lose only a small amount of water with a decrease in temperature and an increase in concentration of external solutes. Morris et al. (1988) investigated the freezing injury in *S. cerevisiae* dependent upon the growth conditions of cells. They used a special microscope to determine shrinkage of cells as well as formation of intracellular ice crystals at precisely controlled rates of temperature change. The cells were, in general, more viable after freezing and thawing if cultivated at higher temperatures. The maximum viability of cells cultivated at 24° C was with the rate of cooling of around 7° C/min. For cells cultivated at 37° C, the highest survival was at the rate of approximately 12° C/min. At the high rates of temperature change (200° C/min), small intracellular ice crystals were observed and the survival levels were very low. Intracellular ice could form at rates of temperature change higher than 10° C/min. At very slow changes in temperature during freezing (around 1–2° C/min), the survival rates were also low, and osmotic effects and cell shrinkage were observed.

Pearson et al. (1990) investigated the stability of genetically manipulated yeasts under different cryopreservation regimes. The authors also used glycerol as a cryoprotectant. The cell viability was found to be very sensitive to cooling rates, and the best results were obtained at the cooling rate of 1° C/min with a rapid decline in viability at cooling rates above 8° C/min. An addition of glycerol improved cell viability at the lower cooling rates, but did not help at the cooling rates higher than 8° C/min.

Zhao et al. (1988) reported on low-temperature preservation of strains of streptomyces and gentamyces. Several freezing procedures were tested. The authors found that cell viability was maintained in a wide range of cooling rates (from 0.6° C/min to 450° C/min). Takahashi and Williams (1983) reported results of investigations of the hemolysis of red cells caused by cold shock. They found that hemolysis increased with an increase in salt concentration and cooling rate.

The optimal cooling rate in a conventional cryopreservation procedure depends on the hydraulic permeability coefficient (Lp), the temperature dependence of Lp, and the characteristic temperature of intracellular ice formation as affected by cryoprotectants (Pitt et al. 1991). Pitt (1992) reviewed the thermodynamics of phenomena of intracellular ice formation. Linear cooling rate protocols were reviewed, and nonlinear cooling protocols were proposed. Cooling in bulk systems was discussed. (Works of Hayes et al. 1988 and Hayes and Diller 1988 were used by Pitt as a background for temperature histories of various large sample locations.)

Silvares et al. (1975) proposed a thermodynamic model for water transport from cells during freezing. The authors' approach was that at slow cooling, the mass transfer dominates, and at rapid cooling, the heat transfer is a dominant effect. Rapid supercooling of an external medium caused more water to remain in the cells with formation of intracellular ice crystals.

Cell Concentration Effects during Freezing

The cell suspensions from fermentation or cell culture may be concentrated prior to freezing using membrane filtration or centrifugation. Cell concentration levels in suspension may affect not only cell metabolism, including membrane transport (Janda and Kotyk 1985), but also phenomena associated with freezing. In general, an increase in cell concentration causes a decrease in transport phenomena, including transport across membranes. Kotyk (1987) showed the effects of yeast suspension density on the accumulation ratio of transported solutes. Transport phenomena significantly declined for yeast cell densities above 10 mg dry wt/mL. Levin et al. (1977) proposed a model for water transport in a cluster of closely packed erythrocytes at sub-zero temperatures. The cells near the cluster center lagged behind with their transmembrane water flux when compared to the cells on the outer cluster boundary. Nei (1981) discussed a mechanism of freezing injury to erythrocytes and reported on the effect of initial cell concentration on a post-thaw hemolysis. He tested various cell concentrations under different cooling rates in the presence and absence of glycerol. The cooling rates were at 1° C/min to final temperatures of –5°, –6°, –8°, and –10° C. In another experiment, the cooling rates were 10^4 °C/min, 10^3 °C/min, 10^2 °C/min, and 10° C/min, with the final temperature reaching –196° C. The slow cooling (1° C/min) gave increasing levels of hemolysis with the final temperature change from –5° to –10° C. There was a hemolysis increase with an increase in cell concentration. At the rapid cooling rate of 10^3 °C/min, the hemolysis levels were the most favorable. The hemolysis level strongly depended on cell concentration, with its value declining from 90 percent to 59.5 percent, when the concentration was reduced to one-fourth of its original level. After adding glycerol at levels of 5 percent, 10 percent, and 20 percent (v/v), the influence of cell density was most significant at 5 percent glycerol level, and at higher glycerol contents tended to decrease. The optimal cooling rate decreased from 10^3 °C/min without glycerol to 10^2 °C/min for 5 percent and 10 percent glycerol and to 10° C/min at 20 percent glycerol. Cells in dilute concentrations at the 20 percent glycerol level retained their shape regardless of cell concentration and were well covered by glycerol. At the cooling rate of 10^3 °C/min, there were some intracellular ice crystals formed. At higher cell concentrations, the cells were closely packed and touching each other. Such mutual contact was suggested as a cause for an increase in cell injury. That effect can be further enhanced by lowering the temperature and noting a resulting increase in membrane fragility. Addition of glycerol can reduce water removal from cells and can also provide a protective coating.

Pegg (1981) investigated a freezing injury to human erythrocytes and its dependence upon cell concentration. Packed cell volumes between 4 percent and 83 percent were studied. The cooling rate during freezing was 35° C/min, and the warming rate during thawing was 5° C/min. The results showed that cell hemolysis was increasing with an increase in cell concentration. A significant increase in cell hemolysis was observed after the cell packed volume level increased beyond 50 percent. Addition of 2.5 M glycerol did not change this pattern of cell damage, although the hemolysis levels were much lower than for nonprotected cells. The author suggested that the densely packed cells located between the ice crystals may be exposed to a different environment than the cells frozen at low concentration, i.e., the water loss and solutes distribution may be affected by the size of cell matrix, and as a result, the solute effects may be augmented and an intracellular freezing may occur.

Cell Damage by Mechanical Stress during Freezing

The mechanical properties of cell walls have been widely investigated (Thwaites and Mendelson 1991). Bacteria and yeast are relatively difficult to break with mechanical stress. During freezing and thawing, these properties may be affected by a low-temperature environment with changes resulting mostly from changes in the properties of lipids. Mazur et al. (1981) reported on relative contributions by a fraction of unfrozen water and salt concentration to the survival of slowly frozen human erythrocytes. The authors suggested that the cell membrane may become brittle below 0° C and that mechanical shear forces may be imposed on the cells located in channels between growing ice crystals. Wolfe and Steponkus (1983) proposed a model for tension in plasma membranes during osmotic contraction. They concluded that the tensions are small (no partitioning will occur) and that they are of a positive sign. Fujikawa and Miura (1986) investigated the plasma membrane structural changes caused by mechanical stress during the formation of extracellular ice as a primary cause of slow freezing injury of *Lyophyllum ulmarium*. The authors concluded that the dehydration and increase in ionic concentration were not the major cause of injury. The damage was done primarily by mechanical stress caused by the formation of extracellular ice. Komatsu et al. (1987) reported results on the membrane injury of yeast cells that were rapidly frozen using liquid nitrogen. Reported survival levels for *S. cerevisiae* were very low if slow thawing followed. This was in accordance with the theory of the damaging effects of intracellular ice, but the authors also decided to investigate possible damage done to the cell membrane. The electron microscopy investigations showed damage done to the cell membrane. Cell dehydration, which occurs during slow freezing due to osmotic phenomena, does not cause such damage. The authors suggested that intracellular ice crystals caused mechanical crushing of cell walls and nuclear membranes.

McGrath (1987) proposed a model describing thermoelastic stress in chilled biological membranes. The model included a cooling rate factor with membrane tension increasing with cooling rate increase. Thom and Matthes (1988) reported on deformation of cell membranes at low temperatures. During freezing, the cells remained between ice crystals and were exposed to a mechanical shear and bending in addition to chemical effects (due to cryoconcentration of solutes). Deformability of cell membranes during freezing can be an essential factor in withstanding the freezing process without a mechanical breakdown. Thom (1988) reported on a deformability of cell membranes using erythrocytes. At about

−20° C, the deformability of these cells reached zero, and the membrane became brittle. The author investigated phenomena of ice formation and deformability of cell membranes and their correlation. He found that the erythrocyte membrane cannot withstand any higher mechanical stress at temperatures below −10° C. Cortez-Maghelly and Bisch (1995) investigated the effects of ionic strength and outer surface charge on the mechanical stability of the erythrocyte membrane. At a constant ionic strength, reduction in the outer surface negative charge increases the membrane stability. Reduction in ionic strength also leads to increased membrane stability. Cells may experience an increase in ionic strength and pH changes in the surrounding medium during freezing, which may lead to reduction in membrane stability. Koerber (private commun. 1993) showed that the advancing extracellular ice crystals in the form of dendrites can impose mechanical stress on cells. The cells can be significantly compressed and deformed in an intercrystalline space.

The use of vitrifying cryoprotectants should reduce stress due to the properties of a cryoprotectant in a glassy or rubbery form and due to the presence of unfrozen water. Mechanical stress during freezing of isotonic solutions in large volumes was studied by Gao et al. (1995). The freezing cylindrical sample had a diameter of 96 mm. Addition of glycerol to the solution showed reduction in mechanical stress. Ice deformation under mechanical stress in the presence of a temperature gradient may cause movement of electric charges with the negative charge moving to the warmer portion of the ice specimen (Takahashi 1983).

Cell Damage Caused by the Freeze-Thaw Cycles

Mammalian cells

Farrant et al. (1977) investigated a two-step cooling procedure and its influence on cell survival following freezing and thawing. The authors froze and thawed Chinese hamster tissue culture cells. The applied thawing procedure was very rapid. The freezing procedures were as follows:

1. Freezing to −25° or −35° C and rapid cooling to −196° C
2. Freezing to −25° or −35° C, holding for 10 minutes, then rapid cooling to −196° C

Rapid cooling to –196° C produced cells with large ice cavities and no viability. Cells held at –35° C showed viability after rapid and slow thawing. Cells held at –25° C could not take slow thawing but were viable after rapid thawing. Cells held at –35° C lost so much water that there were no ice crystals present after cooling down to –196° C. Cells held at –25° C lost less water, and fine ice crystals formed after cooling to –196° C. Ice nuclei were absent while holding the cells at both temperatures (–25° C and –35° C). The ice nuclei formed below –80° C. Such multiple-step procedures may prove valuable during process development, not only for viable cell preservation but also if a selective and controlled processing—for example, inducing certain damage to the cells—is of interest.

McGann (1979) reported results of work on optimizing temperature ranges for control of the cooling rate during freezing of hamster fibroblasts. He used rapid warming and various cooling methods. At a cooling rate of 1° C/min, when the cooling process reached a temperature level of –60° C, cell lethality was high. However, when the slow cooling stopped at –20° C and then the sample was rapidly cooled to –196° C by immersing liquid nitrogen, the lethality level was much lower. These results suggest that use of two-step cooling with a slower cooling followed by a very rapid cooling may be beneficial for cell viability. Griffiths et al. (1979) reported results on freezing and thawing of Chinese hamster ovary (CHO) cells. The cooling rates were 1° and 200° C/min. The slow cooling caused cell shrinking. The rapid cooling caused formation of intracellular ice. Cells without intracellular ice were viable after thawing; cells completely filled with intracellular ice were dead; and cells partly filled with intracellular ice could be rescued by fast thawing (warming rates used for such a rapid thawing were 350° and 450° C/min). During slow cooling, a significant reduction in the cell size occurred in the temperature range between –7° and –12° C. Cryopreservation methods of mammalian cells prior to 1987 were summarized by Armitage (1987).

Kruuv et al. (1988) investigated survival of mammalian cells following multiple freeze-thaw cycles. The cells tested were Chinese hamster fibroblasts. The 2–3 logs of reduction in survival were reached by 3–4 freeze-thaw cycles, with survival rates lower at the rate of temperature change for freezing of 4° C/min than at 8° C/min. The rate of temperature change during thawing was much higher, equal to 485° C/min. The optimum cooling rate for maximum cell survival was found to be around 16° C/min. The same authors (Kruuv and Glofcheski 1990) reported later on an investigation of various cryoprotectants and their effects on cell survival. They investigated cooling rates within the range of 2°–16° C/min. The authors found that cell survival was exponentially dependent on the number of freeze-thaw cycles. The cryoprotectants tested were glycerol, propylene glycol, hydroxyethyl starch, dimethyl-sulfoxide, and glutamine. Shier (1988) investigated mechanisms of mammalian cell killing by a freeze-thaw cycle. He studied nucleated freezing and use of cryoprotectants as ways to prevent cell death. The author concluded that the effective mechanisms of cryoprotectants do not include vitrification phenomena around cells. Neither does the presence of sucrose protect cells by increasing the unfrozen fraction of mixture. He suggested that the solidification and liquefaction of lipid membranes may be responsible for cell death. Lipids may solidify at temperatures ranging between 0° and 12° C. If any ice crystals penetrate cell membranes during thawing, they could melt first and open holes in the membrane that remains solid (lipids may need higher temperatures to melt than ice). Uncontrolled penetration of cell contents or external solutes may occur through these holes in the membrane, killing the cell. Adding agents that lower the melting temperature of lipids and aid in reconstitution of the membrane may save the cell.

McGann et al. (1988) investigated mammalian cell damage after freezing and thawing. They proposed the concept that the osmotic dewatering of cells may negatively affect lysosomes during slow freezing and that the plasma membrane is the main site of injury during rapid freezing.

Singhri et al. (1996) investigated clarification of animal cell culture fluids using depth filtration. They used a suspension of fresh cells as well as suspensions that were freeze-thawed once or twice without cryoprotectants. Cell lysis and aggregation was observed after the freeze-thaw steps, resulting in rapid plugging of microfilters from the twice freeze-thawed suspension and to a lower degree for the once processed suspension. The filtration of fresh cells showed much slower rate of filter plugging.

Escherichia coli

Ray et al. (1976) reported results on cell wall damage due to freezing and thawing of *E. coli* cells. The cells, after a freeze-thaw cycle, can be dead or viable, but even if they survive they develop a very high sensitivity to many agents to which normal cells are resistant. This indicated that the outer cell membrane was damaged. The authors suggested that damage occurred to a liposaccharide (LPS) layer in the cell wall. The cooling and warming rates were about 1° C/min. Kempler and Ray (1978) reported results of investigations on a similar subject. They found that the LPS undergoes a conformational alteration, mostly during freezing, and the structural damage is of secondary importance. The *E. coli* cells were frozen at a rate of 100° C/min and thawed at a rate of 5° C/min. Nei et al. (1968) investigated freezing injury to aerated and nonaerated *E. coli* cells. The viability of nonaerated cells was lower than that of aerated ones, particularly with higher rates of cooling.

Yeast

Obuchi et al. (1990) investigated relationships between freezing-thawing viability of yeast and cell membrane fluidity. The authors used nuclear magnetic resonance (NMR) techniques to investigate membrane fluidity *in vivo*. Introduction of heat shock to yeast cells prior to freezing and thawing enhanced cell viability. The authors suggested that such an improvement may be a result of inhibition of membrane disruption. They proposed a mechanism of intensification of hydrophobic interaction of phospholipids by highly ordered water after the heat shock and suggested that those hydrophobic effects have a role in membrane protection.

Other Cells

Diettrich et al. (1987) investigated an influence of intracellular ice formation on the survival of *Digitalis lanata* cells. The authors found that for cooling rates greater than 4° C/min, the cells were seriously damaged and the borderline of cooling rate was quite defined. Microscopic observations showed formation of intracellular ice crystals between –20°and –25° C. During rewarming, a significant ice recrystallization (growth of ice crystals) could be observed outside and inside the cells. Gazeau et al. (1989) conducted observations of *Catharanthus roseus* cells during freezing and thawing using a cryomicroscope. Serious cell damage occurred after formation of intracellular ice crystals. During thawing of cells with intracellular ice, the authors observed formation of gas bubbles inside the cells. For osmotically shrunken cells, there were no intracellular ice crystals present and the membrane seemed to be impenetrable for external ice dendrites, despite mechanical cell deformation. Unprotected cells could develop internal ice crystals at temperatures as high as –6.8° to –8.7° C. Steponkus and Dowgert (1981) also reported on gas bubble formation inside cells during intracellular ice formation. Bryant et al. (1994) investigated the influence of gas bubbles on intracellular ice formation using differential scanning calorimetry (DSC) technique. No significant effect of gas bubbles on intracellular ice formation was found. Repeated cycles of freezing and thawing can produce declining viability of microorganisms, even if optimal conditions for cell survival have been chosen. Bryant (1995) reported on the effects of multiple freezing-thawing runs on intracellular ice formation.

Warming Rates and Thawing

A well-controlled thawing process may be as important to minimizing cold cell injury as the controlled freezing process. Warming rates during thawing may also affect the size of ice crystals due to a phenomenon of recrystallization. The rate of ice crystal growth and recrystallization is temperature dependent with the maximum recrystallization effect at about 260 K. Recrystallization decreases sharply between 260 K and 273 K on one side of this peak and gradually on the other side, e.g., between 260 K and 160 K. At about 140 K, the ice crystals do not grow. Since the ice crystal growth is possible within the temperature range between 260 K and 140 K, this temperature range should be rapidly passed during the warming process, particularly above 190 K. Close to the melting point in the presence of stress, the growth rate of recrystallization depends on the total strain of the deformed matrix (Ohtomo and Wakahama 1983). In live cells, the melting temperature is depressed, and the recrystallization temperature is increased to about 193 K (–80° C). In untreated biological samples, the recrystallization may not be significant if temperature is maintained at or below 180 to 190 K. In the case of cryoprotected cells, the melting temperature may be lowered (to about 263 K), and the recrystallization point could increase to about 230 K (Robarts and Sleyter 1985).

The recrystallization effects are the most significant when the ice crystals are small and formed during a rapid cooling process under nonequilibrium conditions. Such fine crystals are thermodynamically metastable due to their high surface energies. They tend to minimize their surface to form larger, more stable crystals (Taylor 1987). Such crystal transformation can occur when the frozen mass is being warmed. If the thawing rate is within a range of temperature change that promotes ice recrystallization in the cells, then a cold injury to cells may occur during the thawing process. Ice recrystallization may be hindered by use of antifreeze proteins (Carpenter and Hansen 1992). A typical thawing protocol for the products frozen in bags or bottles is to place the containers in a water bath a with temperature of 37° to 45° C. The containers may be moved in the bath, or the water may be agitated or circulated to improve heat transfer.

Freezing of Protein and Peptide Solutions

Freezing of protein and peptide aqueous solutions may affect the biomolecular structure in a reversible or irreversible way (Wisniewski and Wu 1992, 1996; Wisniewski 1998a). The known phenomenon of cold (low-temperature) protein denaturation is of thermodynamic nature and is, in general, reversible after temperature increases. The irreversible denaturation of proteins may be caused by the phenomena in the liquid phase associated with freezing, e.g., solute concentration, precipitation of buffer components, shifts in pH, and, in certain cases, interactions between the ice surface and protein molecule. In general, the proteins are excluded from the ice crystals and do not bind to them strongly. Hydrogen bonds may form, but since the surface of the protein molecule does not match the ice lattice, the probability of formation of a large number of hydrogen bonds is rather low, that is, typical proteins do not adsorb well to the ice surface. Exceptions are the antifreeze proteins (a small number known to date occur in the cold climate species) that could adsorb to ice (Jia et al. 1996). The cold denaturation of proteins may cause partial unfolding of the molecule and expose normally internal residues to interactions with solutes and the ice surface. The protein interactions with solutes depend naturally on the composition of solution. The final formulations may include relatively few ingredients with an overall composition designed to protect the product molecule activity. The raw or intermediate product may contain a variety of molecules; some of them may be detrimental to product activity under liquid storage and during the freezing processes requiring long periods of time. For example, if the liquid contains certain proteases, the product could be rapidly degraded if kept in the liquid phase or exposed to the liquid phase during long freezing processes. A well-controlled, rapid freezing process that can handle large volumes of product can prevent product degradation since it can rapidly transfer liquid into frozen solid and reach safe temperature levels. The presence of carbohydrates in glycoproteins may protect the proteins from the possible detrimental effects of ice surface-protein molecule and solute-protein molecule interactions. Very rapid freezing (for example, by sudden immersion of product droplets into liquid nitrogen) may produce fine ice crystals with very large crystal surface. A large portion of the product molecules might become exposed to the ice surface. Rapid freezing may also generate electrostatic charges that might adversely affect the protein molecules. The traditional interpretation of ice surface with exposed OH groups and feasibility of hydrogen bond formation has suggested that no large deleterious effects may be anticipated from interactions between the ice surface and hydrated proteins. Recent discoveries regarding the ice surface state (Delzeit et al 1997; Materer et al. 1997) may suggest that the interpretation of molecular phenomena at the ice surface needs to be modified. The strongly vibrational character of the atoms at the surface layer results in a marked disorder in comparison with the core of the ice crystal. It can be compared to a double layer of water molecules with varying cluster ring sizes. The number of dangling active groups is reduced in comparison to the situation wherein a regular lattice is assumed at the surface of the ice crystal.

Laboratory Equipment for Freezing

There are many commercial devices providing controlled freezing and thawing of cell suspensions and tissues on a small scale. They provide precise control over temperature levels, time of freezing at different levels, and temperature change rate. The temperature of the sample also can be monitored, if necessary, at several points. Small samples are typically processed with equipment that uses the application of liquid nitrogen. Cryopreservation systems based on mechanical refrigeration are also available. They offer repeatable cycles of cooling and warming with programmable cooling and warming rate profiles. The cooling and warming rate ranges as well as the lowest temperatures may be, however, limited depending on system design and sample size. Equipment using liquid nitrogen offers a wider range of cooling rates, and the lowest sample temperatures may reach the level of liquid nitrogen boiling. Scheiwe and Rau (1981) reviewed processes and equipment for cryobiology operations. Koebe et al. (1993) reported on the performance of rate-controlled freezing machines. Temperature uniformity within freezing chambers was investigated during various cryopreservation protocols with temperature decreases of 1°, 5°, and 10° C/min. The temperature measurements demonstrated considerable gradients within the freezing chamber, depending on the freezing rate and the probe location. The maximum gradients were between 10.8° and 12.6° C, and they occurred during the period of temperature change from 0° to −20° C and decreased at lower temperatures. This range is crucial for cryopreservation protocols, and small samples placed within the chamber may experience suboptimal conditions. The freezing chamber dimensions were 30 × 30 × 33 cm. Larger chambers may experience even wider temperature gradients, leading to poor cryopreservation results.

When liquid nitrogen is used to produce cold gas as a medium for heat transfer to cool samples, the evaporation process from the liquid nitrogen surface should be well controlled. Beduz and Scurlock (1994) reviewed evaporation mechanisms and instabilities in cryogenic liquids. Evaporation of the cryogenic liquid can be significantly changed (a transient rate increase) by agitating the liquid. This may increase the evaporation rate by an order of magnitude. The stationary cryogenic liquid may also generate a transient increase in evaporation. If the cryogenic liquid container is well insulated (no heat flux through the walls, e.g., no boiling there), the surface heat flux into the liquid causing evaporation is less than 100 $\frac{W}{m^2}$. During undisturbed evaporation, an equilibrium forms involving a thin, thermal conduction layer on the surface of the cryogenic liquid. There is a significant temperature gradient across this layer. Below, a thicker, intermittent convection layer forms with fluid recirculation patterns. When this equilibrium is disturbed, the evaporation rate can rapidly increase. Therefore, when the performance of the cooling equipment depends on the surface evaporation rate of liquid nitrogen, a certain level of liquid agitation should be maintained to stabilize the evaporation rate.

Cryobiology research has been greatly facilitated by use of various design cryomicroscopes. Papers by Cosman et al. (1989),

Table 2.1. Eutectic Temperatures and Concentrations of Aqueous Solutions of Salts.

Salt	Eutectic temp. (K)	Eutectic conc. (%)
NaCl	251.4	23.6
KCl	262.1	19.7
$MgCl_2$	239.6	21.6
$CaCl_2$	218.2	29.8
NH_4Cl	257.4	18.6
KNO_3	270.3	10.9
$MgSO_4$	269.3	19.0

Hayes and Stein (1989), Kochs et al. (1989) and Thom and Matthes (1988) presented examples of such designs. Their use may facilitate research on ice-solutes-cell interactions. The scanning electron microscope provides insight into intracellular phenomena and may aid in determining intracellular ice crystal formation.

Wolfingbarger et al. (1996) recently reviewed the engineering aspects of cryobiology, e.g., the principles of equipment design for freezing, storage, and thawing. Many important practical issues were discussed in this paper; however, the very large-scale problems were only partly addressed.

Storage Conditions

The storage temperature of frozen cells may also influence their subsequent viability. As a general rule, the storage temperature should be below the eutectic temperature of salts present in the solution. The data on eutectics of common salts are given in Table 2.1.

If glassy states are present, the storage temperature is selected below the frozen glass transition temperature T_g' (Table 2.6). Often the storage temperature is selected below the glass temperature at proteins T_g'', e.g. below $-80°$ C. When ice recrystallization is considered, the storage temperatures are usually below $-95°$ C (close to $-130°$ C the ice recrystallization effect disappears).

Frequently, for a long storage period, even the temperature of solid carbon dioxide (around −78.3° C) or of a typical mechanical freezer (from −70° to −86° C) may not be sufficient, and the temperature of liquid nitrogen (−196° C) is needed. The liquid nitrogen temperature is below the point of glass transition of pure water (near −140° C). At such a low temperature, the translational movement of molecules is so limited that the cells could be stored for very long times and strain stability may be preserved. Use of nitrogen vapor permits the storage temperature to be kept near −140° C. A new generation of mechanical freezers permits sample storage at this temperature level.

Ice Nucleation

In pure water, the water molecules form hydrogen-bonded clusters that may become nuclei for ice crystals if water is undercooled. The water trimer, tetramer, pentamer, and hexamer are the typical dominant structures among smaller clusters. The most stable hexamer cluster has a cagelike structure held together by 8 hydrogen bonds (Liu et al. 1996). This configuration possesses the lowest energy when compared to other possible hexamer structures. The cage structure calculated O-O bond length was approximately 2.85 Angstroms, close to the reported results. For the cyclic isomer of this structure, the calculated O-O bond distance was 2.76 Angstroms, very close to the reported value of 2.759 Angstroms for hexagonal ice (Ih) at 223 K. This finding suggests a possible role of such isomers of hexagonal water clusters in nucleation and growth of ice crystals. Cooling the clusters may cause random structural fluctuations, which, in turn, may create short-lived nuclei. Lowering temperature may cause the appearance of nuclei of critical size for crystal formation in multiple water clusters (Kinney et al. 1996).

Ice nucleation in pure water occurs spontaneously at an undercooling temperature within the range of −38° C (MacFarlane et al. 1992) to −44° C (Pitt 1992). The nucleation temperature depends on the size of droplet; for instance, at 1 micrometer diameter this temperature is about −39° C and increases with the droplet diameter (by approximately 2° C with tenfold increase in droplet diameter). Considering the size of cells and the temperature depression caused by solutes, the temperature range of ice nucleation could be close to −38° to −44° C. Practically, in the presence of any solid particles, the nucleation temperature is higher (an intracellular ice nucleation may begin at temperatures as high as –10° to –20° C for many cell types). Molecular structures matching the ice surface lattice may induce ice nucleation at temperatures only slightly below 0° C (Gavish et al. 1990).

Addition of cryoprotectants may significantly lower the intracellular ice nucleation temperature. Rall et al. (1983) reported a decrease in nucleation temperature from the range of –10° to –15° C to the range of –38° to –44° C after addition of 1.0–2.0 M glycerol and dimethyl sulfoxide.

Franks et al. (1983) presented results of work on ice nucleation in undercooled cells using undercooled emulsions. Cells could be undercooled to 242 K (–31° C) before freezing began. General information on freezing emulsions may be commonly found in food technology publications (for example, Keeney 1982). Clausse et al. (1983) reported on ice nucleation in aqueous emulsions. Dumas et al. (1994) proposed models for heat transfer during solidification and melting of emulsion droplets. These processes are not symmetrical—certain temperature hysteresis occurs due to the large undercooling during solidification associated with the small size of emulsion droplets. The ice crystal size depends upon the cooling rates. In general, the slower the cooling rate, the larger the ice crystal size. In large samples, rapid cooling rates are practically impossible to obtain. For small samples; even the rapid cooling rates of about 100 K/sec (6000 K/min) still may give ice crystals of 4–5 micrometers in size (Robarts and Sleyter 1985).

Membrane and Cell Models in Freezing and Thawing

Freezing and thawing of suspensions of liposomes may be considered a simplified model of animal cell freezing and thawing. Liposomes can be damaged by the ice crystals; to reduce the effects, cryoprotecting agents such as mannitol or glycerol may be added (Fransen et al. 1986). Cooling rates might also have an influence on liposome damage. Plum et al. (1988) reported a sudden increase in damage with cooling rates higher than 20°–30° C/min. Talsma et al. (1992) investigated freezing and thawing as a means of liposome preservation. The supercooled water entrapped in liposomes of size 0.2 micrometer and smaller was crystallizing at temperatures below –40° C. Storage of liposomes at temperatures of –25°, –50°, and –75° C showed that for small liposomes (below 0.2 micrometer) the storage at –25° C produced membrane damage due to osmotic forces since the water remained unfrozen within the vesicles. Kristiansen (1992) used a fluorescent marker to study leakage from liposomes under eutectic crystallization of NaCl and internal freezing. The damage to a membrane, as demonstrated by marker leakage, was low for the temperature range between −15° and −35° C and increased significantly in the range between −35° and −55° C. The increase in leakage was associated with two processes: eutectic crystallization of NaCl and internal freezing of undercooled solvent. The presence of uncrystallized eutectic above −35° C did not cause significant membrane damage even in the presence of extracellular ice crystals. The small addition of trehalose reduced damage caused by the eutectic crystallization. Biondi et al. (1992) investigated the permeability of lipid membranes during freeze-thaw processes. They used liposomes with membranes formulated close to biological type. The membrane damage could be induced by a loss of water from the membrane structure. Glycerol was used as a cryoprotectant. The authors concluded that the major barrier to permeation in lipid bilayers would not be the diffusion within the membrane but the transport across the aqueous lipid interface. Johansson et al. (1990) suggested that the addition of antifreeze solvent (glycerol) may stabilize biological membranes in the temperature range 0° to −10° C. The authors drew their conclusion from research work with partitioning in aqueous two-phase systems and synaptic membranes and not with live cells. Liposomes, however, cannot be considered adequate models of microorganisms due to differences in membrane structure.

THE FREEZING PROCESS: PHYSICOCHEMICAL PHENOMENA

Particle Entrapment by an Advancing Freezing Front

A subject related to the freezing of cell suspensions is the entrapment of solid particles by the moving boundary between solid and liquid phases. Bronstein et al. (1981) presented results showing the capture of cells by ice crystals during freezing of aqueous cell suspensions. The velocities of the freezing front were tested within the range of 0–14 μm/sec, and the velocity range for capturing of yeast cells varied from 7 to 14 μm/sec (0.007–0.014 mm/sec), depending on salt concentration in the solution (NaCl and NH$_4$Cl were tested).

Wilcox (1980) discussed the theoretical aspects of a force imposed on a single spherical particle by the freezing interface. Koerber et al. (1985) investigated interactions between particles and a moving ice-liquid interface. In their studies, the authors used a cryomicroscope. Using latex particles, they determined critical interface velocities for different particle radii. Examples of critical velocities were, for a particle with a radius of 4 micrometers, about 0.010 mm/sec and, with a radius of 6 micrometers about 0.006 mm/sec. Lipp et al. (1992) reviewed the interaction of particles and ice-water interfaces during bulk freezing. It was shown that the critical velocity is approximately proportional to $1/R_p$ (R_p is particle radius). Lipp et al. (1994) investigated the encapsulation of human erythrocytes in ice using a moving planar ice-liquid interface. In a solution with 0.85 percent NaCl and a temperature gradient of 15.3 K/mm, the cells were encapsulated at front velocities larger than 1.2 micrometers/sec (0.0012 mm/sec). This result was lower than in other works, in particular, the work by Spelt et al. (1982), in which the encapsulating velocity for erythrocytes was found to be about 88 micrometers/sec (0.088 mm/sec).

Nakamura and Okeda (1976) investigated the phenomenon of coagulation of particles in suspension by freezing-thawing. They found that coagulation can appear for fine particles but does not occur when freezing is sufficiently rapid, e.g., there is no temperature plateau. During slow freezing, the ice crystals could grow large and segregate as well as concentrate the suspended particles. At rapid freezing rates, the ice crystals were small, and the particles remained in their original locations.

These results suggest that, during freezing with controlled ice-liquid interface velocities, the microorganisms might be entrapped by a moving freezing front and the suspension would freeze in a uniform pattern. However, during the freezing of large volumes of suspension with large distances between the heat transfer surfaces, the freezing front velocity may be so low that the cells could be rejected and become concentrated in the liquid phase. Cell settling and local concentration may also occur during the slow freezing of volumetric samples (Wisniewski 1998b). The cell settling depends on cell size, density, and viscosity and density of liquid. These characteristics also depend on cultivation conditions. For example, yeasts such as *Saccharomyces cerevisiae* settle at different rates depending on nutrient composition; cells grown on saccharose and glucose as carbon/energy sources settle faster than cells grown on sucrose with molasses (brown sugar) or on potato dextrose broth (Wisniewski 1996, unpublished results).

Dendritic Crystal Growth

Controlled dendritic ice crystal growth could be a great aid in cell entrapment. After dendrites form and continue to grow, there is a high possibility that the cells may become entrapped in the inter-dendritic space. The freezing conditions can be controlled in such a way that there is a continuous dendritic growth (Wisniewski 1998a,b). Also, the interdendritic spacing could be controlled (i.e., this spacing should be large enough to easily accommodate all cells, even in the presence of branched dendrites). An effort should also be made to control the interdendritic spacing and the dendritic growth to minimize mechanical stress on cell membranes, if cell preservation is a goal. The same parameters could be manipulated differently—i.e., to cause controlled damage to the cell membrane when such damage is required for subsequent processing of cells. Such circumstances could be beneficial in downstream processing when the controlled freezing-thawing damage may facilitate subsequent extraction steps in product recovery.

Published results on dendritic crystal growth, including inter-dendritic spacing were frequently derived from experiments not related to cryobiology, but rather concerned with general solidification phenomena. The interdendritic spacing depends on temperature gradient and freezing front velocity (Somboonsuk et al. 1984). In general, a very rapid and a very slow freezing rate (velocities of the solid-liquid interface) may generate fine ice crystals. The dendritic ice crystal form at moderate freezing rates. The dendrites are crystals of pure or almost pure ice, and their heat conductivity is about 4 times higher than for the aqueous solution. The temperature along the center of a dendrite is lower than the temperature of the dendrite surface. The temperature gradient along the dendrite causes water freezing out from the interdendritic space and concentration of solutes and thickening at the dendrite's base. The heat conduction through the mushy zone is driven mostly by thermal conduction through the dendrites but also includes conduction through the liquid phase. If the thermal diffusivity of the mushy zone is greater than the mass diffusivity of the solutes, then the solutes are not rejected from the mushy zone. The thermal diffusivity of aqueous solutions of salts is almost 100 times higher than the mass diffusivities, and there-fore, the solidification process is dominated by heat conduction and the solute rejection should not occur from the mushy zone. Since there is a density increase in the liquid phase within the mushy zone, some buoyancy-driven flows are expected to occur within this zone. The freezing conditions could be manipulated to accomplish certain desired physicochemical and space/mechanical stress conditions in the cell environment.

During freezing, the extracellular liquid environment is losing water to the growing dendrites, which may cause solute concen-tration between the dendrites. At the same time, the extracellular solution is fed by water leaving the cells due to a process of osmotic equilibration. The growth of dendrites could be slow enough for the cells in the interdendritic space to have time to equilibrate osmotically with the surrounding liquid phase. As already described, cells in the interdendritic space are losing water by osmotic action, decreasing the concentration of solutes in that zone. This water is, however, rapidly frozen-out at the dendrite surface and as a result, the concentration of solutes in the inter-dendritic space increases. As soon as cells are entrapped in the interdendritic space, those phenomena may become localized.

The fundamentals of dendritic crystal growth can be found in papers by Langer (1984) and Glicksman (1989). The growth of a dendrite toward the undercooled liquid can be modeled using as the shape of a dendrite the circular paraboloid of revolution (Kurz and Fisher 1989). The growth velocity can be related to the under-cooling at the dendrite tip. The undercooling is the result of a combination of solute undercooling and thermal undercooling and is also dependent on the dendrite tip radius. The solute and thermal undercoolings are proportional to the square root of solid-liquid interface velocity, V. In the solute undercooling, the relationship between the undercooling and interface velocity also depends on the equilibrium distribution coefficient (solid-liquid concentration ratio), k_c and on the distribution of solute concen-tration in liquid phase at the dendritic tip. For aqueous solutions, solidification with solute rejection, the k_c value, is very small (can approach zero); at the practical undercoolings, the interface veloc-ity, V, is small compared to other solidifying systems (interface velocities in metals can be higher than the aqueous solution by orders of magnitude). If a higher interface velocity is required, conditions of high undercooling should be created and maintained. High rates of undercooling for water and aqueous solutions are frequently obtained using specimens of small dimensions. Studies of supercooled water contained in very small volumes showed that water can be supercooled to about −40° C before solidification begins (Angell 1982). In the freezing of large samples, high rates of undercooling are difficult to achieve, and in practice, one may expect the undercooling to be rather small. Additional undercooling occurs due to the dendrite tip curvature. An increase of undercooling is associated with an increase in front velocity and a decrease in the tip radius. The overall undercooling at the dendritic tip is usually composed of the thermal, curvature, and solute undercoolings.

The thermal Peclet number in the liquid can be described as

$$Pe_t = \frac{V \cdot R}{2\alpha}$$

where V is the growth velocity, R is the dendrite tip radius, and α is the thermal diffusivity of the liquid.

For the typical freezing conditions of aqueous solutions, the thermal Peclet numbers are very small (values much smaller than 1). The solidifying front velocity, V, depends on the thermal undercooling:

$$\Delta T_t = 2 \left(\frac{\Gamma \cdot h_m \cdot V}{\alpha \cdot C_p} \right)^{0.5}$$

The growth velocity, V, can be related to the dendrite tip undercooling (ΔT_t) through the Peclet number:

$$\Delta T_t = \left(\frac{h_m}{C_p} \right) Pe_t \cdot \left(E_1 \left(Pe_t \right) \cdot e^{Pe_t} \right)$$

where h_m is the latent heat of fusion, C_p is the specific heat capacity, Γ is the Gibbs-Thomson parameter, α is the heat diffusivity, and E_1 is an exponential integral function.

The group $Pe_t \cdot E_1(Pe_t) \cdot e^{Pe_t}$ is sometimes called the Ivantsev number.

For small thermal Peclet numbers ($0 < Pe_t < 1$), the integral function $E_1 (Pe_t)$ can be approximated by the series:

$$E_1(Pe_t) = a_0 + a_1 \cdot Pe_t + a_2 \cdot Pe_t^2 + a_3 \cdot Pe_t^3 + a_4 \cdot Pe_t^4 + a_5 \cdot Pe_t^5 - ln(Pe_t)$$

where $a_0=-0.57721566$; $a_1=0.99999193$; $a_2=-0.24991055$; $a_3=0.05519968$; $a_4=-0.00976004$; and $a_5=0.00107857$

These values are for the dendrite shaped as a paraboloid of revolution (Kurz and Fisher 1989). The solid-liquid front velocity, tip radius, and curvature undercooling are interrelated. The model of the dendrite as a paraboloid of revolution is a simplified approach since the coarse, developed dendrites may have an array of side branches. An example: In the aqueous solution with a low concentration of solutes at the front velocity of 40 mm/hour, using the paraboloid of revolution model will give the thermal Peclet number of about $7.6 \cdot 10^{-4}$, the integral function $E_1(Pe_t)=6.6$, and the estimated undercooling $\Delta T_t=0.40$ K. Thermal undercooling is a dominant factor in solidification of the pure substances, whereas in the case of solutions and alloys it becomes a fraction of overall undercooling, which is mostly determined by the solutal under-cooling. The thermal undercooling can be important in the case of growth of equiaxial, free-floating dendrites in the undercooled liquid since the heat flow occurs from the dendritic tip toward the undercooled liquid. In the constrained dendritic growth, e.g., columnar dendrites growing from the cooled wall into the liquid volume, the thermal factor may disappear (thermal undercooling of liquid ΔT_t is close to 0), and the undercooling would depend on solutal distribution (solute distribution between solid and liquid) and dendrite tip curvature (radius).

The local tip curvature undercooling ΔT_r at the dendritic tip (due to the tip curvature) depends on the tip radius, R, and occurs because of the Gibbs-Thomson effect:

$$\Delta T_t = 2 \cdot \frac{\Gamma}{R} = \frac{2\gamma \cdot T_m}{h_m \cdot R}$$

where Γ is the Gibbs-Thomson parameter (solid-liquid interface energy divided by entropy of fusion per unit volume), T_m is the melting temperature, h_m is the latent heat of fusion, and is the interfacial (liquid-solid) tension. When ΔT_r is equal to the difference between the temperature of undercooled liquid and the melting temperature, T_m, the tip radius R reaches the minimum value.

In the systems with solute distribution within the liquid phase close to the dendritic tip (increasing concentration from the liquid bulk toward the dendrite surface) and between the liquid and solid phase during solidification, there is an undercooling associated with the solute distribution. The characteristic solutal Peclet number is then defined as

$$Pe_c = \frac{V \cdot R}{2 \cdot D}$$

where D is the diffusion coefficient of solute in the liquid phase.

The solutal Peclet number can be related to the thermal Peclet number:

$$Pe_c = Pe_t \left(\frac{\alpha}{D} \right)$$

The solidifying front velocity, V, depends on solutal undercooling:

$$\Delta T_c = 2 \left(\frac{\Gamma \cdot k_c \cdot V(T_1 - T_S)}{D} \right)^{0.5}$$

The dendrite tip solutal undercooling (ΔT_c) may be estimated using the equation:

$$\Delta T_c = m \cdot C_p \cdot \left(1 - \frac{1}{1 - (1 - k_c) \cdot Pe_c \cdot E_1(Pe_c) \cdot e^{Pe_c}} \right)$$

using $m = \dfrac{(T_1 - \Delta T_c)}{\left(C_o - C_1' \right)}$ which is a slope of the liquidus line,

where k_c is the equilibrium distribution coefficient (solid/liquid concentration ratio), T_1 is the bulk liquid temperature, T_s is the solid temperature, ΔT_c is the solutal undercooling, D is the solute diffusion coefficient in liquid phase, C_o is the initial solute concentration in liquid phase, C_1' is the solute concentration in liquid at the surface of dendritic tip, and $E_1(Pe_c)$ is an exponential integral function.

The relationship shows not only the importance of the solute distribution solid-liquid (represented by k_c), but also the significance of solute distribution in the liquid at the dendritic tip (represented by the difference between C_1' and C_o).

For small solutal Peclet numbers ($0 < Pe_c < 1$), the integral function $E_1(Pe_c)$ can be approximated by the series:

$$E_1(Pe_c) = a_0 + a_1 \cdot Pe_c + a_2 \cdot Pe_c^2 + a_3 \cdot Pe_c^3 + a_4 \cdot Pe_c^4 + a_5 \cdot Pe_c^5 - ln(Pe_c)$$

where $a_0=-0.57721566$; $a_1=0.99999193$; $a_2=-0.24991055$;
$a_3=0.05519968$; $a_4=-0.00976004$; and $a_5=0.00107857$.

These relationships depend on the value of R, (radius of dendritic tip). In general, the dendrite tip radius is related to the concentration of solute in bulk liquid (C_0), e.g., R is proportional to $(C_0)^{-0.5}$. For small Peclet numbers, the general relationship for R is:

$$R = \frac{\dfrac{\Gamma}{\sigma}}{\dfrac{Pe_t \cdot h_m}{C_p} - 2\left[\dfrac{Pe_c \cdot m \cdot C_o \cdot (1-k_c)}{1-(1-k_c)\cdot\left(Pe_c \cdot E_1(Pe_c)\cdot e^{Pe_c}\right)}\right]}$$

where the stability parameter $\sigma \cdot = 0.025$, according to the marginal stability theory. In practice, the stability parameter for ice, $\sigma \cdot = 0.075$ (the approximate value [Koo et al. 1991]) and for other solidifying systems, $\sigma \cdot = 0.02$ to 0.04 (the range of values). Other symbols are as before.

For small Peclet numbers (less than 1), the following simplified relationship is sometimes used to estimate the front velocity:

$$V = \frac{\left(\Delta T_c \cdot k_c \cdot D \cdot Pe_c^{2}\right)}{\Gamma \cdot \pi^2}$$

where k_c is the equilibrium distribution coefficient (solid/liquid concentration ratio) and Γ is the Gibbs-Thomson parameter (solid-liquid interface energy divided by entropy of fusion per unit volume).

In the freezing of the aqueous solutions, the k_c coefficient is close to zero for salts and many solutes used in cryopreservation and may be larger than zero for solutes that can be incorporated into the ice lattice (see also the section "Use of Cryoprotectants"). Therefore, in freezing of the aqueous solutions, the major undercooling comes form the distribution of solute concentration in the liquid phase in the area adjacent to the dendritic tip. This distribution is described by the difference between the concentration of solute in bulk liquid (C_0) and the concentration of solute at the dendritic tip (C_i'). Availability of the phase diagram may aid in estimates of solutal undercooling and conditions within the mushy zone depending on solute concentration in the liquid phase (Huppert and Worster 1985).

If k_c approaches zero, then to reach high velocity, V, values, very large undercoolings, ΔT, are required. The undercooling caused by solute distribution in the liquid phase is typically small since the formulations may contain relatively low concentrations of solutes.

High values of undercooling are difficult to obtain under typical large-scale freezing conditions, e.g., small solutal undercooling may be anticipated after establishing quasi-equilibrium conditions.

Total undercooling at the dendritic tip is a summary of these partial undercoolings:

$$\Delta T = \Delta T_t + \Delta T_r + \Delta T_c$$

For pure materials, the undercooling consists of thermal and curvature parts:

$$\Delta T = \Delta T_t + \Delta T_r$$

In the practice of directional freezing of the aqueous solutions using the cooling-through-wall principle, the overall undercooling, ΔT, comprises mostly the ΔT_c and smaller ΔT_r parts. The released latent heat of fusion flows from the dindritic tip through the dendrite and solidified material toward the cooled wall. The tip of the dendrite moves through the liquid phase, and in front of it there is an elevated concentration of the solutes, since they cannot be incorporated into ice crystals and cannot diffuse away rapidly enough. This elevated concentration causes a temperature depression in the liquid near the tip when compared to the liquid bulk temperature. The dendritic tip temperature is thus lower than the bulk liquid temperature. Temperature further decreases along the stem of the dendrite.

During volumetric freezing, for example, using the cryogenic fluids (droplets or particles of cryogenic fluid falling into an agitated volume of product), the dendritic growth approaches the equiaxial dendritic crystal growth pattern, that is, with the free crystals floating in undercooled liquid. For the aqueous solution freezing, such as equiaxial crystal growth, the overall undercooling may comprise mostly the ΔT_c and ΔT_t parts with some contribution from ΔT_r. The ΔT_c part may be reduced if there is a vigorous agitation of the crystal-liquid slurry, e.g., the solutal concentration gradient is lowered by mixing in the liquid phase. Intensive mixing can also break away parts of dendritic branches and deform the shape of growing dendrites. Since the heat flow (generated by solidification) is now from the dendritic tip toward the liquid phase, thermal undercooling of the liquid phase must be maintained (Wisniewski 1998b).

More detailed analysis of the conditions near the dendritic tip, its radius, and growth rate can be found in works by Glicksman (1989), Kurz and Fisher (1989), and Langer (1984). Recently, Mullis (1995, 1996) proposed models for growth of nonisothermal parabolic dendrites. Hunt and Lu (1993) proposed a numerical model of cellular and dendritic crystal growth that can predict cellular and dendritic spacings, undercoolings, and transitions of shapes and structures. The cellular crystal structures may form at high and low front velocities, while dendrites form at intermittent velocities. The cellular crystal structures are finger-type with smooth walls. They are of cylindrical shape with a tip approximated as a hemisphere, whereas the dendrites are of a more paraboloidal shape with a smooth tip and side branches along the dendrite. Lower undercoolings are required in the case of dendrite formation than for the cellular crystals. Lower velocities produce wider interdendritic spacing. Schievenbusch et al. (1993) compared techniques used to determine the dendritic spacing.

Ishiguro and Rubinsky (1994) investigated mechanical interactions between the ice crystals and red blood cells during directional solidification. The red blood cells in physiological saline were damaged during freezing with dendritic growth of ice crystals. Addition of 20 percent (v/v) of glycerol protected the cells. Addition of antifreeze proteins apparently changed the ice crystal pattern to a degree that cells were damaged even in the presence of glycerol.

One may anticipate that if an original concentration of solutes in an extracellular medium is low, then the dendrites might occupy a large volume, due to the availability of water, and the cells might suffer more mechanical damage from the dendritic pressure. The eutectic or glassy state volumes between the dendrites might be smaller, and not all of the cells may get exposed to these solutions in a significant way. The cells might also retain more intracellular water due to less intensive osmotic transport. An increase in an original concentration of solutes in extracellular medium might make these interdendritic phenomena more pronounced that is, the eutectic or glassy state volumes would be larger, and the cells could get more exposure to the concentrated solution (the cells would be exposed to concentrated solutes in a liquid phase between dendrites before complete solidification). The cells could be exposed to osmotic transport, and they would occupy reduced volume due to water release and cell shrinkage. There may be no ice crystals present in the interdendritic zones. The cell concentration may not be too high in order to allow for cells to be freely immobilized in an amorphous phase in the interdendritic space. As indicated before, the cells should have sufficient time to attain an osmotic equilibrium with the surrounding liquid. That can be a reason for the cooling rate to be well controlled. Such control of the cooling rate will also control dendritic growth. Therefore, a very careful analysis of the freezing phenomena should be performed prior to a final process design.

A mass balance in the system may be used to estimate: water volume in an extracellular medium, solid mass and water contained in the cells, and anticipated mass of dendrites versus mass of eutectic and amorphous zones in the interdendritic space. The cell concentration could then be selected to ensure that the cells are easily contained in the interdendritic space, and their osmotic transport would not be hindered. The temperature and concentration of solutes in amorphous solidification are parameters to be considered in the process optimization procedure. A subsequent step in process optimization will be the selection of process parameters to control dendritc growth (with the cooling rate as a major factor). When rapid agitation is applied to the liquid phase, it impedes dendritic growth, minimizes cell occlusion, and may be utilized to concentrate the cells in suspension. Flow of liquid, in general, inhibits the solidification process (Chiang and Kleinstreuer 1989; Yoo 1991). At very low freezing front velocities, the interdendritic space diminishes, and the solidifying front resembles a planar one (Kurz and Fisher 1989). This phenomenon might enhance cell rejection. Additional information on the dendritic growth role in freezing and thawing of aqueous protein solutions can be found in works by Wisniewski and Wu (1992, 1996; and Wisniewski 1998a).

The scale-up process involves interpretation of very small-scale results and, in particular, interpretation of the cooling rates obtained on a small scale and recommended for cell preservation. An overall treatment of the large sample volume as a whole may produce changing cooling rates, depending on the position within the sample (Diller 1992). Controlled dendritic crystal growth patterns may create conditions of localized cooling rates for cells in a liquid environment with changing concentrations of solutes and temperature gradients.

Figure 2.4. shows an example of dendritic ice crystal growth, with the free columnar dendrites protuding into the solution and the interdendritic spaces filled with glassy substances or eutectics containing embedded cells. For the sake of simplicity, the side branches of dendrites are not shown. Due to the increasing concentration of solutes between the dendritic tip and its base (where the glass or eutectic temperature is reached), the cells are exposed to changing osmotic conditions and may follow the osmotic equilibrium during this transitional period. An optimized freezing protocol may require

Figure 2.4. Dendritic ice crystal growth with the cells trapped between dendrites in liquid solution and embedded in solidified eutectic and glassy matter. T_g is the glass transition temperature, T_e is the eutectic temperature, T_c is the coolant temperature, T_w is the wall temperature, T_t is the temperature of the dendritic tip, T_l is the temperature of the liquid bulk, C_l is the solute concentration in liquid, C_g is the solute concentration in glassy state, C_e is the solute concentration at eutectic point, L_d is the length of dendrite, V is the moving front velocity, G is the temperature gradient along the dendrites, and x is the interdendritic spacing.

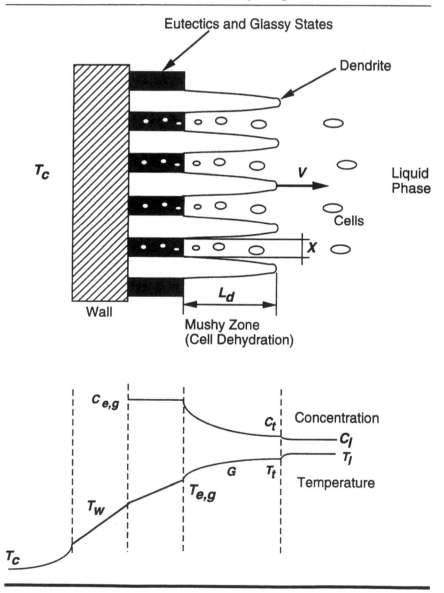

knowledge of the glass temperature, T_g, or eutectic temperature, T_e (e.g., the nature of solutes), solutes concentration (the original C_l in liquid and in glassy state C_g or at eutectic C_e), length of dendrites, L_d, moving front velocity, V, and temperature gradient along the dendrites, G. The temperature of dendritic tip, T_t, is lower than the temperature of liquid bulk T_l (an undercooling required for dendrite to grow). The growth velocity of the dendritic tip depends on the undercooling conditions at the dendritic tip. These undercooling conditions depend in general on the heat removal across the mushy zone (along the dendrites and interdendritic spaces toward the solidified layer and cooling wall) and the temperature gradients there.

The temperature gradient along the columnar dendrites is

$$G = \frac{\left(T_t - T_{e,g}\right)}{L_d}$$

The cooling rate is

$$CR = G \cdot V$$

Cell residence time under conditions of temperature drop and solute concentration increase is

$$t = \frac{L_d}{V}$$

The temperature gradient along a dendrite depends on the $T_{e,g}$ temperature, the T_t temperature and on the free dendrite length, L_d; the dendrite length may depend on front velocity, V (heat flux Q), and also on initial solutes concentration, C_p in the liquid and the concentration at which these solutes solidify, $C_{e,g}$. The cooling rate, CR, which is important for cell survival, depends on the temperature gradient and front velocity (related to heat flux). Relationships between these parameters lead to optimizing the dendritic growth for cell survival or preserving activity of the biological product. The moving front velocity, V, can be controlled by process design, whereas $T_{e,g}$ and C_l and $C_{e,g}$ depend on the original composition of the cell-carrying solution or broth or the formulation of the biopharmaceutical solution. The dendrite growth shown in Figure 2.4 is also important in cryopreservation of protein solutions (Wisniewski 1998a).

An Example

At the solid front velocity of 0.01 mm/sec and the free dendrite length of 0.5 mm, the average residence time of cells between the original suspension state and the dehydrated state embedded into the glassy substance is 50 sec. At the glass temperature of −36° C, liquid temperature of −0.5° C, tip dendrite temperature of −0.9° C, the temperature gradient along the dendrite is 70.2° C/mm and the cooling rate is 0.702° C/sec or 42.1° C/min. If the dendritic growth pattern requires lower temperature gradients than those mentioned, then the medium composition could include solutes with higher eutectic or glass temperatures. For example, if the solidification temperature of solutes is −21° C, then the temperature gradient will be 40.2° C/mm, and the cooling rate will be 0.402° C/sec (24.1° C/min). Further, changing the velocity of the freezing front to 0.005mm/sec would decrease the cooling rate to about 0.201° C/sec (12.0° C/min) and increase the cell residence time between dendrites to 100 sec. Changes in front velocity are associated with changes in dendrite tip radius, liquid undercooling, and interdendritic spacing. All of these variables should be considered in a final analysis.

Cell residence time under changing osmotic conditions might create cell dehydration conditions close to an osmotic equilibrium and permit osmotic cell dehydration under decreasing temperature without the danger of intracellular ice formation. The exposure times are short, and potential damage to cells—due to an exposure to highly concentrated solutes and to potential pH shifts—may be minimized. Dendrite growth (front movement) with approximately stationary cell position gradually exposes cells to increasing concentration and decreasing temperature. Cells are osmotically dehydrated during this process, and with loss of water, they may become less prone to low-temperature-related intracellular ice crystal formation. In an optimized process, the cell might closely follow osmotic equilibrium and after dehydration will be embedded into the solidified glassy or eutectic matter. Dehydration during that transitional stage may be sufficient to permit further temperature decrease in the solidified mass without reaching conditions of any significant intracellular ice crystal formation. Depending on the solution composition, the cooling conditions, and whether horizontal or vertical solid-liquid interface movement takes place, the free dendrites may reach greater lengths than shown in the example (reaching dimensions ranging from several to tens of millimeters). During the large-scale freezing operations (including the simulating runs in the research cartridges), a variety of dendritic lengths are observed—from 200–500 micrometers to about 6–7 mm—depending on freezing conditions and the liquid composition. In the case of the longer dendrites, the temperature gradients along the dendrite could be about 5–6° C/mm for the solidification temperature of –36° C and about 2.9–3.5° C/mm for the solidification temperature of –21° C. The corresponding cooling rates could be 3–3.6° C/min. and 1.7–2.1° C/min. These cooling rates are close to the favorable cooling rates established in very small-scale experiments on the cell survival dependence on the cooling rate. The embedded cells will still be exposed to cooling due to changing temperature within the frozen mass. If the initial osmotic dehydration was insufficient, there is a possibility of intracellular ice crystal formation. Therefore, thermal histories of cells in the frozen state are important and should be considered together with phenomena occurring in the mushy zone prior to freezing (solidification).

During freezing in large volumes cell settling may occur. This may be minimized by freezing cells in relatively thin horizontal layers (cooled from the bottom) with the dendritic ice crystal growth acting in opposite direction to cell settling. However, since many variables are involved (type of the cell, membrane permeability, solution composition, cell cooling rate, parameters of dendritic crystal growth) cells may not experience conditions optimal for preservation of cell viability and integrity, while in the mushy zone. Depending on these variables cells may or may not closely follow the osmotic equilibrium while exposed to changing environment in the mushy zone, where there is a simultaneous change in temperature (cooling rate), concentration of solutes (osmotic conditions) and cell concentration.

The large-scale freezing of biological solutions can also be accomplished under conditions promoting crystal growth in the form of columnar fingerlike cells. The columnar crystal cells do not have side branches as in the case of fully developed dendrites. Such a growth pattern may be beneficial if the protein product is adversely affected by the ice-liquid interface (e.g., by interactions of protein with ice). The ice-liquid interface area for columnar cells can be much smaller than for the fully developed, side-branched dendrites, and the potential damage to protein can be significantly reduced. For microorganisms and animal cells, the interface factor is also of concern, but in addition to the surface area, the spacing among columnar crystal cells or among branched dendrites can be a very important factor in cell protection. The previously discussed model of dendrite (one of paraboloid of revolution) may be applied as well to the columnar cell, but a hemispherical needle crystal shape model provides better approximation.

The spacing among the columnar cells is usually smaller than among the dendrites. The interdentritic spacing depends on the cooling rate and can be related to the solid-liquid interface velocity, that is, the spacing decreases as the velocity increases. The presence of solutes may alter this relationship. For example, in the presence of sucrose, the spacing could change from about 40 micrometers at a front velocity of 10 micrometers/sec to about 15–20 micrometers at a front velocity of 100 micrometers/sec at 5 percent of sucrose, whereas at 35 percent of sucrose, the spacing at those velocities would be 120 micrometers and 15–20 micrometers respectively (Reid 1983). The interdendritic spacing depends on the temperature gradient along the dendrite, the temperature difference across the mushy zone, and the dendritic tip radius. This relationship could also be described as dependency upon the temperature gradient and front velocity and finally simplified to a relationship between the interdendritic spacing and cooling rate, that is, spacing decreases as the cooling rate increases. Rapid cooling (and high front velocity) produces finer dendrites or columnar cells (and a large number of these crystals per unit of area)—the large ice-liquid interface and thin layers of solidified solutes (eutectics, glassy states) in the interdendritic spaces. These two factors (the large ice-liquid interface and a very thin layer of solidified solutes) may be detrimental to cell survival as well as to preservation of protein activity. Therefore, the conditions of dendritic or columnar cell crystal growth should ensure a certain coarseness of crystal formation.

The relationship for approximate estimate of the interdendritic spacing (λ) can be described:

$$\lambda = \left(\frac{3 \cdot \Delta T' \cdot R}{G} \right)^{0.5}$$

where $\Delta T'$ is the temperature difference in the mushy zone (difference between the dendritic tip and eutectic or glassy state temperature), R is the dendritic tip radius, and G is the mean temperature gradient in the mushy zone.

This relationship, simplified for aqueous solutions (for k_c close to 0 and for the growth rates in the range for fully developed dendrites), can be described in a different form:

$$\lambda = \frac{4.34 \cdot \left(\Delta T' \cdot D \cdot \Gamma\right)^{0.25}}{\left(V \cdot m \cdot C_o\right)^{0.25} \cdot G^{0.5}}$$

where the symbols are as described earlier.

The interdendritic space estimates are for the early stages of dendrite formation by growth of perturbations from the flat solid-liquid interface. Later during the process, some dendrites may disappear and others may grow larger, with an overall number of dendrites decreasing. After the initial transition period, the number of dendrites may remain stable. After the dendritic array becomes established, a moderate change in V may not cause any significant change in spacing.

Controlled dendritic crystal growth within the whole volume may provide uniform localized cooling rates across the sample for cells in an aqueous environment at changing osmotic and temperature conditions. The conditions for controlled dendritic crystal growth must be followed, considering the increase in thickness of frozen mass and the increase in conduction heat resistance.

Figure 2.5 shows an approach to controlled dendritic crystal growth. Essential conditions for the controlled dendritic crystal growth are the freezing front velocity, the directional heat flux, and the temperature gradient along dendrites. Since the T_1 and T_e, or T_g, temperatures are not changing, the temperature gradient along the free dendrite remains steady and only the freezing velocity, V, can be variable. The heat conduction through the solidified layer may be described as

$$Q = \frac{k \cdot \left(T_{e,g} - T_w\right)}{s}$$

where s is the thickness of a solidified layer, and k is the thermal conductivity of the solidified material.

Under established cooling conditions, the freezing front velocity will decline with an increase of the frozen layer thickness, s. To maintain a constant freezing front velocity, V, the heat flux, Q, should remain constant. Since the frozen layer thickness increases and the temperatures T_e and T_g are fixed, a constant heat flux can only be maintained when the cooling wall temperature T_w is decreased.

Figure 2.5. Control of the dendritic crystal growth by maintaining constant conditions of crystal growth with the frozen layer thickness (*s*) increase. The constant heat flux, *Q*, and the constant solid-liquid interface velocity, *V* are maintained by the cooling temperature decrease following the frozen layer thickness increase. The temperature gradient slope across the frozen layer remains constant. T_e is the eutectic temperature, T_g is the solidified glassy state temperature, T_w is the wall temperature.

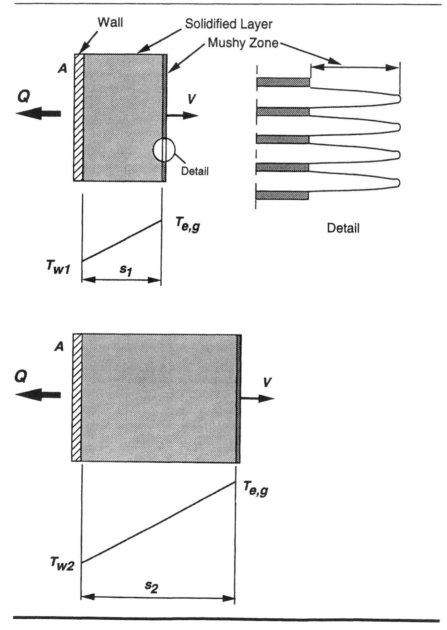

$$Q = \frac{k \cdot \left(T_{e,g} - T_{w1}\right)}{s_1} = \frac{k \cdot \left(T_{e,g} - T_{w2}\right)}{s_2} = const,$$

where $s_2 > s_1$, $T_{w2} < T_{w1}$, and then V = const.

The decrease of wall temperature should be proportional to the increase of frozen layer thickness. A slope of the temperature gradient should be maintained.

A more accurate solution may consider a change in the thermal conductivity of the frozen layer with temperature. A formula for ice thermal conductivity at changing temperature was proposed by Fukusako (1990):

$$k = 1.16 \cdot (1.91 - 8.66 \cdot 10^{-3} \cdot t + 2.97 \cdot 10^{-5} \cdot t^2)$$

expressed with $\dfrac{W}{m \cdot {}^\circ C}$, where t is the temperature in $^\circ$ C.

The nonlinearity of thermal conduction change makes such an estimate complex and requires the position of the freezing front to be tracked. Therefore, for the first approximation, a linear model could be used, in which the slope of the temperature gradient across the solidified layer would be maintained. The ice thermal conductivities for -5° and -30° C are respectively 2.267 and 2.548 $\dfrac{W}{m \cdot {}^\circ C}$. Thermal conductivity of water near 0° C is 0.594 $\dfrac{W}{m \cdot {}^\circ C}$. Thermal conductivities of salt solutions decrease slightly with an increase of salt concentration (as it happens in interdendritic space).

Freezing equipment performance characteristics should provide the feature of changing the coolant temperature according to an increase in the frozen layer thickness. The temperature of gas in blast freezers, liquid in bath, or cooling fluid in contact freezers can be easily controlled to permit optimization of freezing protocols.

Dendritic Crystal Pattern and Freezing of Protein Solutions

Controlled dendritic growth is also important for protein solutions (Wisniewski and Wu 1992, 1996, Wisniewski, 1998a), and the residence time of proteins under potentially adverse conditions of increasing concentration of solutes between the dendrite tip and its base can be estimated as in the example above (page 55–56). Typical residence times of proteins in the mushy zone during bulk freezing may vary from about 30 sec. to a few minutes. Longer times might be experienced if suboptimal freezing conditions are applied. The proteins in the interdentritic spaces may face increased concentration of solutes (including proteins themselves), decreasing temperature below 0° C level, precipitation of solutes, and pH change. The increased concentration of proteins may cause their aggregation and precipitation; the temperature decrease below 0° C in the liquid phase may cause cold denaturation (unfolding) of proteins; the increase in concentration of solutes may destabilize or stabilize the proteins, depending on the nature of the solutes; and the precipitation of solutes and pH change may have a destabilizing effect upon proteins. The initial concentration of protein and other solutes may also influence protein stability in the interdendritic spaces. The interactions with solutes and the ice surface may change along the mushy zone; that is, the decreasing temperature in the interdendritic space in liquid phase may cause cold denaturation of protein and its partial unfolding, exposing many residues to solutes and to contact with ice surfaces. Partial unfolding of protein molecules can cause their aggregation and even precipitation. Exposure of partially unfolded proteins to concentrated solutes may cause a variety of effects, depending on the nature of these solutes. In general, a desirable situation will occur if the proteins can be embedded in the glassy state or solidified with the eutectic state before any serious deleterious effects can take place. The composition of the formulation may be very critical for marginally stable proteins. As for the cells, high eutectic and glass temperatures may be beneficial since they not only affect the storage temperature, decrease cold unfolding of proteins but also can minimize the exposure time to highly concentrated solutes.

The freezing process can be performed to minimize the residence time of molecules in the liquid phase within the mushy zone (freezing conditions should produce a rapidly moving front of short columnar dendrites or cells). The examples of typical residence times in the liquid phase within the mushy zone were demonstrated earlier. In the case of detrimental effects of protein molecule-ice surface interactions, the protein initial concentration and other solutes' character and concentration may influence the protein-ice contact frequency and associated damage. The process of crystal growth itself (thermal conditions in relation to solutes' behavior during freezing) may be conducted in a way that minimizes the ice-protein contact surface area. In such a case, the dendritic crystal growth should be performed to generate the coarse columnar cells with smooth sides, not the dendrites with fully developed side branching (such a branching may create very large solid-liquid contact interface). As described earlier, the pattern of columnar dendritic or cellular growth may be controlled by the thermal environment, but the spacing and crystal shape would also depend on involved solutes, their concentration, and the possible incorporation of solutes into the crystal lattice. During front solidification, there is a release of gas bubbles from the liquid phase. This phenomenon might contribute to detrimental changes in protein structure if the protein is sensitive to the conditions at a gas-liquid interface.

In the equiaxial dendritic crystal growth (freely suspended ice crystals in undercooled liquid), the thermal conditions promoting the continuous crystal growth involve maintenance of sufficient liquid phase undercooling. This undercooling in large-scale operations can be achieved by volumetric cooling of the whole volume—for example, by spraying the cryogenic fluids such as liquified nitrogen or liquified carbon dioxide (with the corresponding transition to the solid phase). The process of such a volumetric cooling with the subsequent formation of frozen granules is described by the author in Chapter 3, page 209, in this volume. In the process, water is volumetrically frozen out at the temperature plateau near 0° C, followed by a rapid formation of frozen granules. The liquid water is being removed by a continuous nucleation of new ice crystals, growth of the formed ice crystals, and maintenance of the liquid phase undercooling. The solutes (including proteins) in the liquid phase are being continuously concentrated due to that water removal process. Since the liquid phase stays at minimum undercooling (close to 0° C) during crystal formation and growth, the protein cold denaturation is limited and may not be as detrimental as in the case of columnar dendritic crystal growth for freezing by cooling through the wall. If there is a detrimental effect of ice surface-protein interaction, protective agents may be required since a large number of ice crystals form during the temperature plateau phase. There is a continuous increase in the concentration of proteins and solutes, and this may be detrimental in certain cases, particularly considering that the whole product mass is agitated and a shear factor is involved (within the fluid and among the ice crystals). The process of freezing and granulation has been designed with agitators shaped to minimize the shear effects within the ice crystal-liquid slurry (Wisniewski, 1998b).

Relatively little is known about the protein-ice interactions (also see Wisniewski, Wu, 1996; Wisniewski, 1998a). Most of the results of such interactions come from the studies of antifreeze proteins that are a very small group among the proteins. A study of the dynamic and binding of antifreeze protein to ice conducted using the molecular modeling approach (Madura et al., 1996) may serve as an example. The close interactions may occur via hydrogen bonding if the water molecules between ice and protein can be replaced and hydrogen bonds formed between the protein residues and ice lattice. At the present time such detailed interactions between the proteins of therapeutic importance and ice surfaces are generally unknown and will require further studies. It can be anticipated that the majority of proteins may not absorb well to the ice surface. The interactions between ice and proteins may disturb the protein structure at the stage of freezing when the hydration shells of the proteins are affected by the ice crystal growth. The presence of other solutes may interfere in these interactions (protect or destabilize the proteins).

Dendritic Crystal Growth and Cooling Rates in Cell Preservation

The cooling rates and corresponding cell survival rates as reported in the cryobiology literature may also be related to the character of dendritic crystal growth. At the slow cooling and corresponding low solid-liquid interface velocities, a solid front or a cellular pattern with almost flat interface forms and the solutes and cells may be excluded from the solidified mass. This can cause a nonuniform solute and cell distribution across the whole frozen mass; that is, cryoconcentration effects may occur and cells can be exposed to elevated concentrations during extended periods of time. At moderate cooling and corresponding solid-liquid front velocities, the fully developed dendrites can grow. In many cases, a coarse dendritic pattern is preferred since it may reduce the surface area of ice-solute interaction. Side branching of dendrites (an increase in ice crystal surface) may not be desirable, and the dendritic growth should be controlled to minimize such a side branching. Freezing conditions with side branch formation may require an increase in concentration of cryoprotectants or other neutral solutes in the original liquid product. At rapid cooling and high front velocities, the ice crystals become small and may develop in a cellular form with an almost flat solid-liquid interface. In this case, the solutes and cells can be excluded from the solidified mass and concentration in the liquid.

At very fast cooling and at volumetric uniform cooling, an equiaxed dendritic crystal formation may occur, with the solutes and cells distributed among a large number of fine crystals. In this case, a very large ice-solution surface forms that may cause detrimental effects in biological solutions and cells due to an extensive solute or cell surface interaction with ice. These fine ice crystals can recrystallize during thawing (within a certain temperature range) and cause mechanical stress and damage to the cells. These ice crystal patterns may play an important role in cell survival and solute distribution during freezing at different cooling rates. These crystal patterns may also aid in explaining cell survival at various cooling rates and solid-liquid front velocities.

Controlled coarse dendritic crystal growth may not be easily accomplished during granule drop-freezing using liquefied gases or in cases of spray cooling or drum or metal belt freezing due to a very rapid solidification. The dendritic growth requires directional heat flux combined with a relatively moderate rate of heat removal. The conditions for freezing cell suspensions with a controlled dendritic growth may exist in thin layers cooled from below with the dendritic crystal growth counterdirected to cell settling. Maintaining optimal conditions for simultaneous cell dehydration and cooling rate, while cell resides within the mushy zone may be difficult to accomplish during freezing of large volumes of cell suspensions, particularly if the process shall ensure appropriate cooling rate of cells embedded into the frozen material. Further cooling after solidification may be conducted in steps, e.g., lowering the temperature with holding periods at certain temperature levels to equalize temperature across the frozen sample.

The dendritic crystal growth pattern and its dependence on the freezing front velocity can be associated with particle and cell entrapment by the freezing front, as was described earlier. The interdendritic spacing may determine which size particles are entrapped between the dendrites and which are rejected. In the case of cellular crystal growth, spacing alone may not determine the critical particle size, since it may be the size of intercellular space (gap between the crystal columnar cells) that determines whether the particle is rejected or entrapped. The intercrystalline space depends not only on the freezing conditions (front velocity and cooling rate) but also on the type and concentration of the solutes. Higher solute concentrations may produce wider gaps between the crystalline cells and allow particle entrapment.

During the thawing process, when the temperature of the frozen mass increases, the cells will remain surrounded by a molten eutectic or a molten glassy-state molecular solution until the temperature reaches a level when dendrites can melt. The cells can remain in an osmotic equilibrium with that liquid phase until the dendrites begin to melt, e.g., until dendrites reach temperature close to 0° C. The melting dendrites release pure water, and the concentration of solutes around the cells decreases. This phenomenon might trigger an osmotic response of cells, which begin to take in some of that released water. It appears that the thawing and osmotic reequilibrium phenomena would be localized until enough dendritic structure melts and cell mobility can be restored.

The freeze-granulation process for large volumes of cell suspensions has been described by the author in Chapter 3, page 209, in this volume. The process is performed in an agitated processor with cryogenic fluids being sprayed over the product. There is an initial volumetric cooling with the subsequent formation of frozen granules. During the initial cooling at a temperature plateau near 0° C, there is formation of a large number of ice crystals and their subsequent growth in undercooled solution. This equiaxial dendritic crystal growth causes volumetric freezing-out of water followed by a rapid formation of frozen granules. The solutes in the liquid phase are continuously concentrated. Since the liquid phase stays near 0° C during crystal formation and growth, the cells are exposed to an increasing concentration of solutes without any cell supercooling. This prevents intracellular ice crystal formation. The cells lose water following an osmotic equilibrium. The decreasing volume of liquid and increasing mass of ice crystals cause apparent mass viscosity to increase to the point at which the whole mass breaks into solidified granules. The agitated slurry of ice crystals and liquid cell suspension generates hydrodynamic shear and physical contact among the crystals. The equiaxial dendrites can break away from ice particles and become nuclei for the new crystals. Interactions among crystals may cause damage to the cells. Therefore, the agitators have been designed to minimize the shear effects within the slurry of ice crystals and liquid with cells. In this process, the equiaxial dendritic crystal growth may be of secondary importance in the cell-dendrite interaction due to the dynamics of the process; the liquid with cells is in continuous movement, the dendritic growth may be suppressed by forced convection in the liquid phase, and there is frequent physical contact between the crystals and breaking off of the dendritic branches. Under such conditions, few cells might be entrapped within the dendritic branches of the crystal, and the majority of cells remain in the agitated liquid phase until the whole mass breaks into granules.

Cell Damage

In summary, three major factors may cause damage to cells: action of concentrated solutes, intracellular ice crystal formation, and mechanical stress. At slow cooling rates, the first mechanism may prevail, and at rapid cooling rates, the second mechanism is a major cause of damage. The mechanical stress from growing external ice crystals in some cases may become a major damaging factor, depending on conditions of external ice crystal growth. Cell dehydration may also have detrimental effects. The intracellular ice formation is commonly thought to be the major cause of cell damage.

If a major process goal is the selective damage to cells, then the parameters listed here can be manipulated in such a way as to cause osmotic as well as mechanical stress on the cells at various levels and different balances between these factors. A change in cell concentration may play an additional role in the cell damage control process. Low cell concentration with a high initial concentration of solutes might cause primarily osmotic damage to cells, but this effect would depend on media composition. Parameters such as dendritic growth velocity, temperature gradient, and the interdendritic space would also be involved. Further research effort is required since there are additional factors involved, i.e., changes in membrane flexibility and fluidity at lower temperatures.

Use of Cryoprotectants in Preservation of Cells and Proteins

Cryoprotective agents have been in use for decades in freezing biological samples (Ashwood-Smith 1987; Meryman 1971). In general, there may be multiple effects of cryoprotectants depending on the compound and the biological material. These effects may include lowering the equilibrium freezing point (T_f), increasing supercooling or lowering the temperature of ice nucleation (T_h), increasing the ice recrystallization level, reducing the critical cooling rate, lowering the eutectic point, forming a glassy (vitrified) solid state, removing unbound water, and removing nuclei for ice crystallization. As has been shown, controlled dendritic crystal growth may create favorable localized conditions for cell dehydration, cooling, and freezing without the danger of intracellular ice

Table 2.2. Some of the Frequently Used Cryoprotectants.

Cryoprotectant	M.W. (Dalton)	Melting Point (K)	Boiling Point (K)
Penetrating			
Glycerol	92.1	293	563
Dimethylsulfoxide	78.1	292	462
Ethylene glycol	62	262	471
Nonpenetrating			
Polyvinylpyrrolidone	>44,000		
Hydroxyethyl starch	450,000		
Human serum albumin	72,300		
Dextran	68,500		

crystal formation. The phenomena associated with controlled growth of dendrites depend on cooling rates, levels of cooling temperature, and composition of solutes (including cryoprotectants).

The cryoprotectants are usually divided into two groups: cell-penetrating (small molecules) and nonpenetrating (large molecules). Table 2.2 shows examples of frequently used cryoprotectants. The penetrating cryoprotectants are glycerol, dimethyl sulfoxide (DSMO), and ethylene glycol. Among nonpenetrating cryoprotectants are polyvinylpyrrolidone (PVP), hydroxyethyl starch (HES), human serum albumin (HSA), and dextran.

Ethanol (small penetrating molecule) is also used as a cryoprotectant. The activity of many cryoprotectants was found to be dependent on the cooling rate and the type of treated biological species.

Large-scale freezing of solutions used for storage and preservation of cells requires understanding how these multicomponent solutions behave during the freezing process. Therefore, phase diagrams (if known) are used to evaluate the freezing effects on cells (Fahy 1981). Fahy (1980) also proposed a set of equations for calculating the ternary phase diagrams: NaCl-DMSO-water and NaCl-glycerol-water. Lin and Gao (1990) developed a heat transfer model for freezing of a ternary system used for cell preservation. Glycerol was added to physiological saline solution, and freezing experiments were conducted in a cylindrical tube. The authors concluded that the addition of glycerol not only reduces electrolyte concentration in the unfrozen part of the solution but also lowers thermal stresses in the frozen solution. Addition of solutes to water may depress both the heterogeneous and homogeneous ice nucleation temperatures. Solutions of sucrose cause a relatively small freezing point depression (0.5° C at 9 % w/w concentration and 1.0° C at 15 % w/w concentration), whereas fructose causes a more significant freezing point depression (1° C at 9.2 % w/w concentration and 1.9° C at 15 % w/w). An addition of 30-% glycerol to water may lower the homogeneous nucleation temperature from about 233 K to about 217 K. At a concentration of glycerol reaching 60–70 %, further cooling may form a glass (vitrified state) at the temperature of about 158 K (–115° C). The critical cooling rate for glycerol solution to reach the vitrified state depends on glycerol concentration: For 49.5 % glycerol the critical cooling rate is about 80° C/min, and for 46-% glycerol it is about 250° C/min. In practice, the low cooling rates may require higher glycerol concentration (above 52–54 %). The presence of sugars and polysaccharides may have an effect on the critical cooling rates and minimal concentrations of glycerol solutions to reach the vitrified state (Sutton 1992). For example, 6 % of PEG400 can reduce the critical cooling rate from about 100° C/min to 30° C/min and the corresponding minimal glycerol concentration from about 49 % to 43.5 %. Bronshteyn and Steponkus (1995) reported on nucleation and growth of ice crystals in concentrated (48–53 % wt) solutions of ethylene glycol using isothermal storage. The example of frequently used protectant ethylene glycol (a small molecule: $HOCH_2CH_2OH$) may serve for analysis of molecular interactions between solute and water. The ethylene glycol interacts strongly with water molecules by forming hydrogen bonds. The ethylene glycol is a flexible molecule and in the hydrated state the intermolecular interactions in addition to interactions with external molecules (water and solutes). The ethylene glycol molecule can form hydrogen bonds with water and with other ethylene glycols. Intramolecular hydrogen formation was also reported (Hayashi et al. 1995). These intramolecular hydrogen bonds were in competition with intermolecular hydrogen bonds. The effects of the molecule internal degrees of freedom were concluded minor. The author conducted molecular simulations of hydrated ethylene glycol using the AMBER force field (Wisniewski, unpublished results, 1996). Intramolecular hydrogen bonds were not observed for ethylene glycol—number of hydrogen bonds with outside water molecules varied from 1 to 4 (in the case of 4 the outside OH groups had 2 hydrogen bonds each, e.g., involving the H and O atoms). Perturbation of water structure around the molecule along the axis was observed. Some of calculated QSAR parameter for hydrated ethylene glycol molecule were as follows: hydration energy –13.04 kcal/mol, logP –0.71, volume 265.59 cubic Angstroms. The distance between oxygen atoms was 3.6283 Angstroms. The angles O-C-C were 109.9° and 109.7° and distances O-C 1.4104 Angstrom and C-C 1.5295 Angstrom. *Ab initio* studies of the peptide-water and peptide-ethylene glycol interactions, showed that the stabilizing effects on peptide induced by hydrogen bonding between peptide and ethylene glycol were stronger than the effects from water-peptide hydrogen (Guo, Karplus, 1994).

The membrane-stabilizing compounds form an important group of cryoprotectants. Egg yolks, milk, lipids, sugars, amino acids, glycerol, 1.2-propanediol, polyvinylpyrrolidone, and bovine serum (BSA) belong to this group (DeLeeuw et al. 1993).

Animal cells grown in monolayers may have their cell junctions affected by the presence of cryoprotectants—including changes in intracellular permeability (Armitage et al. 1995). Le Gal et al. (1995) studied procedures of applying intracellular cryoprotectants before freezing oocytes with a goal of minimizing cryoprotectant cytotoxicity. Glycerol, ethylene glycol, and propanediol were used. Permeability coefficients for water and cryoprotectants were estimated. Procedures for exposure to cryoprotectants were sought to ensure a maximal quantity of cryoprotectant to penetrate the cell within minimal duration of exposure. Use of two cryoprotectants was suggested: one with permeability being only slightly affected by temperature decrease (such as ethylene glycol) and the second with very high initial permeability that might decline with temperature decrease (such as propanediol).

Vitrification and Devitrification

Formation of glassy states has been found to be one of the major means of protecting proteins and cells during freezing. The glassy state forms between the ice crystals after reaching the glass transition temperature or after passing a glass transition temperature zone. It has been assumed that the dynamics of glass formation can be described by the collective motion of molecules over the certain length scale. This length scale increases with decreasing temperature; its increase means slowing down of molecular motion. The molecular dynamics below the glass transition have been investigated only recently using the second-harmonic generation method (Jerome and Commandeur 1997). This approach may be used in further research in the area of protein interaction with substances forming glassy states during freezing. Vitrification of additives during freezing has been found to be one of the more important protective mechanisms for cells in freezing and thawing (MacFarlane et al. 1992). The viscosities of cryoprotectant solutions at glass transition (vitrification) temperatures may reach above $10^{14} \frac{N \cdot sec}{m^2}$ (Chang and Baust 1991a). Molecular diffusion at very high viscosities may be practically neglected, and the vitrified liquid could be treated as a crystalline and maintain its geometric disorder as long as the holding temperature is sufficiently below the glass transition temperature T_g.

Levine and Slade (1988a) published a thorough review on cryoprotection based on control over the physicochemical and thermomechanical properties of solidifying material while controlling the structural state of the freeze-concentrated amorphous matrix between ice crystals. The authors used carbohydrates as cryoprotecting molecules. They also reviewed the thermomechanical properties of aqueous carbohydrate solutions forming glasses and "rubbers" at low temperatures (Levine and Slade 1988b).

Roos and Karel (1991) reported on nonequilibrium ice formation during rapid cooling of fructose and glucose solutions. The rate of ice formation was related to the temperature difference from glass transition $(T–T_g)$. Carrington et al. (1994) investigated ice crystallization for polysaccharide glassy solutions. These temperatures were found to be –42° C for fructose, –36° C for sucrose, and –44° C for glucose.

Ren et al. (1994) proposed a prediction method for vitrification and devitrification of aqueous solutions of cryoprotectants with examples using dimethylsulfoxide, glycerol, ethylene glycol, and PEG. Glycerol solution glass transition temperature is about 168 K (Chang and Baust 1991b). Vitrification of cryoprotectants may depend on pH (Mehl 1995). Her et al. (1994) reported on application of electrical thermal analysis as a complementary technique to the traditionally used differential scanning calorimetry as a means of determining glass transition temperatures for noneutectic, non-crystallizing substances that tend to form glassy states in the frozen material. The technique uses changes in electrical resistance to esti-mate the zone of softening of glassy substances when the frozen mass is being warmed in a controlled way. This technique can also determine eutectics of inorganic salts due to a decrease in electri-cal resistivity after reaching the eutectic temperature in the sample. In the frozen system, the conducting species may have limited mobility. The mobility of these charge-carrying species increases when the glassy state softens and melts and the electrical resistiv-ity decreases. Since the resistivity change transition zone may be smooth for many substances, the authors recommended the addi-tion of a small amount of electrolyte (for example, about 0.1 percent of ammonium salt) to obtain sharper determination of the glass transition temperatures. The design of the measuring cell is impor-tant in this technique, considering such factors as shape and the location of measuring electrodes, the location of temperature sen-sor(s), and the method of controlled heating. Sample freezing may also have an influence on the final result.

Cryoprotectants acting outside the cells (i.e., with molecules too large to penetrate the cell membrane), like modified starch or dex-tran, have been known to provide only limited protection during freezing (Grout et al. 1990). The majority of applied cryoprotectants are of such a molecular size that they can penetrate cell membranes. In the case of gram-negative microorganisms with porins embedded in the outer membrane, the permeability of such cryoprotectants as carbohydrates through porins may depend on the shape of the car-bohydrate molecule. For example, the diameter of raffinose (a linear trimer containing 3 hexose units) is similar to glucose (a hexose), but its diffusion through the OmpF and OmpC porins (solute nonspecific) of *E. coli* is much slower than that of glucose (Jap and Walian 1990). The diameter of hydrated glucose is approximately 8.5 angstroms. The permeability of L-arabinose (a pentose) is half that of glucose. Disaccharides, even if they have the same cross-sectional diameter as glucose, may diffuse much more slowly through the porins.

Sputtek et al. (1987) reported on cryopreservation of human platelets with hydroxyethyl starch and NaCl. The authors used a cooling rate within the range of 5° and 15° C/min with a starch con-centration of 3.6 percent (an optimum cooling rate was found to be between 10° and 15° C/min). The authors also cited an optimum cooling rate for glycerol/glucose as 30° C/min. This may suggest an extracellular influence on the cell osmotic transport by various cryoprotectants. Sputtek et al. (1995) investigated cryopreservation of red blood cells using hydroxyethyl starch (HES). Viability of the cells was 92 percent after the freeze-thaw step. The glass transition temperature for HES extends over a temperature range. The temperature level of about −110° C may be considered a safe storage temperature for biological solutions containing this cryoprotectant (Kresin and Rau 1992).

Brearley, Hodges, and Oliff (1990) reported on the use of low molecular weight and nonpenetrating cryoprotectants as a means of minimizing red cell post-thaw dilution shock. The authors tested several agents, including glycerol, hydroxyethyl starch, betaine, and trehalose, with betaine and trehalose providing the best protection. Protection levels with trehalose were also tested at various NaCl concentrations. Trehalose could protect cells at NaCl concentrations up to 350 mM, with loss of protection above this level. Trehalose performed best at concentrations between 0.2 and 0.4 M. Phase behavior of the trehalose-NaCl-water system was reported by Nicolajsen and Hvidt (1994). Trehalose has been known for a long time as a particularly effective cryoprotectant (Crowe et al. 1985; Green and Angell 1989). Trehalose also shows stabilizing action for isolated proteins such as phosphofructokinase, lactate dehydrogenase, and various restriction enzymes. Trehalose and sucrose provided significantly better protection for red blood cells stored at 4° C in a liquid buffer (phosphate-buffered saline) containing 2.3-butanediol than sorbitol or mannitol (Boutron, Peyridieu, 1994). The concentration of carbohydrate was 4 percent (w/w) in each case. Trehalose and sucrose also showed protective effects for both cell membranes and proteins during lyophilization of *E. coli* and *B. thuringiensis* (Leslie et al., 1995). Aldous et al. (1995) studied transitional states at low temperatures for trehalose and raffinose. The glass transition temperature of trehalose was 304 K (31° C). The sample containing 8 percent moisture was cooled to 230 K at 10 K/min and then heated at 10 K/min. Glass transition was detected by monitoring changes in a heat flux using DSC. The glass transition temperature for raffinose was found to be 313 K (40° C). These temperatures would increase at lower moisture levels. Increasing the moisture levels may lower the transition temperature below 0° C. Table 2.3 shows glass temperatures of some carbohydrates that were reported by Gangasharan and Murthy (1995).

Table 2.3. Glass Temperatures of Carbohydrates (Gangasharan and Murthy 1995)

Carbohydrate	Glass Temperature T_g (K)
D-xylose	274.4
L-arabinose	270.2
D-fructose	277.2
D-galactose	295.9
D-glucose	309
D-sorbitol	267.8

Use of carbohydrates as cryoprotectants for cells and protein solutions requires understanding their behavior at low temperatures (i.e., when they form viscous solutions and reach a glassy state) and their effects on biological membranes. Effects of carbohydrates on biological membranes were investigated by Crowe et al. (1984 a, b, c; 1988).

The interaction of solutes with water is important in the analysis of freezing phenomena. Water can make two hydrogen bonds as a donor and two as an acceptor. Hydrogen atoms and their lone pairs are located along the vertices of a tetrahedron. This tetrahedral structuring is the most visible in ice Ih. The properties of aqueous solutions are determined by mutual interactions and structuring among the solutes and water clusters. The hydrogen bonds of water are relatively strong: Their energies are in the range of 3–6 kcal/mol. The hydrogen bonds play important roles in organized molecular structures, since their energy decreases with distance (above 3.3 angstroms) and with an angle deviating from the aligned O–H–O. The structuring of water molecules can be disturbed by the presence of solutes. These new molecular structures determine the properties of the solute and solution. Hydrogen bonds play a role in the structure and molecular interactions of carbohydrates. The monosaccharides may differ only in their stereoscopic configuration, but their properties in solution may be different. Most of the hydrogen bonding with surrounding water molecules is through hydroxyl groups.

The carbohydrates contain multiple hydroxyl groups that can serve as hydrogen bond donors and acceptors. The sugar molecules are usually considered hydrophilic in nature, but the structures of sugars include both hydrophilic and hydrophobic groups. The hydrophobic surfaces are formed by CH and CH_2 groups, and the hydrophilic surfaces by –OH, –O– and H C̦OH groups (Miyajima et al. 1988). For example, α-D-glucose has an approximate hydrophilic area of 243.3 square angstroms and a hydrophobic area of 97.1 square angstroms. The hydrophilic and hydrophobic characteristics of the molecule determine hydrogen bonding and hydrophobic interactions in aqueous solutions. The maltose may attain a shape with one nonpolar side (with CH groups) and another side with hydroxyl groups exposed (Sundari et al. 1991). Hydrophobic characteristics can be more pronounced in larger molecules such as cyclodextrins. Sorbitol and mannitol are frequently used as protecting agents in pharmaceutical formulations. The molecules vary only by one position of a hydroxy group, but their properties vary (for example, the solubility of sorbitol in water is 3.5 times larger than the solubility of mannitol). Molecular modeling of mannitol and sorbitol in water (Grigera, 1988) showed that sorbitol has produced more significant water structure disruption than mannitol. The author conducted his own molecular simulations of mannitol and sorbitol in aqueous solutions (Wisniewski, unpublished data, 1996). The simulations involved molecular dynamics and geometry optimization of these molecules in periodic water box using AMBER force field. Similar numbers of hydrogen bonds to surrounding water were formed for both molecules (varying from 1 to 5)—simulations showed that mannitol may form larger number of hydrogen bonds than sorbitol. The hydrated molecules with optimized shape were extracted from the periodic water box and their parameters examined. The calculation of QSAR properties showed a difference in hydration energy: for sorbitol it was –22.93 kcal/mol and for mannitol –24.90 kcal/mol. The configurations/dimensions of molecules varied: the distance between C1 and C6 carbons was 6.0474 Angstroms for sorbitol and 5.6057 angstroms for mannitol. For comparison with crystallographic information the bond distance and angles between bonds were checked. The average distances between C-C atoms were found 1.54 Angstrom for mannitol and 1.537 Angstrom for sorbitol, between C-O atoms: 1.409 Angstrom for mannitol and 1.411 Angstrom for sorbitol and the average angles C-C-C for mannitol 113.21° and for sorbitol 113.22°, while angles C-C-O were 109.63° for mannitol and 109.9° for sorbitol. These numbers suggest a relatively good approach to molecule structure by the model used.

The combination of molecular modeling and NMR data showed that the sucrose molecule is extensively bound to water molecules and its first hydration shell may contain on average 25 water molecules (Engelson et al., 1995). The crystalline intramolecular hydrogen bonds were found as short-lasting in the aqueous solution.

Sugars contain hydroxyl and ether groups as well as CH_2 and CH groups all in close proximity. The majority of hydrogen bonding with the surrounding water occurs through the hydroxyl groups (in this hydrogen bonding their behavior is similar to that of water molecules). The past approach to hydrogen bonding between the hydroxyl groups in sugars and water has been that the hydroxyl group imposes a tetrahedral order upon the adjacent water molecules. However, the recent molecular modeling studies (Leroux et al. 1997) pointed out that the most favorable hydration in the sugars may take place when the sugar stereochemistry itself is involved, e.g. when the locations of hydroxyl groups in sugars and the hydrating water molecules are the most compatible as a system (the least interference occurs among such hydrated groups). These hydration patterns may possibly interact with the proteins and might provide some explanation on differences in protein stabilization by various sugars.

Liu and Brady (1996) reported on the structure of pentose sugars and surrounding water molecules in aqueous solutions. Well-defined first and second solvation shells were observed around the sugar molecules. The sugar molecules imposed significant structuring upon the surrounding water. In this structuring, an important role was played by the hydroxyl groups and the conditions for hydrogen bonding formation.

During freezing, carbohydrates become supersaturated in the spaces among growing ice crystals and, finally, vitrified. Hydrogen bonds may also play an important role in the vitrified carbohydrate solutions. After temperatures increase above the glass transition temperature, amorphous carbohydrates may crystallize, reducing the moisture content of the remaining amorphous phases (Aldous et al. 1995).

Figure 2.6 shows the glucose molecule in the "water box" with the hydrogen bonds shown as dotted lines. It is an illustration of a particular state, since the hydrogen bonds between the glucose

Figure 2.6. The glucose molecule in the "water box." The hydrogen bonds are shown as dotted lines.

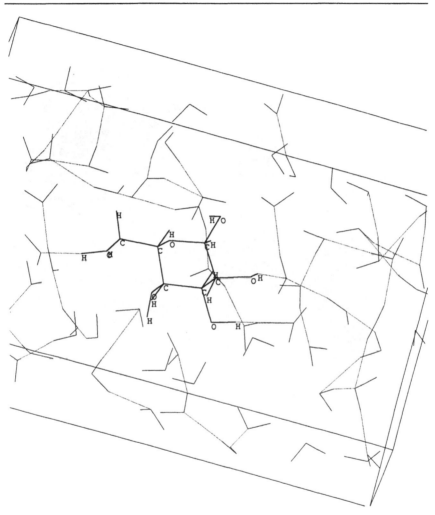

molecules and water molecules are dynamically changing. Figure 2.7 (a and b) shows the trehalose molecule in vacuo (a) and in the "water box" (b).

Views of the molecular structures were created by using the Hyperchem v. 5 molecular modeling software package. In addition to carbohydrates, aqueous media may frequently contain cations

Figure 2.7. (a) The trehalose molecule "in vacuo." (b) The trehalose molecule in the "water box." The hydrogen bonds are shown as dotted lines.

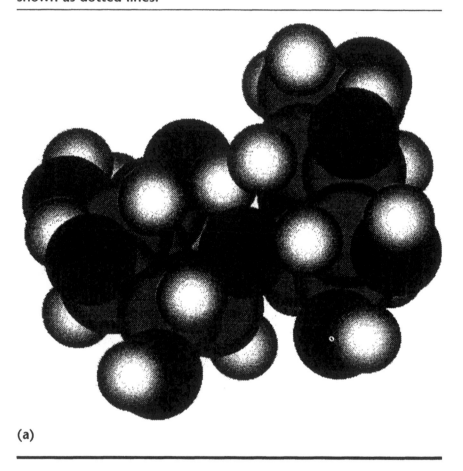

(a)

such as calcium, potassium, or sodium. Interactions of sugars with various electrolytes can be of concern during freezing and thawing of cell-containing solutions compounded with carbohydrates and salts. For example, sugars like ribose, arabinose, xylose, glucose, mannose, galactose, talose, and allose interact only very weakly with KCl. For $CaCl_2$, ribose and talose have shown strong specific interactions (Morel et al. 1988). However, the interactions between

Figure 2.7. Continued.

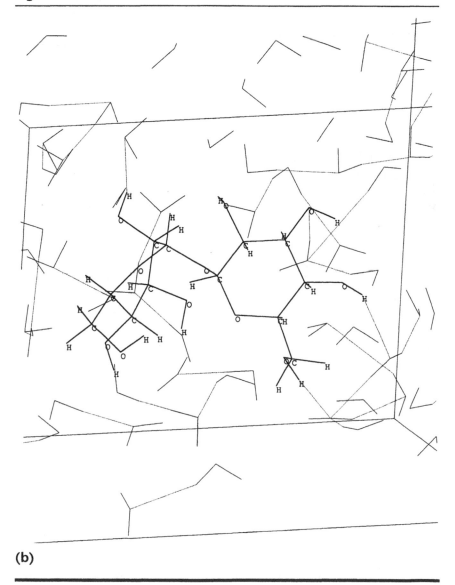

(b)

sugars and cations in dilute aqueous solutions may be very different from those in concentrated solutions or in the solid state.

Ice in its normal form (Ih) consists of a hexagonal arrangement of water molecules that are held together by hydrogen bonds (for a recent review on hydrogen bonding in ice see work by Li [1996]). At the ice-liquid interface, there are exposed hydroxyl groups in the ice hexagonal structure. These hydroxyl groups may form hydrogen bonds with bulk water or solute molecules. Kawai et al. (1992) studied hydration of oligosaccharides, with a particular interest in trehalose hydration. The authors correlated the number of equatorial OH groups for various oligosaccharides with the number of unfrozen water molecules. Trehalose had a relatively large amount of unfrozen water molecules, considering both the OH groups and glucose units. It was suggested that among disaccharides trehalose has the most powerful ability to lower the mobility of surrounding water molecules.

Mashimo, Miura, and Umehara (1991) and Mashimo and Miura (1993), in their studies of water structure, shed additional light on interactions of sugars with water and their possible interface with ice lattice. The authors studied glucose, trehalose, maltose, polysaccharides, and ascorbic acid. Their findings were that glucose, trehalose, and maltose can be incorporated into the ice lattice since their size is comparable with a typical water cluster, whereas larger sugars cannot. Later Mashimo (1994) summarized the earlier work on the structure of water including water in biosystems. His work concluded that water has a structure of distorted ice lattice present in the clusters and that water clusters may consist of about 30 molecules. The molecular model applied by Mashimo showed that the glucose molecule can occupy a site in the ice lattice, and several hydrogen bonds are formed between the molecule and the water cluster structure. These hydrogen bonds keep the cluster structure stable. The glucose molecules can reorient only when the cluster structure is decomposed. Gaffney et al. (1988) estimated the number of hydrogen bonds possible to form for various saccharides. Sucrose and maltose have 4 acceptors and 8 donors each, while trehalose has 3 acceptors and 11 donors. The authors obtained these number of hydrogen bonds by matching the saccharide structure with an ice lattice. Studies with maltose in water revealed that 8 hydrogen bonds form between water and hydroxyl groups of each ring in maltose

(Wang et al. 1996). The hydration shell around maltose and water mobility there were not significantly affected when compared to bulk water. This might suggest a similar characteristic of hydrogen bonds between maltose and water and among water molecules in water clusters. More on hydrogen bonding in carbohydrates in aqueous solutions can be found in works by Brady and Schmidt (1993) and Jeffrey and Saenger (1991). Molecular modeling using the maltose and trehalose molecules showed that the number of hydrogen bonds between the molecules and water molecules included in the water box may vary (Wisniewski unpublished results 1996), as occurs among the water clusters. Molecular modeling of maltose and trehalose molecules showed that the number of hydrogen bonds between these molecules and water may vary, as it occurs with water molecules in clusters. (Wisniewski, unpublished results 1997).

Figure 2.8 (a and b), shows a molecule of maltose (a) and a molecule of sucrose (b) in the "water boxes."

Maltose, trehalose, and sucrose have similar overall formulas ($C_{12}H_{22}O_{11}$, MW 342.30), but their characteristics vary. For example, some of the QSAR (Quantitative Structure-Activity Relationships) parameters estimated using HyperChem v. 4 (Wisniewski, unpublished results 1997) are shown in Table 2.4.

There is an intramolecular hydrogen bonding within many carbohydrate molecules. These bonds may be mostly pronounced in the crystallized state but are also retained in aqueous solutions. For example, the sucrose molecule may lose one of the two intramolecular hydrogen bonds after dissolution but may still retain its shape from the crystallized state (Jeffrey and Saenger 1991). Molecular dynamics modeling of sucrose molecules in the periodic water box demonstrated the existence of an intramolecular hydrogen bond ($O***H$) between two rings of sucrose (Wisniewski, unpublished results 1997). Several hydrogen bonds among sucrose atoms and surrounding water molecules were also observed.

In biological systems, water can be divided into two main groups: free water and weakly bound water. Studies with DNA showed that during cooling the free water freezes between 0° and −10° C, and the bound water freezes within the range of −25° −65° C. In hydrated DNA, thirteen water molecules are bound to A-DNA and nine water molecules to the Z-DNA. This weakly

Figure 2.8. (a) The maltose molecule in the "water box." The hydrogen bonds are shown as dotted lines. (b) The sucrose molecule in the "water box." The hydrogen bonds are shown as dotted lines.

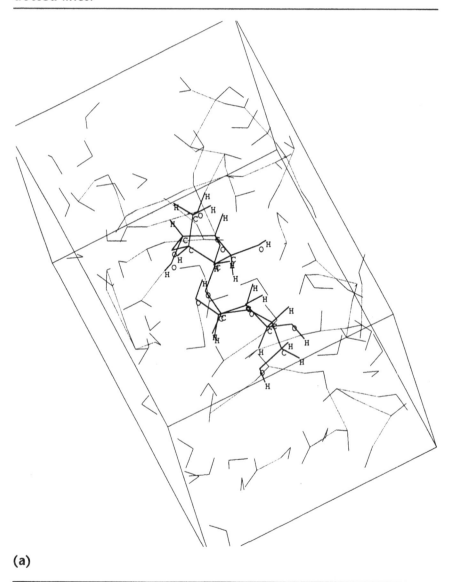

(a)

Figure 2.8. Continued.

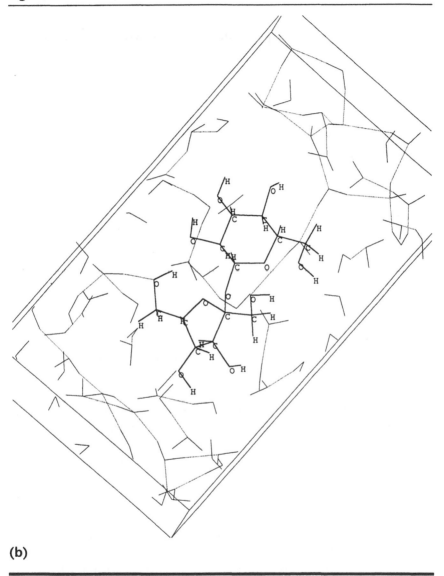

(b)

Table 2.4. Estimated QSAR Parameters of Maltose, Trehalose, and Sucrose.

	Maltose	Trehalose	Sucrose
Volume ($Å^3$)	844.13	859.34	831.91
Surface ($Å^2$)	358.76	377.35	351.25
Polarizability ($Å^3$)	28.25	28.25	28.25
Refractivity ($Å^3$)	68.34	68.34	68.77
log P	–2.98	–2.98	–2.47
Hydration energy (kcal/mol)	–26.76	–27.25	–26.27

bound water does not freeze easily and has no definite freezing temperature. Its structure and properties (specific heat, freezing point) are different from the bulk water. As temperature decreases, these water molecules are gradually taken into the ice structure. This amount of bound water is independent of the amount of bulk water. Mashimo (1994) suggested that some tightly bound water may not even freeze down to the –100° to –200° C levels.

Interaction of carbohydrates and proteins has been studied as a general topic (Sears and Wong 1996), but studies in the cryopreservation field are mostly concerned with such interactions as protective means against protein denaturation caused by phenomena occurring during freezing and thawing. During chemical denaturation of proteins by chaotropic agents, the protection mechanisms by carbohydrates may involve binding the carbohydrate to protein via hydrogen bonding (Figlas et al. 1997). The protective binding of carbohydrate (D-glucose) to protein was found to be entropy-driven, using differential scanning calorimetry (DSC) in investigations of thermal denaturation of yeast hexokinase B (Catanzano et al. 1997). Further research is needed in the area of carbohydrate (and other solutes) interactions with proteins during freezing and thawing processes. This should involve effects of low temperature, protein unfolding, and formation of glassy states from solutes other than the product biomolecules.

Biopharmaceutical Formulations of Protein Drugs for Lyophilization and Freeze-Thaw Steps

Carbohydrates are used as stabilizers for final formulations of protein drugs that undergo lyophilization (pass through the freezing processing step). Such formulations frequently are treated with the freezing and thawing process as a means of cryopreservation in the bulk phase prior to filling and finishing operations. In addition to using the carbohydrate as a protein stabilizer, the formulation contains the bulking agent and a buffering substance. Frequently, a nonionic surfactant is added to prevent protein aggregation. Traditionally, mannitol has been used as a bulking agent and the sodium phosphates (mono- and bi-) as buffering substances. The approved biopharmaceuticals in many cases reflect the state of the art in protein formulation at the time the drug was developed. Table 2.5 shows examples of the formulations approved for lyophilized pharmaceuticals. The liquid formulations are also shown for comparison.

An introduction to formulation selection for lyophilized protein products was outlined by Pikal (1990). He pointed out such important facts as the changing concentration of solutes during freezing (for example, the concentration of NaCl may increase 40-fold before reaching eutectic) and solidification of solutes in a crystalline or amorphous/glassy state (the last with a substantial amount of water). The importance of component crystallization during concentration and cooling processes introduces a possibility of sudden shifts in pH, which may adversely affect the proteins. The importance of glassy state solidification is that the solidified mass may contain a substantial amount of water that has to be removed during lyophilization (this glass formation temperature $[T_g']$ is associated with the collapse temperature during lyophilization and may determine the approach to the lyophilization procedures). The importance of the glass transition temperature on product processing and storage was emphasized, e.g., what the processing and storage should undergo below the glass transition temperature. Recently, Pikal (1997) summarized and updated information on formulations for lyophilized biopharmaceuticals. He recommended that the buffer components should be selected to remain amorphous (no crystallization during freezing) and the buffer quantity should be reasonably small. The list of recommended stabilizers included

Table 2.5 Formulations for Approved Protein Biopharmaceuticals in Lyophilized and Liquid Form. (Summarized from the PDR 1997)

Product and Manufacturer	Description	Form	Formulation*
Humegon Organon	Gonadotropins (human origin)	Lyophilzed	per vial with 2 mL Sodium Chloride Injection: 10.5 mg lactose 0.25 mg sodium phosphate (mono) 0.25 mg sodium phosphate (di-)
Koate-HP Bayer	Factor VIII (human origin)	Dried concentrate	(all NMT) Heparin 5 units/mL 1500 ppm PEG 0.05 M glycine 25 ppm polysorbate 80 5 ppm tri-n-butyl phosphate 3mM $CaCl_2$ 1 ppm Al 0.06 M histidine 10 mg/mL albumin (HSA)
Kogenate Bayer	Factor VIII (recombinant glycoprotein)	Dried concentrate	10–30 mg/mL glycine 500 µg imidazole/1000 IU 600 µg polysorbate 80/1000 IU 2–5 mM $CaCl_2$ 100–130 mEq/L Na 100–130 mEq/L Cl 4–10 mg/mL albumin (HSA)
Epogen Amgen	Erythropoietin (recombinant glycoprotein, MW 30,400)	Liquid	*Single dose:* 2.5 mg/mL albumin (HSA) 5.8 mg/mL sodium citrate 5.8 mg/mL NaCl 0.06 mg/mL citric acid pH 6.9 ± 0.3 Multiple dose (preserved): 2.5 mg/mL albumin (HSA) 1.3 mg/mL sodium citrate 8.2 mg/mL Na/Cl

Product and Manufacturer	Description	Form	Formulation*
			0.11 mg/mL citric acid 1% benzyl alcohol pH 6.1 ± 0.3
Neupogen Amgen	Granulocyte colony stimulating factor (recombinant, MW 18,800)	Liquid	300 µg/mL G-CSF 10 mM sodium acetate (0.59 mg/mL) 50 mg/mL mannitol 0.004% Tween 80 pH 4.0
Procrit Ortho Biotech	Erythropoietin (recombinant glycoprotein, MW 30,400)	Liquid	*Single dose:* 2.5 mg/mL albumin (HSA) 5.8 mg/mL sodium citrate 5.8 mg/mL NaCl 0.06 mg/mL citric acid pH 6.9 ± 0.3 Multiple dose (preserved): 2.5 mg/mL albumin (HSA) 1.3 mg/mL sodium citrate 8.2 mg/mL NaCl 0.11 mg/mL citric acid 1% benzyl alcohol pH 6.1 ± 0.3
Orthoclone OKT3 Ortho Biotech	Murine MAb to CD3 antigen of human T cells-immuno-suppressant (recombinant IgG_{2a} heavy chain MW 50,000, light chain MW (25,000)	Liquid	1 mg/mL MAb-CD3 0.45 mg/mL sodium phosphate (mono-) 1.8 mg/mL sodium phosphate (di-) 8.6 mg/mL NaCl 0.2 mg/mL polysorbate 80 pH 7.0 ± 0.5
Proleukin Chiron	Interleukin-2 (recombinant, MW 15,300)	Lyophilized, reconstituted	IL-2 1.1 mg/mL 50 mg/mL mannitol 0.18 mg/mL SDS 0.17 mg/mL sodium phosphate (mono-) 0.89 mg/mL sodium phosphate (di-) pH 7.5 ± 0.3

Continued on next page.

Product and Manufacturer	Description	Form	Formulation*
Roferon-A Roche	Interferon alfa-2a (recombinant, MW 19,000)	Liquid	133.3 µg/mL Interferon-α (max.) 7.21 mg/mL NaCl 0.2 mg/mL polysorbate 80 0.77 mg/mL ammonium acetate 10 mg/mL benzyl alcohol
		Lyophilized, reconstituted	22.2 µg/mL Interferon-α 9 mg/mL NaCl 1.67 mg/mL albumin (HSA) 3.3 mg/mL phenol
Actimmune Genentech	Interferon gamma-1b (recombinant, dimer 2 x MW 16,465)	Liquid	200 mcg/mL Interferon-γ 40 mg/mL mannitol 0.72 mg/mL sodium succinate 0.11 mg/mL polysorbate 80
Activase Genentech	Tissue plasminogen activator (recombinant, glycosylated, 527 amino acids)	Lyophilized, reconstituted	1 mg/mL tPA 35 mg/mL L-arginine 10 mg/mL phosphoric acid 0.11 mg/mL polysorbate 80
Pulmozyme Genentech	Deoxyribonuclease 1 (recombinant, glycosylated, MW 37,000)	Liquid, inhalant	1.0 mg/mL rhDNase 0.15 mg/mL $CaCl_2 \cdot 2H_2O$ 8.77 mg/mL NaCl pH 6.3
Nutropin Genentech	Growth hormone (recombinant, MW 22, 125)	Lyophilized, reconstituted	1.0 mg/mL hGH 9 mg/mL mannitol 0.08 mg/mL sodium phosphate (mono-) 0.26 mg/mL sodium phosphate (di-) 0.34 mg/mL glycine 0.9% benzyl alcohol pH 7.4
Nutropin AQ Genentech	Growth hormone (recombinant, MW 22, 125)	Liquid	5 mg/mL hGH 8.7 mg/mL NaCl 2.5 mg/mL phenol

Product and Manufacturer	Description	Form	Formulation*
			2 mg/mL polysorbate 20 5 mM sodium citrate pH 6.0
Protropin Genentech	Growth hormone (recombinant, MW 22,000)	Lyophilized, reconstituted	1 mg/mL hGH 8 mg/mL mannitol 0.02 mg/mL sodium phosphate (mono-) 0.32 mg/mL sodium phosphate (di-) pH 4.5–7.0
Genotropin Pharmacia- Upjohn	Growth hormone (recombinant, MW 22,124)	Lyophilized, reconstituted	5 mg/mL hGH 41 mg/mL mannitol 1.93 mg/mL glycine 0.28 mg/mL sodium phosphate (mono-) 0.27 mg/mL sodium phosphate (di-) 0.3% m-cresol pH 6.7
Humatrope Eli Lilly	Growth hormone (recombinant, MW 22, 125)	Lyophilized reconstituted	1 mg/mL hGH 5 mg/mL mannitol 1 mg/mL glycine 0.26 mg/mL sodium phosphate (di-) 0.3% m-cresol 1.7% glycerin pH 7.5
Humalog Eli Lilly	Insulin (recombinant, MW 5,808)	Liquid	100 U/mL insulin lispro 16 mg/mL glycerin 1.88 mg/mL sodium phosphate (di-) 3.15 mg/mL m-cresol 0.0197 mg/mL zinc pH 7.0–7.8
Ceredase Genzyme	Beta-glucocere- brosidase (recombinant, glycosylated, MW 59,300)	Liquid	80 U/mL or 10 U/mL alglucerase sodium citrate buffer (53 mM citrate 143 mM sodium) 1% albumin (HSA)

Continued on next page.

Product and Manufacturer	Description	Form	Formulation*
Cerezyme Genzyme	Beta-glucocere-brosidase (recombinant, glycosylated, MW 60,430)	Lyophilized, reconstituted	29.2 mg/mL mannitol sodium citrates (9.8 mg/mL trisodium c., 3.4 mg/mL disodium hydrogen c.) 0.1 mg/mL polysorbate 80 pH 6.1

*These published formulation compositions for the lyophilized protein drugs (the right column) are shown after reconstitution and may not reflect the exact concentrations (and composition) of the bulk prior to the lyophilization manufacturing step.

glycine, sucrose, and nonreducing disaccharides. The stabilizers should not crystallize during freezing. Mannitol and glycine also typically serve as bulking agents (although mannitol could crystallize in the process and lose its protecting properties). The presence of collapse temperature modifiers may also be considered (examples: dextran, hydroxyethyl starch (HES), ficoll, and gelatin). Isotonicity adjustment of the formulation is typically done by including NaCl or glycerol.

Modern formulation design is based on better understanding of the molecular interactions of protein-stabilizer-bulking agent-buffer-surfactant occurring during cooling, freezing, thawing, and lyophilization, as well as during subsequent storage. The classical approach to protein stabilization in aqueous solutions states that all structure stabilizing agents are excluded from the protein domain, i.e., the proteins are preferentially hydrated (Timasheff 1993). The denaturing agents are considered to be preferentially bound to proteins (ibid.). Timasheff summarized in a comprehensive form the typical patterns of preferential interaction of proteins with precipitating and stabilizing cosolvents. Several mechanisms have been proposed:

    I. Exclusion resulting from increase in surface tension,

    II. Surface tension increase compensated by binding,

    III. Binding that can be overcome by surface tension increase,

    IV. Binding,

V.   Solvophobic effect, and

VI.   Steric exclusion.

For the preferentially excluded protein stabilizers (sugars, most nonhydrophobic amino acids, and good salting-out salts [NaCl, $(NH_4)_2SO_4$]) the preferential hydration does not depend on the concentration or pH, whereas for other types of molecules (such as $MgCl_2$, PEG), the preferential hydration depends on concentration or/and pH, and their effect on protein stability may vary. Agents influencing the surface tension are important in protein stability and denaturation issues. Many protein stabilizers such as sugars, nonhydrophobic amino acids, and many salts increase the surface tension of water. Recent work by Cioci (1996) demonstrated an increase of protein stability with the surface tension increase. However, there are exceptions; for example, urea increases the surface tension but destabilizes proteins, whereas glycerol or betaine may stabilize proteins despite the fact that they lower the surface tension. Other factors than surface tension are also involved in protein stability in the presence of solutes (Timasheff 1993).

An example of urea as the protein destabilizing agent may be used to demonstrate the approach to understanding of solute interactions with proteins in aqueous solutions. This example is used since urea has been widely used as a protein unfolding agent and its mechanisms were subject to numerous studies. Urea is a polar, planar molecule with high solubility in water. The protein denaturing characteristics of urea occur at high urea concentrations, typically, urea concentrations above 6M are used for protein unfolding (Pace 1986). Despite the fact that the mechanism of protein destabilization by urea has been widely investigated, there may still be no clear agreement on all the phenomena and interactions involved. Breslow and Guo (1990) discussed the model of urea as an agent that destabilizes the proteins by decreasing the hydrophobic effect. They concluded that urea may not act as a simple water structure breaker and may change the solvation effects, possibly affecting the protein conformation.

Makhatadze and Privalov (1992) proposed an explanation of protein destabilization by urea using a simple binding model based on calorimetric studies. The concept of the binding model was earlier proposed by Prakash et al. (1981). The behavior of the urea molecule in water may provide a key to urea interactions with proteins, e.g., better understanding of urea hydration and possible interaction of concentrated urea solutions with water, hydration shells of proteins, and proteins themselves. The molecular simulations of urea molecules in water using the Monte Carlo method have revealed that each urea molecule may coordinate 5 molecules of water in its tight first hydration shell (Tanaka et al. 1984). The hydration shell was closely attached to urea and did not affect significantly the surrounding bulk water. Molecular simulations by Kuharsky and Rossky (1984) showed that very little difference in properties was noticed between the bulk water and the urea hydration shell. At high concentrations of urea where the protein unfolding begins, the 5 water molecule hydration shells might lead to an estimate that the number of water molecules coordinated this way around urea is approaching most of the "available" water, e.g., water that is not in the hydration shells of protein molecules. Lee et al. (1995) investigated the urea molecule, urea dimer, and complexes of urea and water molecules using density functional theory. The authors concluded that urea readily forms hydrogen bonds with water and, because of this fact, no urea dimers could be stable in aqueous solution. The author's own molecular dynamics simulations showed that the number of hydrogen bonds between urea and surrounding water can be between 1 and 5 with some of them being a multiple type (Wisniewski, unpublished results 1996).

Kuramoto and Ishikawa (1995) used an ultrasonic relaxation method to study experimentally the urea-water interactions. They concluded that urea has an hydrogen bond breaking effect on water structure and changes the water bulk structure to the denser one. Recent studies of urea molecule by Tsai et al. (1996) showed that addition of urea to water increases the number of hydrogen bonds (not by an increase in water-water hydrogen bonds, but because of an increase in water-solute hydrogen bonds). Urea was found not to aggregate in solution and be well mixed with water even at high concentrations (high solubility of urea in water has been a well-known fact). Urea at high concentrations seemed not to disturb O–O distribution in water (this distribution was not much different from pure water). Other Y-shaped analog molecules (similar to urea) tended to aggregate and decreased the number of hydrogen bonds, pointing out the uniqueness of urea. The authors did not find any tendency for urea molecules to aggregate into dimers (urea surrounds itself with water molecules). The urea molecule did not disturb the surrounding water structure (was evenly distributed). The analysis showed that at the hydrophobic surfaces, a urea molecule was able to replace 3 water molecules. This effect may be of significance since one might expect limited interaction of urea with hydrophobic surfaces due to the high capability of urea to form hydrogen bonding with the surrounding water. Such a release of water may lower the free energy of cavity formation for hydrophobic solutes. Urea in aqueous solution could minimize the amount of disruption caused by apolar, hydrophobic solutes. The authors proposed a mechanism in which urea destabilizes proteins by decreasing the free energy required to solvate hydrophobic residues of protein molecules. Further studies of urea interactions with peptides and proteins with changing numbers of exposed hydrophobic and hydrophilic residues are required to validate this model. This example of multiple interpretations of the issue of protein stability depending on solute-water-protein interactions involving such a simple molecule as urea shows that there is still need for better understanding of protein stabilization by solutes of higher complexity than urea and that detailed analyses of molecular interactions should be conducted.

Advances in neutron scattering instrumental techniques are bringing new insight into the hydrophobic effect (Finney et al. 1993). In that work, a disordered structure of water clusters around nonpolar molecules has been found to be even less structured than typically thought in explanation of the hydrophobic effect. Considering the generally accepted importance of the hydrophobic effect for stability of biomolecules, such experimental studies bring additional perspective to the understanding of the nature of this effect. Recent studies of water structure around hydrophobic amino acids, including neutron scattering experiments and molecular dynamics simulations, showed that the water structure around the hydrophobic amino acids deviated from the bulk water; instead of bent and strained hydrogen bonds, the hydrogen bonds are more linear and unstrained (Pertsemlidis et al. 1996). The authors reported that the geometric water structures of supercooled water and hydration water around hydrophobic groups are similar. A good agreement was reported between the neutron scattering experiments and molecular dynamics simulations.

Behavior of carbohydrates in aqueous solutions is essential in protein stabilization, as already discussed. Molecular simulations, similar to the previously described approach to urea, can be conducted since there are computational methods developed for studies of carbohydrates (Reiling et al. 1996).

An introductory review of the relationship between formulation design and drug delivery for biopharmaceuticals was published by Cleland and Langer (1994). Factors such as formulation composition, pharmacokinetics, toxicity, drug delivery route, and clinical indications were discussed, and the process of formulation development was outlined. It includes studies of physicochemical properties of the drug, information on its *in vivo* actions, selection of initial formulations, investigation of drug activity degradation, solving the potential degradation problems (modification of formulation, change of formulation from liquid to lyophilized, etc.), processing (including problems associated with lyophilization), and final stability testing. Most of the information collected in this process may be used in the design of the freezing and thawing processes.

An approach to designing a modern formulation may begin with considering sucrose as a protein stabilizer, glycine as a bulking agent, citrate or histidine as a buffering substance, and a nonionic surfactant as an inhibitor of protein aggregation. Sucrose may be eventually replaced by trehalose in the case of protein stability problems in the sucrose-containing formulation. If possible, human serum albumin (HSA), frequently used in the past, may not be considered since its origin is human blood.

The glass transition temperature (T_g) of the formulation is critical to the lyophilization procedure and subsequent storage conditions of the product. Even at a very low water content, the actual glass transition temperature may be lower than the reported temperature for anhydrous substances (T_g declines with an increase of water content). Since after lyophilization there is a certain level of water remaining in the product, the actual glass temperature should be determined experimentally rather than using the literature data on glass transition temperature, particularly if they are listed without the corresponding water content. As a general rule, the storage of lyophilized product should be at temperatures below the actual product glass temperature.

During freezing, the formulation becomes more and more concentrated in the interdendritic space and at the same time its temperature decreases. This process involves an increase in viscosity and slowing down of the molecular motion. At a certain concentration and temperature (T_g') the viscosity becomes so high that the formulation behaves first as a rubbery substance and then becomes a glass. This temperature can be estimated within a range and may vary depending on formulation composition. There is a certain water level included in the glassy state at this stage (water concentration, W_g'). At very high viscosity, the water cannot diffuse through the liquid toward the surface of the dendritic ice crystal and is trapped inside the glassy state at this residual level. Typically, the protein concentration is low while the concentration of protective carbohydrates is much higher than that of protein. The solidification process may, therefore, involve only a limited exposure of protein to ice surfaces, e.g., the majority of proteins may be embedded within the solidifying carbohydrate. Growth of relatively coarse dendrites might minimize this protein-ice contact. The temperature of liquid between dendrites is higher than the dendrite surface, and, because of this solid undercooling, the dendrites grow thicker toward their bases and branches may form on their side surfaces. As a result, the solute concentration occurs along the dendrite length (freezing out of water) together with its cooling (due to the temperature gradient on the dendrite surface toward its base). Very rapid freezing (such as spraying the liquid on precooled surfaces or droplets plunging into liquid nitrogen) may generate very fine ice crystals (their number will be very large per unit of volume with correspondingly very large ice-solute interface). Under such conditions, more protein molecules will be exposed to contact with ice and possible product degradation might occur. While growing fine ice crystals, one may try to increase the content of carbohydrate, but it has to be kept in mind that rapid cooling may trap water within the carbohydrate itself since the water might not be able to diffuse through rapidly cooled viscous carbohydrate toward the ice crystal surface. Lowering the temperature of the solution at high water content would cause reaching the glass temperature corresponding to this water content (lower than for small water contents) and in practice would require lower storage temperature of the product. If the storage temperature is established using slower freezing rates then the product may experience degradation. In general, all bulk freezing is anticipated to be slow, e.g., the solute concentration in the intercrystalline space would follow closely equilibrium conditions until the solutes became so viscous that water molecules could not migrate toward the ice crystal surface any longer. The controlled dendritic crystal growth may create conditions of simultaneous cooling of solutes combined with their concentration, e.g., the water can diffuse toward dendritic crystal surfaces until the solutes become viscous and solidify between the dendrites, as was described earlier in this chapter. Controlled dendritic crystal growth can also be very important during freezing prior to lyophilization. Relatively slow growth of parallel dendrites may ensure conditions close to equilibrium with water migrating toward the ice crystals and minimum water being entrapped within the solidified glassy state, e.g., with relatively high glass temperature. Developed dendritic structure would also facilitate ice sublimation and provide an extended network of pores helpful in the secondary drying. Rapid freezing causes more water being trapped in the glassy state. This corresponds to lowering of the frozen glassy state temperature T_g'. With T_g' temperature decrease the required storage temperature also has to be lowered to be maintained below T_g'.

Therefore, very rapid freezing not only creates unstable fine ice crystals, prone to recrystallization, but also lowers T_g' and increases W_g'. These changes may affect the storage of granulated products (ice crystal growth and storage temperature above T_g' may lead to fusion of granules) and may create conditions of product deterioration (high water content combined with ambient temperature above the real T_g'). Low T_g' and high W_g' are very undesirable for lyophilization (the primary drying has to be conducted at lower temperature and the lyophilization process will last much longer).

The data shown in Table 2.6 represent a compilation from various sources. Estimates of T_g and T_g' should carefully consider the water content in the glassy state (an increase in trapped water may cause lowering the glass temperature). In estimates of T_g' and W_g' the results may also depend how the freezing was conducted (rapid freezing would produce higher W_g' and lower T_g', while slow freezing may generate conditions closer to equilibrium). Water entrapment in the glassy state may also depend on initial concentration of glass-forming solute(s). At high concentration, the migration distance for water molecules in the viscous liquid phase towards the ice crystal surface is longer and the obtained W_g' may be higher.

Table 2.6 shows approximate values of the glass transition temperature T_g, the transition temperature T_g' occurring at cooling and cryoconcentration during freezing, and the water concentration associated with T_g' (Franks 1991; Franks et al. 1992).

Table 2.6. Glass Transition Temperatures (°C) of Carbohydrates and Derivatives. (Franks 1991; Franks et al. 1992; Levine and Slade 1988b)

	T_g	T_g'	W_g' (percent H_2O)
Glycerol	–93	–65	46.0
Ribose	–10	–47	33.0
Sorbitol*	–3	–43.5	18.7 (23)
Xylose	10	–48	31.0
Fructose	13	–42	49.0 (15)[†]
Glucose	30	–43	29.1 (17)[†]
Maltose	43	–29.5	20.0
Sucrose	56	–32	35.9 (16)[†]
Cellobiose	77	–29	
Trehalose	110 ° C	–29.5	16.7
Polydextrose	56		
Glucopyranosyl-mannitol	57		
Glucopyranosyl-sorbitol	60		
Inulin	81		
Dextran	83		
Maltohexose	173		

[†]Values obtained in the specially designed freeze-drying runs in F. Franks's laboratory.
*Values of the glass temperature for sorbitol at the water content 0 percent to 50 percent can be found in work by Quinquenet et al. (1988).

The differences in values reported in Tables 2.3 and 2.6 were probably caused by product origin (purity levels) and measurement methodology. The glass transition temperatures may be considered not as sharp points but rather as ranges around certain temperature levels.

The good cryoprotectant trehalose can be distinguished by high T_g and high T_g'. Frequently, sucrose is being used as a cryoprotectant due to its availability, low cost, and regulatory approval. Both T_g and T_g' values for sucrose are sufficiently high to ensure good cryoprotection during freezing, storage, and thawing as well as during lyophilization and subsequent storage. Glycerol, on the other hand, may work as a cryoprotectant during freezing and thawing, but may not be a good agent for lyophilization formulations because its T_g' and T_g temperatures are low (lyophilization may require special equipment to maintain very low condenser temperature and storage of lyophilized product below the glass temperature. Due to its low T_g' and T_g temperatures, sorbitol may not be well suited for the lyophilized formulations either.

Table 2.7 lists glass transition temperatures for various biopharmaceutical formulations (Chang et al. 1996).

This approach has led to an optimized formulation for Interleukin-1 receptor antagonist (rhIL-1ra): 100 mg/mL rhIL-1ra, 2 percent (wt/vol) glycine, 10 mM sodium citrate (pH 6.5), and 10 percent (wt/vol) sucrose (Chang et al. 1996). The lyophilized product was stored for 56 weeks at 30° C without any detectable damage, and when stored for the same time at 50° C the loss was only 4 percent (due to deamidation). In many cases, the optimized formulation may consist of less than 10 percent sucrose and still provide adequate protein protection.

Frequent use of glycine may pose a question of how many water molecules are associated with this molecule (this can be of interest during concentration and solidification of solutes in the intercrystalline space). The form of glycine in aqueous solutions and the crystalline state is zwitterionic within the pH range of 2 to 10. The isoelectric point of glycine pI is near 6. The surface tension of glycine solution reaches maximum at the isoelectric point and depends on concentration: $\Delta\sigma/\Delta C_M = 0.92$ (Greenstein and Winitz 1961). Jensen and Gordon (1995) analyzed the hydration of glycine and reported that the minimum number of water molecules to stabilize glycine is most likely two. Glycine has a unique property among amino acids

Table 2.7. Glass Transition Temperatures (T_g) of Biopharmaceutical Formulations. (°C)

Stabilizer	Bulking Agent	Buffer	T_g (°C)
1% sorbitol	5% glycine	Sodium citrate	20
1% sucrose	4% mannitol + 2% glycine	Sodium phosphate	26
	4% mannitol + 2% glycine	Sodium citrate	46
1% alanine	5% glycine	Sodium citrate	47
1% sucrose	4% mannitol	Sodium citrate	47
1% trehalose	5% glycine	Sodium citrate	50
0.5% sucrose	5% glycine	Sodium citrate	50
1% sucrose	4% mannitol + 2% glycine	Sodium citrate	51
1% sucrose	5% glycine	Sodium citrate	54
1% sucrose	2% glycine	Sodium citrate	56

in that it can assume various conformations, unlike other residues in the protein structure (Branden and Tooze 1991). This flexibility may minimize disruption of protein structure. The presence of sodium chloride (a frequent component of formulations) may increase the solubility of glycine. The glass temperature (T_g') of glycine during freezing of its aqueous solutions is low: about –37° C (Carpenter and Chang 1996); it is lower than sucrose, and the freezing may need to proceed down to about –40° C. The glycine also forms a metastable glassy state during freezing. During thawing, the glycine glass may recrystallize at around –25° C, and further warming may cause it to pass through a eutectic melting point (ibid.). The phase equilibria water-glycine-sucrose were investigated by Shalaev and Kanev (1994), and the phase equilibria water-sucrose-NaCl were investigated by Shalaev et al. (1996). Sodium citrate has glass temperature during freezing (T_g') at about –40° C, and mannitol has two glass temperatures (T_g'): at around –37° C and at –57° C due to doubly unstable glass (Chang and Randall 1992). The data indicate that the freezing of formulations that include these components should proceed at least at –40° C, and the subsequent storage should be maintained at least at this temperature or below. Nonionic surfactant Polysorbate 80 is the pharmaceutical version of the well-known Tween 80 surfactant. It is an oleate ester of sorbitol and its anhydrides: copolymerized 20 moles of ethylene oxide for each mole of sorbitol and sorbitol anhydrides. The hydrophilic part includes free hydroxyl and oxyethylene groups.

The glass temperature during freezing (T_g') for many proteins has been found near –10° C (Chang and Randall 1992). Norberg and Nilsson (1996) reported results of studies on glass transition in DNA. The molecular dynamics results agreed with experiments establishing the glass transition temperature within the 223 K to 234 K range (–50° to –39° C)—the experimental result was 238 K (–35° C) at a water content of about 32 percent. The definition of glass transition involved the number of hydrogen bonds to water; this number reached maximum at the glass transition temperature (the origin of the glass transition was interpreted as the melting of hydrogen bonds of bound water molecules on the surface of DNA).

The formulations are designed to optimize the lyophilization process; both glass temperatures (T_g' and T_g) are important, and the goal is to obtain a formulation with sufficiently high T_g' and T_g temperatures.

In addition to estimating the behavior of salts during freezing and thawing (considering a possibility of precipitation and crystallization and resulting changes in solution buffering capacity and pH shifts and estimating the glass transitions), one should also consider the interactions of ions formed from those salts, including effects of their hydration shells and possible interruption to the bulk water clusters (Marcus 1985). For example, the hydration water of K^+ may weaken the water structure. The hydrogen bonds of the hydrogen sphere of the Na^+ ion are stronger than in bulk water, i.e., it strengthens the hydrogen bonds of water (Luck 1991). Such ionic interactions have been classified as water structure "breaking" or "making." Ca^{++} and Mg^{++} are other examples of structure making ions, whereas Cl^- may belong to the structure breaking ions (Marcus 1985). Both $H_2PO_4^-$ and HPO_4^{--} ions can be considered structure making (ibid.).

Increasing concentration of salts in the interdendritic space can lead to the well-known phenomena of salting-in and salting-out of proteins and other biological macromolecules. The phenomena of salting-out or -in are explained by competition between dissolved salt and protein for water of hydration. For example, at high concentration, the Na^+ ion can be a more active protein precipitant than NH_4^+ ion, and SO_4^{--} ion is a better precipitant than Cl^- ion. A simple approach based on those interpretations of ionic-hydration may consider that the Na^+ and SO_4^{--} ions are water structure makers and NH_4^+ and Cl^- ions are water structure breakers (Marcus 1985). Recent interpretation of the salting-in and -out phenomena has been based on interaction of ion pairs and on partial molar volumes of ions (Leberman and Soper 1995). The proposed mechanism is related to the density increase in water shells adjacent to the ions, as compared to bulk water. The NH_4^+ and Cl^- ions have partial molar volumes close to the one for water (they may fit relatively well into the bulk water structure), and the Na^+ and SO_4^{--} ions have significantly smaller partial molar volumes than water, i.e., they rearrange the normal tetrahedral coordination of water molecules. Recent approaches to the ion hydration and ion pair interaction problems may be studied by following the example of studies of aqueous solutions of NaCl or Cl^- ion alone behavior in the aqueous solution. The effect of NaCl solution concentration increase may affect differently the Na^+ and Cl^- ions. In the case of Na^+ ion, the number of water molecules in the hydration shell may decrease, since some of the water may be replaced by Cl^- ions, whereas for the Cl^- ion the number of water molecules may remain the same. The Na^+ ion may not substitute for the water molecules in the hydration shell of the Cl^- ion (Lyubartsev and Laaksonen 1996). Recent controversy about hydration of Cl^- ion (Jorgensen and Severance 1993; Perara and Berkowitz 1991, 1992, and 1993; Stuart and Berne 1996) may suggest that ionic hydration effects and interactions of hydrated ions with proteins require further investigations.

The addition of surfactants may prevent protein aggregation, but it may also lower the surface tension. The surfactants are typically used in low concentrations. In general, an increase in the hydrophobic tail length of a surfactant molecule may enhance surfactant action at low concentration and may reduce surface tension (Smit et al. 1990).

The role of cryoprotectants also includes creation of an environment for active product biomolecules during freezing, thawing, lyophilization, and storage, which promotes maintenance of the native state of protein. Certain deleterious reactions may even occur in the glassy states, and the formulation composition should minimize these effects. An analysis of protein and solute hydration and molecular interactions among protein-water-solute-ice is of great importance in the understanding of protein protection in the frozen and lyophilized state.

The general trend toward designing formulations with high T_g' temperatures may also have an influence on freezing with the dendritic ice crystal growth, e.g., the temperature gradient along the free dendrite may be less steep (smaller temperature difference between the dendritic tip and the glassy/eutectic temperature at the free dendrite base). The cooling rate in the interdendritic space may be affected as well as the dendritic spacing, as was discussed in the section on the dendritic ice crystal growth. These two variables alone may determine the conditions and effects of the freezing/thawing process (important for protein and biological cells) as well as the conditions of lyophilization (ice sublimation and secondary drying) and the resulting porous product structure.

The formulations of modern biopharmaceutical drugs are developed in most cases with the final lyophilization process in mind. These formulations are designed to provide drug stability during freezing, first-stage drying (ice sublimation), second-stage drying (residual moisture removal), and subsequent storage. Such formulations, if correctly designed, should also provide protection during the bulk freezing and thawing steps. However, the freezing step of the bulk may vary from the final freezing prior to the lyophilization. The bulk freezing is conducted in volumes as large as possible for economy and logistics reasons with an intensified heat transfer, whereas the freezing prior to lyophilization takes place in thin layers (usually thinner than 20 mm) with one side heat transfer (shelf-vial-product) hindered by thermal resistances of the shelf and vial walls and of the air gap between the shelf and vial bottom. In the freezing prior to lyophilization, the dendritic ice crystal growth may be conducted in a pattern to grow parallel dendrites in the direction vial bottom-upwards across the layer. The growth conditions may promote dendrites with the fine side branches to develop a network of delicate ice crystals. Such a pattern may aid ice sublimation during primary drying and water diffusion through the product texture toward the pores left by sublimed ice during the secondary drying. Since there is no prolonged storage of product in the frozen state, such an extended product-ice interface may cause only limited or no product deterioration if the formulation is properly designed. It is believed that the protection of protein by sugars (such as sucrose or trehalose) in the lyophilized state is possible by substitution of water bound on protein surface by these sugars. Recent review of formulation optimization for lyophilized products (e.g., those that are most likely to be encountered in the large-scale bulk freezing and thawing process) was published by Carpenter and Chang (1996).

In the case of bulk freezing with a long storage time for the product in the frozen state, the extended interface product-ice that occurs for dendrites with developed side branches or with columnar dendrites/cells with very fine spacing may not be desirable if the contact product-ice is detrimental for the product. This factor may become even more important if the frozen product storage temperature is higher than the solute's eutectic or glassy state temperatures, e.g., if there is a liquid phase present between ice crystals. In general, the recommended storage temperature of the frozen product should be below the solute's eutectic and glassy state temperatures.

A new trend in formulation design is to adapt the formulation to the novel drug delivery methods even closer than in traditional liquid and lyophilized/reconstituted products. Cleland and Jones (1996) described the design of stable formulations for recombinant human growth hormone (rhGH) and interferon-gamma (rhIFN-γ) for the specific drug delivery method, e.g., using microencapsulated biodegradable microspheres. The formulation's composition had to consider the manufacturing process of obtaining microspheres. The highly concentrated proteins (>100 mg/mL) must not lose activity after being dissolved in organic solvents (such as methyl chloride and ethylene acetate) prior to production of microspheres and subsequent storage. Trehalose and mannitol were found to be successful protein stabilizers.

Freezing of Cell Suspensions and Fermentation Broths

Large-scale cell suspension freezing processes can involve the cell culture and fermentation broths. The freezing process design should consider medium composition as well as cell by-products present in the broth. For example, a defined (simple) growth medium for growing *Escherichia coli* consists of carbon and nitrogen sources and inorganic salts. The medium may be composed of glucose (a carbon and energy source), NH_4Cl or $(NH_4)_2SO_4$ (a nitrogen source), salts ($MgSO_4$, KH_2PO_4, Na_2HPO_4), and salts of iron, manganese, and copper. Complex media contain all necessary ingredients for growth of microorganisms, but their exact composition is not always known. Some of their components are acid or enzymatic digests of meat tissue, casein, or yeast cells. They are a source of polypeptides, amino acids, vitamins, and minerals. General media formulations for bacteria and yeast can be found, for example, in books by Atlas and Parks (1993) or by Cote and Gherna (1990), in the Difco Manual (1985), in the ATCC Catalogues (1992), and in textbooks on microbiology (Brock and Madigan 1988). Carbohydrates present in the broths also can be cryoprotectants for cell freezing.

The media for animal cell culture are more complex than media for bacterial fermentations and contain glucose as an energy source, salts (with the following ions: Na^+, K^+, Mg^{++}, Ca^{++}, Cl^-, SO_4^{--}, PO_4^{---} and HCO_3^-), vitamins, amino acids, and serum. For example, the popular Eagle's Dulbecco's modified medium contains 4.5 g/L of glucose, 6.4 g/L of NaCl, 0.084 g/L of arginineHCl, 0.584 g/L of glutamine, and a variety of other components in small quantities (Freshney 1994).

The freezing process itself and the ice-solute, ice-cell, and unfrozen solute-cell interactions may depend on the growth medium and cell metabolic products present in the medium. If antifoams are added during the fermentation process, their presence may also affect these interactions.

Effects of Sample Size and Shape

Most of the reported work on freezing and thawing of suspensions of microorganisms and cells pertains to a small scale, i.e., samples are frozen in vials, straws, capillaries, slides, and other small containers of sizes from a few microliters to several tens of milliliters. Even for small samples cooled at low overall cooling rates, there may be differences in cooling rates in time and at the different sample points. The necessary condition of obtaining rapid and uniform cooling rates across the sample volume is to minimize sample size. For larger samples (with volumes 0.5–1.0 cubic millimeter or more), there is a difference between cooling rates at the sample surface and in its interior. Larger samples may require cryoprotectants to reduce the effects on cells caused by differences in the cooling rate.

Bald (1993) investigated temperature histories in a 1.5-mL vial cooled at 1° C/min. Cooling rate was found to change in time. Hartmann et al. (1991) investigated the methodology of determination of cooling rates in large samples, e.g., finding which temperature measuring points may provide a representative cooling rate for the sample to ensure reasonable cell survival rate at various points.

Reports on freezing cell suspensions in quantities of several hundred milliliters or several liters are rare from fields other than blood preservation. Information on freezing batches of several tens to several hundred liters are scarce from areas other than the food industry. Large-scale operations frequently use the approach of filling blood/serum bags with suspensions and freezing them in cabinet or blast freezers. Large batches of several hundred liters can be subdivided into numerous bags with a capacity from a few hundred milliliters up to a few liters each. The advantage of using spherical and cylindrical samples as compared to cubes, rectangular prisms, or small slabs is due to lack of, or limited, edge effects and, as a result, more consistent freezing conditions.

Controlled Cell Damage Using Freezing and Thawing

A controlled process of freezing and thawing of microorganisms may result in selective damage (lysis) to the cell membrane and also to the product located between the outer and inner membranes in gram-negative microorganisms. Carefully applied freezing and thawing procedures may release the product from the intermembrane space into surrounding liquid without release of internal contents of the cell. Such a step may facilitate subsequent product purification steps. However, the procedure may require a very precise control of freezing and thawing processes. Under controlled cooling rates, the extracellular ice crystals may reach a predetermined size. This phenomenon may be used to inflict limited damage to cells if controlled release of the product of interest is a goal. This field, however, requires further investigation.

ESTIMATE AND ANALYSIS OF FREEZING AND MELTING PROCESSES

Estimate of Freezing Rates

The freezing process has been investigated to provide answers regarding freezing times of materials of various shapes and volumes. Figure 2.9 shows a typical thermal history of points located in the

Figure 2.9. Freezing histories of the sample surface and the center for large and small samples. *T* is the temperature, and *t* is the time.

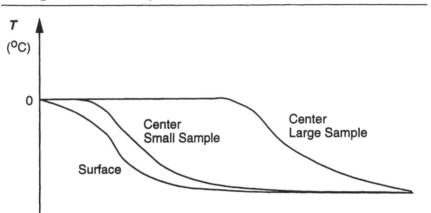

freezing material (for example, a tissue) near the cooled surface and at the center when a large sample in the form of a block is being frozen.

With a decrease in sample size, the curve for the central point approaches the curve for the surface and the temperature plateau almost disappears. With a small enough specimen, points near the surface and near the center may experience very similar thermal history during freezing.

Initially, the formulas by Plank could be used (Geankoplis 1983) to calculate approximate freezing times. Those formulas may provide an indication of anticipated dimensions of process equipment based on a freezing time. The formulas, however, do not apply if there is a temperature change on the cooling agent side and may not be used for final process specifications. Freezing and thawing are the phase change problems (or moving boundary problems), and they have received wide attention in recent years. The need for process estimates has arisen in order to find solidification rates and total solidification and cooling times (depending on anticipated sizes of samples of solidifying material, cooling agent temperature, and its temperature change rate). An example of an introductory review of freezing and melting is the work by Yao and Prusa (1989).

The general problem of freezing with latent heat release and moving solid-liquid boundary was first approached by Neumann and Stefan in the 19th century. To honor their work, the solidification problem is often called the Stefan-Neumann problem. Analytical solutions for solidification and melting for simple geometries were summarized by Carslaw and Jaeger (1959). Crank (1984) presented an overview of general moving boundary problems. Fukusako and Seki (1987) reviewed fundamentals of analytical and numerical methods for freezing and melting heat transfer problems. A newer summary of analytical and some approximate solutions were given in a work by Lunardini (1991). Saitoh (1991) provided a summary of numerical methods in freezing and melting with comparisons to the analytical approach. In this work, computer models were used to estimate freezing times when the temperature of the cooling agent was changing. The models permitted changing the cooling temperature and the cooling rate. This temperature may continuously decrease, and the temperature gradient between the freezing material and the cooling agent may change (Wisniewski, 1998a). The purpose of these investigations was to provide control over the freezing rate (by controlling the cooling rate) as well as to affect the freezing pattern and damage done to microorganisms.

Analytical and Approximate Methods

The analytical approaches that provide exact solutions to freezing and melting problems are developments of the original Neumann solution. The Neumann solution uses the Gauss error function (erf) for calculating temperatures in the freezing or melting media. The equations for solid and liquid phases can be solved using a trial and error approach. A summary of exact solutions for simple geometries such as slab, cylinder, sphere and semi-infinite space is given in a book by Carslaw and Jaeger (1959); the reader may look there for further details. Analytical solutions for freezing and melting in a wedge configuration with various angles were proposed by Budhia and Kreith (1973).

The Plank Approximate Solution and Quasi-Steady Approximation

The approximate solution for freezing time of a sphere was proposed by Plank (Geankoplis 1983):

$$t = \frac{h_m \cdot \rho}{T_m - T_f} \cdot \left(\frac{a}{6 \cdot U} + \frac{a^2}{24k} \right)$$

where t is freezing time, T_m is solidification temperature, T_f is final temperature of frozen solid, h_m is latent heat of phase change, ρ is density, a is diameter of a sphere, U is heat transfer coefficient at the cooled surface, and k is thermal conductivity of a frozen mass. Plank's formula estimates times for freezing until the sample center reaches the temperature of 10° C below the initial freezing temperature. The changing water content factor may be included by multiplying the latent heat of phase change of ice (333 kJ/kg) by the fraction of water content. A granule 5 mm in diameter with an external temperature of -30° C and a heat transfer coefficient of approximately 100 $\dfrac{W}{m^2 \cdot K}$ would freeze in about 1 min 41 sec, according to this formula. The Plank approximate solution for freezing times of slab, cylinder, and sphere shapes was developed to estimate freezing times of meat and fish tissue for refrigerated storage. This approximate method may, however, give a wide margin of error. For example, the calculated freezing times for a 50-mm (2-inch) thick slab in a blast freezer can be approximately 5 hrs 5 min using the Neumann approach or 5 hrs 36 min using the Plank equation (Charm 1981). The Plank equation may give shorter freezing times than computer simulations (Heldman 1983).

The preliminary approach to granule freezing can be done by a quasi-steady approximation using Stefan, Biot, and Fourier dimensionless numbers.

The Stefan number is described as

$$Sn = \frac{h_m}{C_{p1} \cdot \left(T_m - T_f\right)}$$

and is the ratio of the latent heat of phase change $H_m = m \cdot h_m$ to enthalpy of temperature change (heat removed during cooling

$$H_m = m \cdot h_m$$

$$H = m \cdot C_{p1} \cdot \left(T_m - T_f\right)$$

The Stefan number is also described in literature as the reciprocal of the above

$$\frac{C_{p1} \cdot \left(T_m - T_f\right)}{h_m}$$

where T_m is the freezing/melting temperature, m is the mass of freezing liquid, T_f if the final temperature of frozen mass, h_m is the latent heat of phase change, and C_{p1} is the specific heat of the frozen mass.

The quasi-steady approximation uses the Fourier and Biot numbers and their relation to the Stefan number to derive the approximate freezing time. The Fourier number is

$$Fo = \frac{k}{\rho \cdot C_p} \cdot \frac{t}{r^2}$$

where k is the solid thermal conductivity; ρ is the solid density; C_p is the specific heat; t is the time; and r is the characteristic dimension.

The significance of the Biot number is the ability for transitional heat transfer from the surface into the object interior, i.e., considering cooling of the solidified layer during freezing or warming of the solid core during thawing. The Biot number is

$$Bi = \frac{r \cdot h}{k}$$

where r is the characteristic dimension, h is the heat transfer coefficient on a surface, and k is the solid thermal conductivity.

Samples at $Bi < 0.1$ undergo rapid cooling (thermal histories of different points of samples, e.g., cooling rates may be similar), whereas at $Bi > 0.1$ the cooling is slower and thermal histories (cooling rates) of different parts of the sample may vary. Due to low thermal conductivity of ice, heat transfer by conduction in a frozen sample has significance and has to be considered along with the external heat transfer. The coolant temperature (coolant-freezing temperature difference) as well as the heat transfer coefficient at the sample surface have influence on the cooling rate. The geometry of a specimen and the area of heat transfer surface also have an influence on thermal histories of different sample points and their cooling rates. The Biot number meaning for a transitional heat transfer of granule solidification is at the end of freezing when only a small volume remains unfrozen near the center and most of the granule is frozen and being cooled through the surface. The Biot number of solidified droplet of aqueous solution 2 mm in diameter cooled in liquid nitrogen (with the external heat transfer coefficient of about 150 $\frac{W}{m^2 \cdot {}^\circ C}$) is approximately 0.0225.

The quasi-steady approximations (Grigull and Sandner 1984) are applicable for aqueous solutions and water-containing solids since in the range of Stefan numbers between 2 and 5.6 (Stefan number for water is 5.6), the error to estimate the freezing time is between 4 and 8 percent. For smaller Stefan numbers, the error increases (for a Stefan number of 1, the error may reach 14–15 percent). Using this method, the calculated freezing time of a sphere (droplet of aqueous solution) with a diameter of 3 mm suddenly immersed in liquid nitrogen (temperature of –196° C; heat transfer coefficient of approximately 150 $\frac{W}{m^2 \cdot °C}$) is about 6.2 seconds. Fine droplets of liquid (in the size range from a few to tens micrometers) would freeze very rapidly, and water supercooling and formation of very small ice crystals may be expected. This method gives the freezing time of a slab 20 mm thick and cooled on both sides by an agent at a temperature of –50° C and a heat transfer coefficient of 20 $\frac{W}{m^2 \cdot °C}$ of about 52 min 46 sec and at heat transfer coefficient of 200 $\frac{W}{m^2 \cdot °C}$ of about 9 min 10 sec. Conti (1995) compared the quasi-steady approximation with a numerical method for a finite slab with results for freezing front position in time closely following each other at shorter times and diverging at longer times.

Comparison of the Freezing Times Calculated Using Different Methods

Use of the exact solution based on the error function to find the final solidification time involves finding the root of the equation where the sought variable is included in the error function erf. The representation of the error function by a series (Lunardini 1991) may facilitate this step (a solver in any mathematics software package can be used to solve an equation with such a series representation). The erf function can be approximated as follows:

$$erfx = 1.12838 \cdot \left(x - \frac{x^3}{3} + \frac{x^5}{10} - \frac{x^7}{42} + \frac{x^9}{216} - .... \right)$$

The 4-component series gives relatively accurate values for the error function when $x<1$, while 5- and 6-component series give acceptable results up to $x=1.3–1.4$. The series with 5 components was used for the calculations of the freezing times of the slab in the following discussion. Since the use of error function for the exact solution may be too cumbersome for a rapid estimate, other methods may be considered. The exact solution can be closely approximated by the heat balance integral method based on enthalpy (Voller 1987). The author's experience showed that the quasi-state approximations give different solidification times than the exact solution (using an error function). The difference is caused by the fact that the quasi-state approximations use an external heat transfer coefficient; consider a possible heat flux limitation on the cooling agent (heat removal) side, whereas the exact solution uses a boundary at a certain cooling temperature and the solidified layer grows from this boundary toward the liquid side. At low heat transfer coefficients, the solidification times calculated using the quasi-stationary method are longer than those obtained using the exact solution. Only at very high heat transfer coefficients can these results match the exact solution or can the times become even slightly shorter (however, special techniques should be used to obtain such high heat transfer coefficients, such as nitrogen boiling with moving heat transfer surfaces, etc.). Size of the sample may also affect the solidification time differences calculated using these two methods.

Following are the results of calculations for freezing slabs of various thicknesses using the exact solution based on an error function, the heat balance method based on an enthalpy (Voller 1987), and the quasi-stationary method (Grigull and Sandner 1984). The same cooling temperature (233 K) is used for all cases, and various heat transfer coefficients $U\left(\dfrac{W}{m^2 \cdot K}\right)$ are used with the quasi-stationary method.

(a)   Slab thickness 25 mm, $T_c = 233$ K

Freezing times: exact solution $t_f = 306.46$ sec; heat balance integral $t_f = 306.73$ sec

Table 2.8 shows the freezing times calculated using the quasi-stationary approximation at different heat transfer coefficients.

Table 2.8.

Heat tr coeff U	20	40	60	80	100	200	500	1000	2000
Time, sec	5,083	2,675	1,872	1,471	1,230	749	460	363	315

The high heat transfer coefficients are possible to obtain using a boiling cryogenic fluid with a moving heat transfer surface, but cannot be achieved using traditional cooling techniques (such as conventional wall-liquid heat transfer using a low-temperature heat transfer agent).

(b)   Slab thickness 100 mm, $T_c = 233$ K

Freezing times: exact solution $t_f = 4,903.35$ sec; heat balance integral $t_f = 4,907.70$ sec

Table 2.9 shows the freezing times calculated using the quasi-stationary approximation at different heat transfer coefficients.

Table 2.9.

Heat tr coeff U	20	40	60	80	100	200	500
Time, sec	23,543	13,911	10,701	9,096	8,133	6,206	5,051

(c)   Slab thickness 200 mm, $T_c = 233$ K

Freezing times: exact solution $t_f = 19,613$ sec; heat balance integral $t_f = 19,630$ sec

Table 2.10 shows the freezing times calculated using the quasi-stationary approximation at different heat transfer coefficients.

Table 2.10.

Heat tr coeff U	20	80	150	250
Time, sec	55,647	26,753	22,259	20,204

These results indicate that for the larger samples the quasi-stationary approximation may approach the exact solution at practically achievable heat transfer coefficients.

Numerical Methods

Numerical methods use the following main approaches: finite difference, finite element, and finite volume techniques of solving the set of equations describing the system with a phase change. These methods may also allow using a pre-programmed temperature change outside a freezing granule, e.g., to introduce additional variables into freezing dynamics.

There are numerous publications describing numerical small-step methods for phase change problems based on solving moving boundary problems in various geometries. At negligible convection, the numerical approaches may use fixed or variable grid methods. Fixed grid methods have been more popular due to simpler algorithms than those needed for variable grids. Basics of the fixed and adaptable grid approaches can be found in work by Crank (1984). Lacroix and Voller (1990) demonstrated that the fixed grids could produce comparable accuracy with the transformed grids.

The finite difference, finite element, and enthalpy methods had initial difficulties with accurate tracking of moving interface. Numerous solutions for handling this problem have been proposed (for review of those, see Fukusako and Seki 1987; Saitoh 1991; and Samarski et al. 1993). Voller and Cross (1981) proposed an improved enthalpy method to solve these problems. The improved enthalpy method has become one of the popular numerical approaches to the moving boundary problems (Voller 1985). The reason for its frequent use is the fact that a continuously moving, liquid-solid boundary does not need to be tracked over a numerical grid. However, the accuracy of this method may be influenced by the position of the interphase front.

Viswanath and Jaluria (1993) compared the enthalpy method with fixed grid to the transformed grid method with immobilization of moving boundary via transformation. The enthalpy method, although it may sometimes give worse prediction of the moving interface, was found suitable for analysis of the flat interface and mushy (dendritic) regions. Voller and Swaminathan (1993) described approaches to treatment of discontinuous thermal conductivity in phase change problems when using fixed space grids. Vicks and Nelson (1993) proposed to use the boundary element method for freezing and melting with the purpose of simplification of handling the moving solid-liquid interface. Sutanto et al. (1992) reported on solidifying front velocities in the freezing plate, cylinder, and sphere using an adaptive finite element analysis of axisymmetric freezing. After initial acceleration, the front velocity stabilized for the plate. For the cylinder, the front velocity initially accelerated, reached a short plateau, and further increased while approaching the center. For the sphere, the velocity profile was similar to the cylinder, but the velocity continuously increased (no significant plateau existed).

Voller (1997) presented a thorough review of numerical methods for solving phase change problems. The fixed and deforming grid methods were compared as well as the hybrid schemes (node jumping, local tracking, and deforming enthalpy methods). In addition to basic approaches, the more complex problems were also reviewed: phase change with discontinuous thermal properties, phase change with fluid flow, and problems with mushy zones. Information on the construction and performance of various schemes was provided. The simple cases of the flat solid-liquid front were examined.

The computational methods for mushy zones, problems with changing enthalpy, and problems with the coupling of thermal and solute fields remain active areas of developmental work. The numerical methods are tested by checking their results against known exact analytical solutions for a few standard geometries. Usually at the small-step sizes, the reported results of numerical methods closely approach those obtained by analytical methods. The author has developed a method with pre-programmed external temperature change with a constant or variable grid for tracking the moving front toward the center of a sphere or slab. The method included an option to allow the changing freezing temperature and changing composition of liquid phase following the phase diagram to account for possible cryoconcentration effects and, as a result, to reach eutectic freezing in the center of a sphere or slab. The heat transfer coefficient on the cooling agent side as well as the container wall thickness and its thermal conductivity could be entered for more accurate representation of the real conditions. For example, the 25-mm slab freezing between two stainless steel plates 5 mm thick and at heat transfer coefficients on the cooling liquid side of $100 \dfrac{W}{m^2 \cdot K}$ and the cooling fluid temperature of $-40°$ C gives the freezing time of eutectic solution (including the solidification of a eutectic at $-21°$ C) of 1,498 sec. The analytical solution and the heat balance integral solution (both without considering wall parameters, external heat transfer coefficient, and eutectic) gave about 306 sec freezing time, while the quasi-stationary solution (with the heat transfer coefficient, but with no wall effect, and no specific, low-temperature eutectic) gave a freezing time of about 1,230 sec. The numerical solution with a eutectic near $-11°$ C gave a freezing time of about 1,172 sec. This solution is close to the quasi-stationary solution, which is based on the principle that the center of the slab has to be cooled below the initial solidification temperature of 0° to –0.5° C by about 10° C. The analytical and heat balance integral solutions provide the freezing time when the solid-liquid interfaces meet, but without further lowering of temperature, and do not consider thermal resistance at the boundary (heat transfer coefficient on the cooling agents side, wall thermal conductivity). Therefore, shorter freezing times may be anticipated when using these methods. These methods were also tried for an estimate of the freezing time of freezing droplets and granules (see Chapter 3 "Large-Scale Cryopreservation: Freezing and Thawing of Large Volumes of Cell Suspensions, Protein Solutions, and Biological Products" by the author). For the granule formed during the agitated mass freezing, the limits of this approach have been recognized, since there was no provision for the presence of gas bubbles and previously nucleated ice crystals within the freezing granule.

Updated numerical approaches to the phase change problems are being frequently published, and readers may follow such publications as, for example, International Journal of Heat and Mass Transfer, or Numerical Heat Transfer Journal for recent developments.

SLAB, CYLINDER, AND GRANULE FREEZING

Slab/Cube Freezing

A slab is a simple geometry, but slab freezing is a complex heat transfer problem with the latent heat of phase change and moving boundary factors. Slab freezing may be conducted with one- or two-sided cooling. The simplest case is a single-side cooling of a slab of infinite length. In reality the slab position may influence the outcome of the freezing process. In a vertical slab position the natural convection effects in the liquid phase will be more pronounced than in the horizontal position. In the horizontal position, there is a difference in solidification depending on which surface is being cooled (the upper or lower one). Often the lower surface is cooled for practical reasons since the container with product may be placed on a cooled plate. Cube freezing is a particular case of slab freezing with corner effects. These effects cause more rapid freezing in sample volumes adjacent to the corners. Practical applications of the slab freezing problem often occur in the food industry where slabs of meat or fish are frozen. Freezing of liquids in flat bags may be modeled using slab freezing.

Spherical Granule Freezing

Freezing Phenomena

Granule freezing is typically approached by letting liquid droplets fall into a low-temperature bulk liquid such as brine, ethanol, or liquefied gas (such as nitrogen) or spraying the liquid into an atmosphere of cold gas. The freezing process is very rapid, and its completion depends upon droplet size and water content of the solution. In the case of liquid nitrogen, the temperature difference is high considering the freezing point of aqueous solutions at near 0° C and the liquid nitrogen boiling point of –196° C (77 K). Therefore, the surface heat flux is very high. The nitrogen boiling phenomenon on the granule surface and formation of the gaseous phase may lower the heat flux since the gaseous phase may locally separate the bulk of liquid nitrogen from the solid surface. Formation of liquid droplets may become a critical factor in uniformity of freezing conditions because the droplets should have uniform diameters (droplet size distribution should be as narrow as possible).

The spray freezing method may generate very high cooling rates if fine, uniform droplets are developed. At rapid cooling rates, the limiting thickness of the sample surface layer is less than 0.1 mm (100 μm) with a layer of uniform fine crystals occurring at a depth of about 20 micrometers. Therefore, the required size of droplets should be below 120–180 micrometers. Atomization of aqueous solutions into droplets of 50–100 (or even 10–50) micrometers is technically feasible. The spray may be directed into a pool of liquid nitrogen (kept at 77 K), propane (cooled to about 85 K), or ethanol (cooled to about 170 K) or may be injected into a cold gas chamber where the liquid nitrogen is simultaneously injected.

Freezing of droplets in the cold gas involves heat and mass transfer phenomena in the cloud of droplets that vary from the heat and mass transfer using the single droplet approach. In addition to extensive empirical and semiempirical descriptions of such phenomena, numerical models have been proposed. These models are based on solving the Navier-Stokes equations with heat and mass transfer at low and intermediate Reynolds numbers using fine meshing for arrays of particles (Dwyer et al. 1994).

Heat flux during granule freezing may decline due to a decrease in the area of solid liquid interface while approaching the granule center as well as due to an increase in the heat conductive solidified layer. This effect can be offset by maintaining a high heat transfer coefficient at the surface and by a continuously decreasing external temperature of bulk environment (e.g., if freezing at the solid-liquid interface occurs at declining temperature, the temperature difference between the phase change interface and granule surface may increase and counteract the increase in thermal conduction resistance through the solidified layer).

Droplet Formation

Information on droplet formation can be found in works related to liquid fuel combustion and spray drying—two areas in which uniform droplet size is of great importance (Beer and Chigier 1983; Masters 1979). When evaluating an atomization technique for cell suspension spraying, one must consider the effects of hydrodynamic shear generated during atomization. Typically, formation of droplets may be accomplished by a rotary atomizer or a single- or multiple-fluid nozzle. The rotary atomizers are used for high capacity installations in the spray drying application. High shear rates occur in the disks of these atomizers, since the rotating speeds may reach 20,000 RPM. The cooling/solidification and freezing applications may use lower speeds since larger droplet sizes can be acceptable.

Nozzle spraying requires higher energy input and higher pressure than rotary atomization and can be used from very low to medium capacities with a scale-up approach using an increasing number of identical nozzles. In the spray drying applications, the single-fluid nozzles operate under pressures of 20–330 bars (330–5000 psig). Solidification applications may use coarse nozzles at low operating pressures. Two- and multiple-fluid nozzles use compressed air/gas to disperse the liquid into droplets. These nozzles are used for low-capacity installations due to the high cost of compressed air. The air pressure can be within the range of 1.3–6.5 bar (20–100 psig). Turbulence parameters of the gas jet may affect the quality of liquid atomization (Shavit and Chigier 1996). For spray freezing application in the cold nitrogen gas atmosphere, the compressed dry nitrogen gas can be used for atomization to minimize product oxidation. These nozzles offer a wide range of capacity without significantly affecting the droplet size distribution. However, the droplet size distribution is wide. Luo and Svendsen (1996) proposed a model for drop breakup in turbulent dispersions. The drop oscillations leading to breakage depend on the size of the turbulent eddies and energy dissipation. Unlike earlier models, this model does not have any unknown parameters. Panchagnula et al. (1996) investigated the stability of swirling liquid sheets at varying atomizing gas velocities. Gas-generated atomization typically produces high local shear forces.

Nozzle spraying typically provides a wide particle size distribution. Average and mean droplet diameters may not be representative of completion of the solidification process; therefore, the maximum droplet diameter should be used for system design parameters estimates. Large droplets may contain most of the liquid volume. For example, for droplets formed by a nozzle with the droplet diameter range 10–600 micrometers, the average Sauter diameter was found to be 331 micrometers and droplets larger than this diameter had 49.4 percent of the total droplet surface and 71.6 percent of the total droplet volume (the Sauter average diameter was used since it involves surface and volume of droplets). The droplet volume is of particular importance during solidification due to a need for latent heat removal. The droplet surface determines external cooling heat transfer. Considering the volumetric distribution of droplets, the largest diameter (d 95 percent) may be estimated as approximately 2 times the d 50 percent diameter for the single-fluid centrifugal nozzles. Droplet size distribution may be estimated by freezing the spray in a pool of liquid nitrogen and examining the sizes of frozen droplets.

Due to a relatively wide range of droplet size distribution, traditional spraying nozzles may not be used for repeatable biological freezing applications. Narrow diameter bore tubes could be applied with two possible approaches: using a continuous pulseless low rate pumping and a liquid jet breakup and droplet formation by hydrodynamic instability, or using a hanging drop formation and a subsequent droplet shaking off of the tube tip. In the second case, a pulsating flow may be required with an initial drop formation and a final shake, e.g., a special dosing pump may be used. The naturally unstable flow of a liquid jet would produce some droplet size distribution, while controlled drop formation at the tip may produce more uniform spheres. Since the capacity of a single nozzle is low (particularly in the second case), an array of nozzles may be required.

This type of flow belongs to the surface tension driven flows (Keller and Miksis 1983) under the condition that there is no initial velocity imposed upon the liquid thread. Eggers and Dupont (1994) proposed one-dimensional approximation to drop formation using an equation derived from the Navier-Stokes equation. Results were compared to experiments on the breakup of a liquid jet. Computed profiles coincided with experiments. Theory of droplet formation was later reviewed by Eggers (1995). The free flowing vertical column of liquid develops surface instabilities leading to column breakup (Papageorgiou 1995a,b). The presence of surfactants (including biomolecules with surfactant properties) would affect the droplet formation and its size and contribute to the stability of liquid thread (Awati and Howes 1996) and its breakup into droplets. The jets ejected from circular orifices under unsteady inflow conditions are expected to produce main size droplets and smaller satellite droplets (Hilbing and Heister 1996). The size of the droplets can be affected by changes in perturbation wavelength, perturbation magnitude, and the Weber number.

$$We = \frac{\rho \cdot a \cdot U^2}{\sigma}$$

where ρ is liquid density, U is mean inflow velocity, a is orifice radius, and σ is liquid surface tension. The perturbations may be purposely introduced (randomly or periodically by the droplet-forming equipment) to affect the main/satellite droplet diameter ratio. The Weber number increase may cause a decrease in the size of the satellite droplets.

The formation of satellites is a nonlinear phenomenon, and the satellites form in the last period of liquid thread breakage (Mansour and Lundgren 1990). An extensive description of satellite and subsatellite drop formation in capillary breakup of a very long fluid filament was published by Tjahjadi et al. (1992). Droplet/particle uniformity is important considering the requirement of identical cooling and freezing conditions for all the product. Small satellite droplets are thus undesirable and an effort should be made to generate droplets of uniform diameter.

The large droplets of uniform diameter in the range of 1–3 mm can plunge into liquid nitrogen and rapidly solidify in 77 K environment. Heat transfer coefficient at the droplet surface is very high due to the boiling of liquid nitrogen. Boiling produces gas bubbles or even a gaseous envelope, which hinders heat transfer (gas cushioning). Rapid movement of freezing granule through liquid nitrogen may reduce this effect and maintain a high heat transfer coefficient. The values of heat transfer coefficient with gas cushioning may be as low as about 200 $\frac{W}{m^2 \cdot K}$ and much higher than 1000 $\frac{W}{m^2 \cdot K}$ at fully developed boiling.

Figure 2.10 shows the dependence of time of freezing of such large droplets on the heat transfer coefficient U and droplet diameter (range 0.5 to 3 mm shown) under conditions of plunging into liquid nitrogen (at 77 K). Low U values correspond to conditions close to stationary with a gas envelope around solidifying particles. High U values correspond to moving particle with reduced effects of gas phase presence. Freezing time strongly depends on droplet diameter at low heat transfer coefficient. At higher U values, the freezing times level off and are closer together for various droplet sizes.

Other types of spraying devices include electrostatic, acoustic and ultrasonic atomizers. They typically produce very fine droplets, which may not be required in the freezing process. For example, the ultrasonic atomizer may produce droplets within the range of 0.5–7 micrometers at a frequency of 2500 kHz at low-flow capacities (up to 10 mL/min) at a power input of less than 750 W. Larger droplets are produced using lower frequencies; for example, droplets of 10–30 micrometers can be obtained using 100 kHz and droplets of 30–90 micrometers using a frequency of about 50 kHz. The process of drop formation through the orifice may be modified by application of an electric field (Byers and Perona 1988). Drop diameter can be affected by voltage; drop volume decreases with an increase in voltage. Recent results on droplet formation from a capillary in the presence of an electric field were published by Zhang and Basaran (1996).

Spray Cooling and Freezing

A possible way to change the droplet solidification into a well-controlled process is to pass the droplet rapidly through a cold zone using liquid nitrogen (in the form of a film, spray, or vapor) to solidify the surface and continue subsequent freezing at a slower pace. Heat removal from the surface must be maintained since the interior of freezing droplets releases heat, and, at the same time, the droplet surface must remain hard enough to prevent formation of drop aggregates and lumps. The drops may partly freeze by a free fall in a tower in an environment of cold gas, and, after their surface hardens sufficiently, they may land in a fluidizing or vibro-fluidizing bed to complete cooling. Bricard (1991) reviewed designs of direct-contact heat exchangers including solidification towers. He described a tower to form ice drops to be used later for surface cleaning as an ice particle in compressed air jet. Liquid nitrogen was

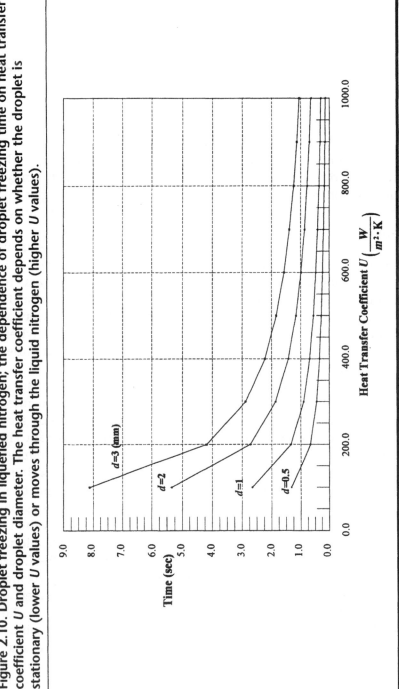

Figure 2.10. Droplet freezing in liquefied nitrogen; the dependence of droplet freezing time on heat transfer coefficient U and droplet diameter. The heat transfer coefficient depends on whether the droplet is stationary (lower U values) or moves through the liquid nitrogen (higher U values).

sprayed concurrently with water at the tower top. Droplets of water solidified during free fall in the nitrogen vapor and were removed at the bottom by a screw conveyor. The temperature of gaseous nitrogen in the heavily insulated tower was given as −196° C. This concept may be applied for freezing droplets containing cells, proteins, or other biological materials. Droplet solidification during free fall might not be as rapid as in the case of plunging it into liquid nitrogen. Larger droplets formed by dispensing orifices may require long fall time to sufficiently harden. Final freezing in a bottom nitrogen pool may provide a means for solidification of droplet cores. There are no good models available to account for mixed density sprays of aqueous solution and liquid nitrogen. Some limited analogies might be drawn from the high-temperature models of mixed sprays of vaporizing fuel and water droplets (Kleinstreuer et al. 1993).

Mizuno and Wakahama (1983) described experiments with water droplets of diameter 100–300 micrometers sprayed into air kept at temperatures ranging from −8° to −23° C. The droplets traveled in cold air for several seconds before hitting the ice surface. At temperatures below −14° C, the droplet ice crystals were very small. Above −10° C, the droplets froze as a single crystal, whereas below −14° C there were polycrystals present with the number of crystals increasing with temperature decrease.

Epstein and Fauske (1993) analyzed kinetic and heat transfer controlled solidification of highly supercooled droplets using integral profile computation. The surface heat flux was sustained by a combined convection and radiation. Erukhimovitch and Baram (1994) analyzed the solidification of fine droplets in the spray atomization process. Buchlin (1995) included the subject of droplet freezing and melting in a spray in his von Karman Institute lecture. During droplet freezing in a stream of very cold gas, the droplet undergoes phase change simultaneously with convective heat transfer. When the droplets freeze, after exchanging the latent heat of solidification, the heat transfer (cooling) continues at temperatures below the freezing point. For small freezing droplets, the temperature is approximately even across the droplet volume (depending on Biot number). The heat transfer during freezing consists of the latent heat of solidification being removed convectively through the droplet surface into the surrounding gas. The temperature level in the freezing tower may be controlled by metered injection of liquid nitrogen.

The heat transfer between small particles and surrounding gas depends on the particle diameter. For a nonstationary heat transfer during solidified droplet cooling, if the Biot number is smaller than 0.1, then at all Fourier numbers the temperature gradient across the particle will be negligibly small and, therefore, the heat transfer solid gas will depend on convection around the particle. Heat transfer between the particle and still gas is governed by

$$Nu = 2.0$$

where Nu is the Nusselt number

$$Nu = \frac{h \cdot d_p}{k_g}$$

and h is the heat transfer coefficient, d_p is the particle diameter, and k_g is the gas thermal conductivity.

The terminal velocity (in m/sec) of free fall for a spherical particle/droplet can be estimated using the following equation (Masters 1979):

$$U_f = \frac{0.225 \cdot d_p^{1.14} \cdot g^{0.71} \cdot \left(\rho_p - \rho_g\right)}{\rho_g^{0.29} \cdot \mu_g^{0.4}}$$

where d_p is the droplet/particle diameter, g is the gravitational constant, ρ_g is the gas density, ρ_p is the particle density, and μ_g is the dynamic gas viscosity.

The heat transfer under relative velocity conditions can be described by equations (Boothroyd 1971):

$$Nu = 2 + 0.6 \cdot Re^{0.5} \cdot Pr^{0.33},$$
$$for \ \ 1 < Re < 450 \ \ at \ \ Pr < 250$$

and

$$Nu = 2 + 0.6 \cdot Re^{0.5} \cdot Pr^{0.33},$$
$$for \ \ 450 < Re < 1040 \ \ at \ \ Pr < 250$$

where the Reynolds number

$$Re = \frac{d_p \cdot u \cdot \rho_g}{\mu_g}$$

and the Prandtl number

$$Pr = \frac{C_p \cdot \mu_g}{k_g}$$

where u is the gas velocity relative to particle, μ_g is the gas viscosity, ρ_g is the gas density, and C_p is the gas specific heat.

The concentration of suspended particles in a gas stream may affect the average heat transfer between gas and cloud of particles (Gorbis and Kalenderyan 1975):

$$\frac{Nu_c}{Nu} = 0.033 \cdot \beta^{-0.43},$$

$$for \quad 0.35 - 0.45 \times 10^{-3} < \beta < 2 - 2.5 \times 10^{-3}$$

where β is the particle concentration (by volume), and Nu_c is the Nusselt number for highly concentrated suspensions.

The heat transfer coefficient between droplet (particle) and gas may be estimated from these equations. The freezing time of a spherical droplet can be estimated using the quasi-steady approximation or the numerical approach. Height of the freezing tower may then be estimated using the terminal velocity of free fall for the droplet/particle.

Figure 2.11 shows the dependence of droplet freezing time on heat transfer coefficient U at droplet surface and droplet diameter (given range: 50–500 micrometers) at liquid nitrogen boiling temperature (77 K). Low values of heat transfer coefficient U are for cooling of droplets in gas phase (product spraying into the cold gas atmosphere), and high U values correspond to freezing in the liquid nitrogen (boiling with gas cushioning effects). The range of droplet size corresponds to the typical droplet sizes obtained from high-pressure fine spraying nozzles.

The estimated tower height should be adjusted (increased at downward spray) considering droplet deceleration after leaving the atomizing device. The thoroughly frozen droplets can be removed from the tower using belt, screw, or vibrating conveyor.

Final cooling also could be accomplished in the liquid nitrogen pool or in a fluidized bed located in the bottom of the tower. The drops may be transported on a fluidized bed conveyor using cold fluidizing gas to finish cooling and provide final handling. A conveyor with directed jets could move granules independently before they reach the denser bed. The temperature of gas leaving the dense fluidized bed is approximately the same as the surface temperature of bed granules. Therefore, monitoring of cooling can be performed by temperature measurements above the bed, although the temperature sensor immersed into the bed may provide more accurate information since there is no gas mixing factor above the bed. The granule removal from the fluidized bed may be performed using a

Figure 2.11. Small-diameter droplet freezing at the temperature of liquefied nitrogen boiling point (77 K); the dependence of droplet freezing time on heat transfer coefficient U $\dfrac{W}{m^2 \cdot K}$ and droplet diameter d (mm). The lower values of U are for freezing in the gas phase, and the higher values of U are for freezing in the liquid phase under liquid nitrogen boiling conditions.

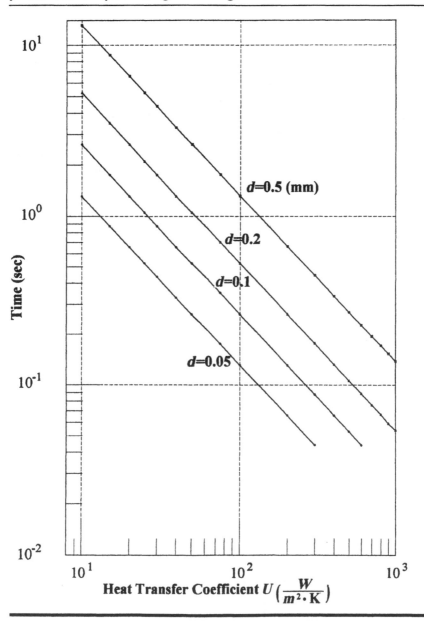

traditional weir with overflow/underflow approach or using a screw conveyor or an air lock. The fluidization processing and conveying may be accomplished using a vibratory fluid bed apparatus.

Immersion (Dip) Cooling

Slower cooling also may be accomplished by dropping the cooled granules into a low-temperature liquid other than liquefied nitrogen, which is kept at a higher temperature (e.g., NaCl or KCL brine) at a lower heat transfer coefficient (no boiling involved). This approach may create conditions for a gradual temperature decrease and for a dendritic crystal growth inside granules in a part of the volume other than the rapidly solidified crust. Droplet size should be large prior to freezing to minimize the volumetric relationship of crust/unfrozen interior, e.g., if product is sensitive to a rapid freezing, then only a small percentage will be affected by a rapidly solidifying surface crust. The crust will be then frozen in a more or less uncontrolled way, while the interior may be frozen under more controlled conditions with a temperature change slope as required. Due to limitations regarding the initial droplet size and minimum thickness of crust to provide drop rigidity, control over freezing rates is limited to the unfrozen part of the drop left after crust formation (part of product may be exposed to the detrimental effects of rapid freezing).

Atomization by disks, single- or two-fluid nozzles, or ultrasonic nozzles may produce droplets too fine to be able to conduct such extended controlled freezing, and other types of droplet-forming devices may be needed. Fine atomization also develops large liquid-gas interfaces, which may be detrimental to some products (such as proteins). In practice, in order to avoid any possible fluid-product interaction, product encapsulation may precede the freezing. This kind of freeze-granulation resembles freezing of small specimens, which is frequently used for cryobiology studies.

Granule Freezing in Agitated and Fluidized Beds

Fluidized beds are frequently used in the food industry for freezing berries or small fruit. The particles are kept in suspension by a vertically moving gas or by vibration of the fluidizer. The gas-solid heat transfer in fluidized beds is well researched and can be estimated in a straightforward way (Zabrodsky and Martin 1988). For spherical particles, the heat transfer gas-solid is very intensive and the heat transfer coefficient only slightly changes with changes in fluidizing gas velocity. It can be estimated using the formulas:

$$Nu_p = 2 + \left(Nu^2{}_{lam} + Nu^2{}_{turb}\right)^{0.5}$$

where

$$Nu_{lam} = 0.664 \cdot Re^{0.5} \cdot Pr^{0.33}$$

$$Nu_{turb} = \frac{0.037 \cdot Re^{0.8} \cdot Pr}{1 + 2.443 \cdot Re^{-0.1} \cdot \left(Pr^{0.66} - 1\right)}$$

and other symbols are as before.

The heat transfer coefficient in fluidization cooling of granules of 6 mm diameter using air may reach values of about 180–220 $\frac{W}{m^2 \cdot K}$.

The fluidized bed devices may freeze solids with high moisture content or encapsulated biological products (liquids, pastes, or suspensions). Heat transfer coefficients solids-wall between the fluidized bed and internal heat transfer surfaces can be very high and reach 700–800 $\frac{W}{m^2 \cdot K}$ or more. A dense fluidized state also may be achieved in the agitated bed of granulated material. The granulated state may, however, not easily be reached in the machine itself. One possible way is to spray cryogenic fluids (such as liquefied nitrogen or carbon dioxide) into agitated liquid until the whole mass breaks into granules, which are further cooled and frozen through. Such a process was developed by the author and is described as a case study 3.1 in Chapter 3. Freezing granules either immersed in a coolant or as part of an agitated bed with coolants delivered directly to the bed can be approximated by a spherical particle with uniform surface cooling. Freezing of a sphere involves a moving solid-liquid interface from surface toward the center with heat removal through the surface.

Gas-fluidized granules might become cores for a layered growth freezing with spraying the product onto a fluidized bed. The thin layer of sprayed product would rapidly freeze after contacting cold granules with additional cooling provided externally by fluidizing gas. The cryogen may also be sprayed into the bed to increase freezing rate. Cooling capacity of the bed would increase in time due to an increase in size (and mass) of granules and an increase in gas flow to maintain original intensity of fluidization. The bed temperature may be constant or decreasing following the present setpoint data. The cooling control may be maintained by the temperature of fluidizing gas, the product delivery rate, and the rate of cryogen sprayed into the bed. The process starts with the seed granules and there is a possibility of particle lumping unless a strict control of temperature and product delivery is applied.

Cylinder Freezing

Cylinder freezing may be evaluated for "infinite" length or including the end effects. The end effects would cause freezing intensification in volumes adjacent to the edges. Cylinder freezing involves simple geometry and is approached in a way similar to the sphere freezing. Cylinder freezing can refer to freezing in small vials and straws, bottles, and large cylindrical containers. Large-size cylinders may experience more significant cell settling, and natural convection effects.

Scale-Up of Freezing and Thawing of Cell Suspensions and Biological Solutions

The approach to the freezing scale-up of biological samples must consider that for larger samples the cooling rate can vary significantly throughout the sample volume. Since the cooling rate may strongly affect cell survival, the survival may vary, depending on cell position in the sample. Only limited information from the reviewed literature can be directly applied to the large-scale freezing of cells and microorganisms. The specimen geometries used in laboratory experiments cannot be easily duplicated on a large scale because of large volumes of processed products.

Scaling-up would require an individual approach. The following estimates may lead to a choice of method of freezing and to a selection of favorable freezing conditions.

1. Temperature histories in time within the freezing specimen must be evaluated in advance and local cooling rates estimated.
2. Movement of freezing fronts in time should be estimated and distribution of cells and solutes evaluated.

3. Information on eutectics of salts and glassy states of cryo-protectants and other substances must be known and their distribution within a specimen evaluated, including possible cryoconcentration phenomena, changes in pH, precipitation effects, etc.

Heschel et al. (1996) discussed possible applications of directional solidification in cell preservation scale-up. They proposed to enclose the sample between 2 heat transfer surfaces at different temperatures to create controlled temperature gradient between the plates. The cooling process involved lowering the temperatures of these 2 blocks with control of temperature difference. This approach was a scale-up concept of the Bridgman technique used in studies of very small and microscopic samples.

During freezing of cell suspensions in large volumes using cooled heat transfer surfaces, two additional phenomena may be involved that could cause significant deviations in freezing patterns and cell entrapment by the freezing front. These phenomena are natural convection and cell settling. With the increased height of cooled surfaces, natural convection currents in the liquid phase can interfere with patterns of dendritic growth. As a result, fewer particles might be entrapped by advancing freezing fronts and a process of particle concentration in the liquid phase might occur. Cell settling may cause excessive cell concentration at the bottom of the container. This will cause nonuniform freezing conditions across the container and changing composition of the frozen mass. An attempt to maintain cells in suspension by agitating the liquid phase will cause cryoconcentration effects for cells and solutes and will also result in changing freezing conditions for cells. Cell settling may be avoided by cell immobilization in gels or beads or by an entrapment in porous materials.

Freezing and thawing of porous materials or packed beds has attracted attention in a variety of fields (Eckert and Pfender 1978; Konrad 1991; Ohrai and Yamamoto 1991). Kececioglu and Rubinsky (1989) proposed a model for propagation of phase-change fronts in porous media. The problem of freezing and melting of salt solutions in porous media was also investigated. Yang et al. (1993) investigated solidification of a binary mixture (water/NH_4CL) in a packed bed of spheres. Glass and steel beads were used. The two solution concentrations used were below (5 percent) and above

(25 percent) the eutectic point. Large temperature gradients were observed in the glass bead matrix resulting in convective flows in the liquid phase. The convection effects were much smaller for the steel beads due to the small temperature gradients (high thermal conductivity of the steel aids in equalizing matrix temperature). Porous structures filled with aqueous solutions or cell suspensions may freeze rapidly if the porous material has high thermal conductivity and good contact with the main cooling surface. Biological material may be exposed to relatively uniform freezing conditions within the pores if the major heat transfer and removal of latent heat of solidification occurs through the porous matrix. The packed beds of regular spheres may have limited use since the bed thermal conductivity is low and affected by the small contact points among spheres. Three-dimensional extruded matrices can be designed with the purpose of enhanced freezing and thawing of entrapped materials. Such matrices, if made of highly conducting materials, may permit rapid freezing of large volumes.

Thawing would also be facilitated by use of these porous structures with latent heat of melting being delivered through the matrix structure to the frozen product embedded in pores or within the matrix. At high thermal conductivity of the matrix, rapid thawing of large volumes can be accomplished. Porous or extruded matrice structures could be considered a further development of extended heat transfer surfaces used by the author in freezing and thawing of bulk protein solutions (Wisniewski and Wu 1992, 1996; Wisniewski 1998a). Such matrices may prevent any significant cell settling. Sanitary and aseptic characteristics of such porous structures and the system ability to clean- and sterilize-in-place must be critically investigated. Thawing of large volumes of biologicals included in porous matrices also can be accomplished by applying an electric current to the matrix and heating the matrix material. Heat generated within the matrix would cause the product to melt. Resistivity of the matrix material and current parameters should be carefully selected to maximize heat delivery without overheating the product.

Thawing of large frozen blocks of material may cause very nonuniform environmental conditions for cells. This is dependent on the cell location (center or surface of the block), since only a limited temperature difference can be applied (usually, the heating agent temperature cannot far exceed the temperature of the cell culture). After some warming, the cells at the center of a block may be exposed to liquefied concentrated solutions from between the ice crystals and be affected by the phenomenon of ice recrystallization. This phenomena may limit applications of large containers with cooled walls and internal heat transfer surfaces for a cell suspension freezing and thawing process.

Hayes et al. (1988) presented a work on prediction of cell survival during freezing in a cylindrical container. The model, based on a finite element method, was used for freezing a cylindrical specimen 8 cm in diameter. Survival rates for cells at various locations in the container (exposed to different freezing conditions) were evaluated. A salt/water phase diagram was included in the model. Results of that simulation may be considered an argument against freezing of high volumes of cell suspensions in a single large container. In addition, the subsequent thawing cannot be performed in a rapid mode, and an ice recrystallization could occur, causing further damage to cells. The paper was a conclusion of earlier works by Hayes et al. on the computer modeling of phase change phenomena and its application to freezing cell suspensions (Hayes et al. 1984; Hayes et al. 1986). The freezing and thawing times were long, and thermal and osmotic histories of cells could vary significantly, depending on their location within the specimen. Temperature in the liquid phase would remain close to the freezing point, and temperature in the frozen phase would gradually decrease with time. Long freezing times would also allow cell sedimentation and, as a result, cause nonuniform freezing conditions along the height of the specimen.

In summary, the thermal and osmotic history of cells will vary if a large volume is frozen (freezing in large bags, containers, or tanks). Times of exposure for some cells to concentrated solutions can be very long, and their cooling rates and thermal histories may vary. In addition, the cell sedimentation phenomenon will create significantly different freezing conditions along the height of the container. To prevent cell sedimentation in the liquid phase during freezing, agitation of liquid can be applied. However, if any convectional currents are introduced into the liquid phase, the dendritic growth will be impeded, cell occlusion by the solid front significantly limited, and cryoconcentration phenomena intensified (Wisniewski and Wu 1996; Wisniewski 1998b). If the freezing volume can be reduced, then overall freezing times are shorter and individual cells are not exposed to significantly differing conditions before the freezing process can be completed. An additional benefit of freezing small volumes is that the rate of temperature change can be better controlled within the volume. Containers may be divided into compartments composed of heat exchange structures. The design of such compartments should overcome negative effects at cell setting, e.g., distances shall be short and freezing relatively rapid.

Thawing of large volumes also means slow temperature changes within a frozen mass and possible recrystallization of ice crystals, with subsequent damage to cells. Small frozen volumes can be thawed in a controlled way with a rate of temperature increase similar for almost all cells included in the thawed volume. Consequently, an optimum thawing rate can be achieved and an ice recrystallization process can be minimized.

SUMMARY

Development of freezing and thawing in large-scale processes involving biological cells requires a thorough knowledge of cell structure, physiology, sensitivity to substances that may be present in the environment, osmotic characteristics, transport phenomena across the cell wall, and cell response to the cold shock.

The approach to the process design should consider whether any cryoprotectants are going to be introduced into the cell suspension or whether a raw product is going to be treated "as is." In the case of freezing with cryoprotectants, the freezing process may be affected by the cryoprotectant type (penetrating or nonpenetrating) and by the molecular structure of the particular cryoprotectant.

The reported mechanisms of cell damage may all occur during large-scale freezing of cell suspensions. Since the cooling rates can be relatively low, damage by intracellular ice crystal formation can possibly be avoided, unless the droplets of cell suspension are immersed directly into a cryogenic fluid and the resulting cooling rate is high. The damaging mechanism may also depend on the freezing method used; e.g., in the case of a large sample size the time of freezing can be long and the cooling rate may vary depending on the location within the sample and time. In the large volume, cell settling may occur causing differences in cell concentration. In addition, a cryoconcentration of solutes may occur as well as a growth of extracellular ice crystals (at different rates depending on the location in the sample). Since a uniform cooling rate in large volumes is difficult to realize, division of the sample into small volumes may create conditions for more uniform cooling and freezing.

The effects occurring during freezing of biological cell suspensions are typically described as follows: Freezing based on the principle of sample immersion into liquefied gases (giving a very high rate of temperature change) can produce very small ice crystals outside and within the cells. If the temperature changes slowly, as during freezing in bags in a cabinet freezer, then the ice crystals can grow to a large size. There is a good chance that they will form outside the cells depending on the chosen rate of cooling. If the ice crystals form inside the cells, then cell cold injury or intracellular damage may depend on the size of the ice crystals. During typical freezing procedures employing slow cooling, the ice crystals first form only in the extracellular medium. Slow cooling not only affects the growth of crystals but also provides conditions allowing cells to be in an osmotic equilibrium with the extracellular medium. At relatively slow cooling rates, the generation of ice crystals outside cells causes the phenomenon of cryoconcentration, or increase in concentration of solutes in the liquid phase. Such a process can concentrate solutes to levels at which they can adversely affect the cell membrane. At slow cooling, this effect can cause damage to the cell membrane if the exposure time is long. Due to a partial permeability of cell membranes, osmotic pressure tends to remove the water from the cell in order to reach an equilibrium between the cell and environment. Thus, at a controlled rate of cooling, water content inside the cell gradually declines.

Freezing and sample cooling may cause mechanical stress within the sample and result in cell damage. When a liquid solidifies at its eutectic point, or as an amorphous glassy phase within the certain temperature range, the cell interior may be at its solidification point or may still be above it and yet the ice crystals may form inside the cell. These crystals may be very fine if the cooling rate is very high (for example, during pouring of small droplets of cell suspensions into a pool of liquid nitrogen). Formation of intracellular ice can be more lethal to cells than an increase in the concentration of solutes during slow cooling. The character of cell damage during cooling and freezing, and warming and thawing processes may depend on the cell type, thermal change protocols, and the type and concentration of solutes (small and large molecules, cryoprotectants).

Optimum freezing procedures for cells may require strictly controlled freezing and cooling rates. These parameters might control not only the osmotic conditions in the cell but also the structure of ice matrix surrounding the cells as well as intracellular ice formation.

Cell membrane characteristics may be one of the most important factors to be considered during the process development and selection of freezing and thawing techniques. Membrane composition, its mechanical strength, presence of pores and their size, and membrane transport properties are a few examples. Osmotic effects and effects of concentrated solutes and pH change on the cell may depend on those membrane characteristics. The same applies to possible mechanical damage by the extracellular ice. Some types of cells may be more resistant to these effects. Cell membrane characteristics may be different for wild strains and cell mutants.

Slow freezing rates are considered when the cooling rates are between 1° and 10° C/min, and rapid freezing rates are when the cooling rates are above 100° C/min. Faster freezing rates may reduce cell exposure to concentrated solutes and to possible pH shifts, or salt denaturation of cell surface proteins. A limit on freezing/cooling rate is usually determined by intracellular ice formation and associated phenomena of dehydration and osmotic equilibrium.

The survival of different cells may vary depending on the cooling rate and the type of cell. Considering practically reachable cooling rates for yeasts, the highest survival occurs at cooling rates below 0.2 K/sec (12 K/min), and the lowest survival is for cooling rates of 1.0–10.0 K/sec (60–600 K/min). The stem cells have the highest survival rate within the cooling rates of 0.01–0.1 K/sec (0.6–6 K/min), with hamster cells at the rage 1–5 K/sec (60–300 K/min) and red blood cells at 20–50 K/sec (Robarts and Sleyter 1985).

If concentrated cell suspensions directly from the fermentation step are going to be frozen, the osmotic conditions in cell environment during freezing may be more complex than those found in literature reports on freezing and thawing of cells and microorganisms. Transport phenomena can become complicated due to the presence of a variety of ions and molecules from the fermentation broth.

The size and shape of a sample may affect localized damage to the cells. Freezing and thawing of cell suspensions in large isolated volumes may involve problems associated with different thermal histories (cooling rates) and osmotic and mechanical conditions. In the case of cooling through a stationary wall of a container, the cooling rate changes due to an increase in the thickness of frozen material (increasing resistance to heat transfer by conduction). At the same time, the solutes may concentrate in the liquid phase and the cells may be exposed to a mechanical stress, if they are trapped between advancing ice crystals.

The majority of results found in literature cannot be directly applied to freezing or thawing of large volumes, since the laboratory work is typically done using very small specimens.

Large-scale freezing and thawing operations involve heat transfer for large single volumes or multiple smaller containers depending on the process chosen. The heat transfer coefficients for cooling by contact or immersion in brine, heat transfer fluid, or liquid nitrogen can be approximately 8–10 times higher than for cooling by forced convection in air. The contact freezers may be considered feasible for large-scale freezing under the condition that the problem of material handling is successfully solved.

The thawing process can also be a critical procedure, and cells may be damaged by mechanical stress caused by recrystallization of ice. An additional effect may be caused by dissolved solutes after liquefying the eutectic or glassy state (vitrified) substances in the intercrystalline spaces during the warming process. As a result, the cells after thawing may show varying viability, changes in membrane structure, and intracellular damage, depending on where these cells were located in the freezing bulk volume and what freezing and thawing regimes were applied.

Dividing the large volume of product into smaller portions may significantly improve uniformity of freezing and thawing conditions. Such a division can aid in freezing of the entire cell mass simultaneously under well-controlled conditions since the freezing parameters change uniformly in the whole volume. Bulk freezing combined with granulation leads to a controlled freezing and cooling rate in the whole volume, i.e., the entire cell mass can be exposed to a similar freezing regime. Such a uniformity may assure that the majority of cells will have similar characteristics after the freeze-thaw cycle.

Due to the relatively low heat conductivity of ice, the cooling rates of large biological samples are dominated by the internal heat conduction, and the heat transfer coefficient at the sample surface plays a less important role.

Granulation by immersion of cell suspension droplets into liquid nitrogen may not be an optimal way of freezing since the cooling rate is difficult to control. After reaching the surface of cryogen, a rapid freezing occurs with a possible oscillatory drop movement due to the nitrogen boiling around it and generation of a gas cushion. This period is very brief, and the encrusted granule may immerse into the liquid nitrogen. In the case of freezing of granule interior, there is still a significant latent heat of phase change removed from the freezing water in addition to the cooling of solidified mass. This would make the nitrogen boiling last longer than in the case of cooling only. The solidification is rapid, and very fine ice crystals form. Such a crystallization pattern develops a very large total surface of ice crystals. The solutes not entrapped within the ice crystals are spread between the crystal surfaces into very thin layers, and a very large interface of solute-ice forms. Such a pattern may require higher concentrations of cryoprotectants to create layers of eutectics and glassy states between the ice crystals to embed product (cells or proteins) to make the ice-product contact less significant. The glassy states formed during very rapid solidification may contain more trapped water than glassy states formed during slow freezing. This lowers the glass temperature T_g' and requires lower temperature for storage and for primary stage of lyophilization. The dendritic growth of relatively large crystals with eutectics and glassy states of solutes between them typically provides limited exposure to ice crystal surfaces. However, in the case of rapid droplet solidification, fully developed dendritic growth may not occur.

Controlled dendritic ice crystal growth might provide localized conditions favorable for a gradual cell dehydration prior to freezing. These conditions may occur within the mushy zone, e.g., in the interdendritic space of the moving solidification front. The concentration of solutes increases along the free dendrite length while the temperature decreases. These conditions provide the cell aqueous environment with a defined cooling rate accompanied by changing osmotic conditions. When optimized, these conditions may create gradual cell dehydration and temperature decrease without formation of intracellular ice crystals. However, such optimization may not always be possible, since many variables are involved. The cells and their environment may finally solidify at the eutectic or glassy state temperature, at an intracellular water content low enough to prevent cell damage. If the controlled dendritic growth can be maintained across the large volumes, the cell thermal histories prior to the solidification can be essentially similar at various sample points. After solidification, the cells embedded in the frozen material may experience various temperature histories, depending on cell location in the frozen mass, configuration of cooling surfaces, sample shape, and cooling protocol. Cell cooling in the frozen mass may cause intracellular ice formation, particularly if cells passed the mushy zone rapidly, far from reaching an osmotic equilibrium. The freezing configuration should be such to minimize or eliminate cell settling effects. Cell settling may interfere with control over the local cell environment during the cell cooling, dehydration, and freezing (cell movement in the interdendritic liquid should be minimized or eliminated). This concept works better for the freezing of protein solutions since there are no problems associated with the cell settling effect.

Controlled dendritic ice crystal growth can be also very important for freezing of protein solutions (pharmaceutical formulations or solutions between the processing steps). The key phenomena are increasing concentration of solutes in the interdendritic space along the dendrites (freezing out water on sides of dendrites) and simultaneous temperature decrease (due to the temperature gradient along the dendrite). The solutes concentration increase and temperature decrease in the liquid phase lead to formation of glassy states between the dendrites composed of the formulation solutes. The proteins may be embedded in those glassy states and protected from degradation. The process operational parameters are critical, but the composition of solutes in formulation is also of great importance. The eutectic points and ranges of glass temperatures for glassy state-forming solutes should be known before approaching the process design.

The reported cell survival rates, depending on the cooling rates, may correspond to the solid-liquid front velocities and associated front and ice crystal shapes. At very low front velocities (associated with slow cooling) flat front or cellular ice crystals form with an almost flat solid-liquid interface. At very high front velocities (associated with rapid cooling), again a flat front forms with a fine cellular ice crystal pattern. In both cases, the cells and solutes may be excluded from the solidified frozen mass and concentrated in the liquid phase. Such conditions may be detrimental to the cell survival. At moderate front velocities and corresponding moderate cooling, the ice crystals grow in the form of fully developed dendrites with a distinctive mushy zone. Such conditions may be favorable for cell survival and uniform solute distribution across the frozen mass. The earlier described pattern of gradual cell dehydration among dendrites may occur in the mushy zone. At very rapid cooling the equiaxed, very fine ice dendrites form with the solutes and cells distributed among them. Such a pattern may create very high ice crystal surface (increasing the ice-molecule and ice-cell surface interaction effects), and ice crystals may mechanically affect the cells (if the amount of protective solutes is small). Very fine crystal structure may also mechanically affect the cells during thawing due to the ice recrystallization, which may become very significant due to the large number of very fine crystals present.

The control of the dendritic ice crystal growth is associated with maintaining certain heat flux and the solid-liquid interface velocity. The increase of the frozen layer thickness causes increase of resistance to heat transfer by conduction. This decreases the heat flux and slows the movement of the solid-liquid interface. To maintain the heat flux and velocity constant, the cooling agent temperature should be lowered with an increase of the solidified layer thickness. The temperature gradient slope across the solidified layer should be maintained.

The current practice of freezing small samples often uses the two-step process with first cooling the samples in liquid nitrogen vapor (controlled cooling rate) and then finally plunging the sample into the liquid nitrogen. Such a procedure may not be easy to duplicate for large product volumes. Spraying the cryogenic agent into a cooled product provides uniform volumetric freezing in large-mass granule formation and a subsequent cooling at a controlled rate. (See case study 3.1 Chapter 3.)

Freezing and thawing of cell suspensions in large volumes poses more challenges than freezing and thawing of large volumes of proteins. In the case of proteins, the phenomena of importance are cryoconcentration, cooling/freezing rate, and molecular interactions such as protein-protein, protein-ice, protein-solute, and solute-ice (Wisniewski and Wu 1996; Wisniewski 1998a). The protein stability may be affected by thermodynamics (cold unfolding of protein), by pH shifts in the liquid phase prior to freezing, by an increase in concentration of solutes, and by protein-ice surface interactions. Some proteins may be more susceptible than others to adverse effects of freezing and thawing. In general, the multimetric molecules might be more labile than smaller molecules. Protein glycosylation may make proteins more stable during the freezing and thawing steps. Some proteins may demonstrate a certain degree of activity loss after the freezing and thawing cycle, and in such a case a careful optimization of the freezing conditions in large scale should be performed. Addition of cryoprotectants may also be considered. Freezing and thawing processes should be well controlled across the whole product volume. The freezing and thawing of cells includes all of these issues plus additional factors of cell physiology, cell cultivation conditions, cell membrane behavior during freezing and thawing, osmotic transport and cell water content in the environment of freezing and thawing, intracellular ice formation, action of penetrating versus nonpenetrating cryoprotectants, cell capture or rejection by the freezing front, ice recrystallizaiton, and cell settling. Therefore, conditions of freezing and thawing of cells in large volumes have to be very carefully evaluated, including temperature histories of various parts of the sample, movement of freezing fronts, size and shape of forming ice crystals, cryoconcentration effects, freezing and thawing thermal regimes (sample and cooling medium temperature change in time), selection of cryoprotectants, etc.

Thawing of large samples using the heat transfer surfaces such as cooled walls may involve varying temperature histories at different sample points. The liquid-solid interface will be at near 0° C (ice melting temperature), whereas the frozen solid undergoes local temperature changes from its original low temperature of storage through increasing temperatures toward the near 0° C level. Histories of particular sample points depend upon configuration of heating surfaces and the shape of the sample. At certain temperature levels, the following events will take place: eutectics between ice crystals will melt, and glassy (vitrified) solids will soften and liquefy; and ice crystals will recrystallize (small crystals may dissolve and large may grow). Since the cells are distributed within the intercrystalline spaces, these events may have detrimental effects on them, e.g., the cells might be exposed to elevated solute concentration and might be mechanically squeezed by the growing ice crystals. For large volumes, shortening the time the sample zones are exposed to these temperature conditions may be impossible and cell damage would occur. A possible solution to avoid such thermal thawing damage occurring in large samples is to divide large samples into multiple smaller ones and thaw them as independent entities. This can be accomplished by a simultaneous process of freezing and granulation or by freezing and thawing using extended heat transfer surfaces (fins). Individual granules or compartments may then be rapidly thawed, and their temperature histories will include rapid passing through the critical temperature ranges.

Another method of large-sample thawing without subdivision into small volumes is to thaw the whole sample volumetrically, i.e., heating up the whole volume rapidly to the melting temperature of ice. Such a volumetric heating will allow rapid passage through the critical temperature zones of ice recrystallization and melting of eutectic and glassy substances. In practice, such a melting may be accomplished using microwave irradiation, dielectric, or resistance methods. The thawing process by microwave irradiation needs to be carefully analyzed since the shape of a sample may affect temperature distribution (for example, any extending corners and edges may be heated faster than the rest of the sample volume). A combination of these methods with thermal conduction/convection can also be applied.

The quest for better understanding of phenomena occurring during freezing and thawing of cell suspensions leads to detailed investigations of molecular interactions between cell membranes and water and solutes, molecular phenomena (including molecular transport) within membranes under nonequilibrium conditions, interaction of water with cryoprotectants, interactions between ice and free and bound water molecules and between ice and solutes, etc. Computer-based molecular modeling finds an increasing use in such studies. The freeze-thaw step is often used for cell disruption, but the conditions of the process are not always optimized and may not be specified in the publications.

Large-scale cryopreservation (freezing and thawing operations) of biological materials is becoming increasingly important as modern biotechnology industries are introducing the idea of a multiproduct facility operating on a campaign basis. Preservation by freezing may be applied at various stages of the manufacturing process. The product generated in a short campaign may be then frozen and stored for 1–2 years. The frozen product can be then thawed and moved through the fill and finish steps according to market demands. In such a way, the manufacturing process can become decoupled and the overall product life extended. These two factors provide great flexibility in manufacturing operations (including contract manufacturing at multiple locations, since the frozen product can be easily shipped between manufacturing facilities) as well as in marketing strategies (including an ability to respond rapidly according to market demands).

The phenomena of stabilization of proteins depend mainly on molecular interactions protein-water-solute-ice and an effort should be made to understand these interactions prior to the process design. Experimental work combined with molecular simulations is the current approach. In the cryopreservation of cells a properly designed experimental work still remains the best approach due to the combination of molecular, microscopic, and macroscopic effects occurring during the freezing and thawing of cell suspensions. The experiment design should consider the subsequent scale up issues (the sample size, cooling rates, ice crystal growth phenomena, concentration of solutes and osmotic effects, etc.).

REFERENCES

Aldous, B. J., A. D. Auffret, and F. Franks. 1995. The crystallization of hydrates from amorphous carbohydrates. *Cryoletters* 16:181–186.

Angell, C. A. 1982. Supercooled water. In *Water—A comprehensive treatise,* vol. 7, edited by F. Franks, 1–82. New York: Plenum Press.

Armitage, W. 1987. Cryopreservation of animal cells. In *Temperature and animal cells,* Symp. Soc. Experimental Biology 41, edited by K. Bowler and B. Fuller, 379–393.

Armitage, W., B. Juss, and D. Easty. 1995. Differing effects of various cryoprotectants on intracellular junctions of epithelial (MDCK) cells. *Cryobiology* 32:52–59.

Ashwood-Smith, M. 1987. Mechanism of cryoprotectant action. In *Temperature and animal cells,* Symp. Soc. Experimental Biology 41, edited by K. Bowler and B. Fuller, 395–406.

ATCC Catalogue of Bacteria and Bacteriophages. 1992. Rockville, MD: ATCC.

Atlas, R. M., and L. C. Parks. 1993. *Handbook of microbiological media*. Boca Raton: CRC Press.

Awati, K., and T. Howes. 1996. Surfactant induced stationary modes on a cylindrical fluid jet. *J. Colloid Interface Sci.* 181:344–346.

Bader, J., and B. Berne. 1996. Solvation energies and electronic spectra in polarizable media: Simulation tests of dielectric continuum theory. *J. Chem. Phys.* 104:1293–1308.

Bald, W. B. 1993. Real cooling and warming rates during cryopreservation. *Cryoletters* 14:207–216.

Ballou, C. E. 1983. Yeast cell wall and cell surface. In *The molecular biology of the yeast* Saccharomyces. *Metabolism and gene expression,* edited by J.N. Strathern, E.W. Jones, and J.R. Broach, 335–360. Cold Spring Harbor: Cold Spring Harbor Laboratory.

Beduz, C., and R. Scurlock. 1994. Evaporation mechanisms and instabilities in cryogenic fluids. *Adv. Cryogen. Eng.* 39:1749–1757.

Beer, J. M., and N. A. Chigier. 1983. *Combustion aerodynamics*. Malabar, FL: R. Krieger.

Benz, R. 1988. Structure and function of porins from gram-negative bacteria. *Ann. Rev. Microbiol.* 42:359–393.

Benz, R., and K. Bauer. 1988. Permeation of hydrophilic molecules through the outer membrane of gram-negative bacteria. *Eur. J. Biochem.* 176:1–19.

Biondi, A. D., G. A. Senisterra, and E. A. Disalvo. 1992. Permeability of lipid membranes revised in relation to freeze-thaw processes. *Cryobiology* 29:323–331.

Boothroyd, R. G. 1971. *Flowing gas-solid suspensions*. London: Chapman and Hall.

Boutren, P., and J. F. Peynedieu. 1994. Reduction of toxicity for red blood cells in buffered solutions containing high concentrations of 2,3-butanediol by trehalose, sucrose, sorbitol and mannitol. *Cryobiology* 31:367–373.

Brady, J., and R. Schmidt. 1993. The role of hydrogen bonding in carbohydrates: Molecular dynamics simulations of maltose in aqueous solution. *J. Phys. Chem.* 97:958–966.

Brearley, C. A., N. A. Hodges, and C. J. Oliff. 1990. The use of low molecular weight nonpenetrating cryoprotectants to minimize red cell post-thaw dilution shock. *Cryoletters* 11:295–304.

Breierova, E., and A. Kockova-Kratochvilova. 1992. Cryoprotective effects of yeast extracellular polysaccharides and glycoproteins. *Cryobiology* 29:385–390.

Bricard, A. 1991. Direct-contact heat exchanger. In *Industrial heat exchangers*, edited by J.M. Buchlin. Lecture series 1991–2004, Rhode Sâint Genese, Belgium: von Karman Institute for Fluid Dynamics.

Brock, T. D., and M. T. Madigan. 1988. *Biology of microorganisms.* 5th ed. Englewood Cliffs, NJ: Prentice Hall.

Bronshteyn, B., and P. Steponkus. 1995. Nucleation of ice crystals in concentrated solutions of ethylene glycol. *Cryobiology* 32:1–22.

Bronstein, V. L., Y. A. Itkin, and G. S. Ishkov. 1981. Rejection and capture of cells by ice crystals on freezing of aqueous solutions. *J. Cryst. Growth* 52:345–349.

Bryant, G. 1995. DSC measurement of cell suspension during successive freezing runs: Implications for the mechanisms of intracellular ice formation. *Cryobiology* 32:114–128.

Bryant, G., I. Heschel, and G. Ray. 1994. The influence of gas bubbles on intracellular ice formation. *Cryoletters* 15:113–118.

Brzobohaty, B., and L. Kovac. 1986. Factors enhancing genetic transformation of intact yeast cells modify cell wall porosity. *J. Gen. Microbiol.* 143:3089–3093.

Buchlin, J. M. 1995. Thermohydraulic modeling of liquid sprays. In *Two-phase flows with phase transition*, edited by C.H. Sieverding. Lecture series 1995–2006, Rhode Sâint Genese, Belgium: von Karman Institute for Fluid Dynamics.

Budhia, H., and F. Kreith. 1973. Heat transfer with melting and freezing in a wedge. *Int. J. Heat Mass Transfer* 16:195–211.

Buechner, M., A. H. Delcour, B. Martinac, J. Adler, and C. Kung. 1990. Ion channel activities in the *Escherichia coli* outer membrane. *Biochim. Biophys. Acta* 1024:111–121.

Byers, C., and J. Perona. 1988. Drop formation from an orifice in an electric field. *AIChE J.* 34:1577–1580.

Cabib, E., R. Roberts, and B. Powers. 1982. Synthesis of the yeast cell wall and its regulation. *Ann. Rev. Biochem.* 51:763–793.

Calcott, P. H. 1985. Cryopreservation of microorganisms. *CRC Crit. Rev. Biotech.* 4:279–297.

Calcott, P. H., and A. H. Rose. 1982. Freeze-thaw and cold shock resistance of *Saccharomyces cerevisiae* as affected by plasma membrane lipid composition. *J. Gen. Microbiol.* 128:549–555.

Calcott, P. H., J. D. Oliver, K. Dickey, and K. Calcott. 1984. Cryosensitivity of *Escherichia coli* and the involvement of cyclopropane fatty acids. *J. Appl. Bacter.* 56:165–172.

Calcott, P. H., and K. N. Calcott. 1984. Involvement of outer membrane protein in freeze-thaw resistance of *Escherichia coli*. *Can. J. Microbiol.* 30:339–344.

Calcott, P. H., and R. A. MacLeod. 1974. Survival of *Escherichia coli* from freeze-thaw damage: A theoretical and practical study. *Can. J. Microbiol.* 20:671–681.

Calcott, P. H., and R. A. MacLeod. 1975a. The survival of *Escherichia coli* from freeze-thaw damage: Permeability barrier damage and viability. *Can. J. Microbiol.* 21:1724–1732.

Calcott, P. H., and R. A. MacLeod. 1975b. The survival of *Escherichia coli* from freeze-thaw damage: The relative importance of wall and membrane damage. *Can. J. Microbiol.* 21:1960–1968.

Cannon, W., M. Pettitt, and A. McCammon. 1994. Sulfate anion in water: Model of structural, thermodynamic and dynamic properties. *J. Phys. Chem.* 98:6225–6230.

Carman, G., and G. Zeimetz. 1996. Regulation of phospholipid biosynthesis in the yeast *Saccharomyces cerevisiae*. *J. Biol. Chem.* 271:13293–13296.

Carpenter, J. F., and T. N. Hansen. 1992. Antifreeze protein modulates cell survival during cryopreservation: Mediation through influence on ice crystal growth. *Proc. Natl. Acad. Sci. USA* 89:8953–8957.

Carrington, A. K., J. E. Sahagian, H. D. Goff, and D. W. Stanley. 1994. Ice crystallization temperatures of sugar/polysaccharide solutions and their relationship to thermal events during warming. *Cryoletters* 15:235–244.

Carslaw, H. S., and J. C. Jaeger. 1959. *Conduction of heat in solids.* Oxford: Clarendon Press.

Cascales, L., H. Berendson, and G. de la Torre. 1996. Molecular dynamics simulation of water between two charged layers of dipalmitoylphosphatidylserine. *J. Phys. Chem.* 100:8621–8627.

Catanzano, F., A. Gambuti, G. Graziano, and B. Barone. 1997. Interaction with D-glucose and thermal denaturation of yeast hexokinase B: A DSC study. *J. Biochem.* 121:568–577.

Cevc, G. 1990. Membrane electrostatics. *Biochim. Biophys. Acta* 1031–1033:311–382.

Chandrasekaran, M., and R. E. Pitt. 1992. On the use of nucleation theory to model intracellular ice formation. *Cryoletters* 13:261–272.

Chang, Z., and J. Baust. 1991a. Physical aging of glassy state: DSC study of vitrified glycerol systems. *Cryobiology* 28:87–95.

Chang, Z., and J. Baust. 1991b. Further inquiry into the cryobehavior of aqueous solutions of glycerol. *Cryobiology* 28:268–278.

Charm, S. E. 1981. *Fundamentals of food engineering.* 3d ed. Westport, CT: Avi.

Chiang, H., and C. Kleinstreuer. 1989. Solidification around a cylinder in laminar cross flow. *Int. J. Heat and Fluid Flow* 10:322–327.

Cid, V., A. Duran, F. del Rey, M. Snyder, C. Nombela, and M. Sanchez. 1995. Molecular basis of cell integrity and morphogenesis in *Saccharomyces cerevisiae*. *Microbiol. Rev.* 59:345–386.

Clausse, D., L. Babin, F. Broto, M. Aguerd, and M. Clausse. 1983. Kinetics of ice nucleation in aqueous emulsions. *J. Phys. Chem.* 87:4030–4034.

Cocks, F. H., and W. E. Brower. 1974. Phase diagram relationships in cryobiology. *Cryobiology* 11:340–358.

Conti, M. 1995. Planar solidification of a finite slab: Effects of the pressure dependence of the freezing point. *Int. J. Heat Mass Transfer* 38:65–70.

Cortez-Maghelly, C., and P. Bisch. 1995. Effect of ionic strength and outer surface charge on the mechanical stability of the erythrocyte membrane: A linear hydrodynamic analysis. *J. Theor. Biol.* 176:325–339.

Cosman, M. D., M. Toner, J. Kandel, and E. G. Cravalho. 1980. An integrated cryomicroscopy system. *Cryoletters* 10:17–38.

Cote, R., and R. Gherna. 1990. Nutrition and media. In *Methods for general and molecular bacteriology*, edited by P. Gerhardt, R. Murray, W. Wood, and N. Krieg, 155–178. Washington, DC: Am. Soc. Microbiol.

Crank, J. 1984. *Free and moving boundary problems*. Oxford: Clarendon Press.

Cronan, J. E., R. B. Gennis, and S. R. Maloy. 1987. Cytoplasmic membrane. In Escherichia coli *and* Salmonella typhimurium. *Cellular and molecular biology*, edited by F. C. Neidhardt et al., 31–55. Washington, DC: Am. Soc. Microbiol.

Crowe, J., M. Whittam, D. Chapman, and L. Crowe. 1984b. Interactions of phospholipid monolayers with carbohydrates. *Biochim. Biophys. Acta* 769:151–159.

Crowe, J., L. Crowe, and D. Chapman. 1984c. Infrared spectroscopic studies on interactions of water and carbohydrates with a biological membrane. *Arch. Biochem. Biophys.* 232:400–407.

Crowe, J., L. Crowe, J. Carpenter, A. Rudolph, C. Wistrom, B. Spargo, and T. Anchordoguy. 1988. Interactions of sugars with membranes. *Biochim. Biophys. Acta* 947:367–384.

Crowe, L., R. Mouvadian, J. Crowe, S. Jackson, and C. Wormersley. 1984a. Effects of carbohydrates on membrane stability at low water activities. *Biochim. Biophys. Acta* 769:141–150.

Crowe, L., J. Crowe, A. Rudolph, C. Womersley, and L. Appel. 1985. Preservation of freeze-dried liposomes by trehalose. *Arch. Biochem. Biophys.* 242:240–247.

Csonka, L., and A. Hanson. 1991. Prokaryotic osmoregulation: Genetics and physiology. *Ann. Rev. Microbiol.* 45:569–606.

Daldrup, N., and H. Schoenert. 1988. Excess enthalpy and excess volume in ternary aqueous solutions with sucrose-glucose, sucrose-glycerol, and glucose-glycerol at 298.1 K. *J. Chem. Soc. Faraday Trans.* 84:2553–2566.

Davies, P., and C. Hew. 1990. Biochemistry of fish antifreeze proteins. *FASEB J.* 4:2460–2468.

Daw, A., J. Farrant, and G. J. Morris. 1973. Membrane leakage of solutes after thermal shock or freezing. *Cryobiology* 10:126–133.

De Leeuw, F. E., A. De Leeuw, J. Den Daas, B. Colenbrander, and A. Verkleji. 1993. Effects of various cryoprotective agents and membrane-stabilizing compounds on bull sperm membrane integrity after cooling and freezing. *Cryobiology* 30:32–44.

De Loecker, R., and F. Penninck. 1987. Biochemical and functional aspects of recovery of mammalian systems from deep sub-zero temperatures. In *Temperature and animal cells,* no. 41, edited by K. Bowler and B. Fuller, 407–427. Symp. Soc. Experiment. Biology.

Delzeit, L., K. Powell, N. Uras, and P. Devlin. 1997. Ice surface reactions with acids and bases. *J. Phys. Chem.* 101:2327–2332.

Delzeit, L., M. Devlin, B. Rowland, J. Devlin, and V. Buch. 1996. Adsorbate-induced partial ordering of the irregular surface and subsurface of crystalline ice. *J. Phys. Chem.* 100:10076–10082.

de Nobel, J. G., F. M. Klis, T. Munnik, J. Priem, and H. van den Ende. 1990a. An assay of relative cell wall porosity in *Saccharomyces cerevisiae, Kluveromyces lactis,* and *Schizosaccharomyces pombe. Yeast* 6:483–490.

Devlin, J. P., and V. Buch. 1995. Surface of ice as viewed from combined spectroscopic and computer modeling studies. *J. Phys. Chem.* 99:16534–16548.

Diettrich, B., U. Haack, B. Thom, G. Matthes, and M. Luckner. 1987. Influence of intracellular ice formation on the survival of *Digitalis lanata* cells grown in vitro. *Cryoletters* 8:98–107.

Difco Manual, 10th ed. 1985. Detroit: Difco Laboratories.

Diller, K. R. 1992. Modeling of bioheat transfer processes at high and low temperatures. *Adv. Heat Transfer* 22:157–357.

Diller, K. R., and J. F. Raymond. 1990. Water transport through a multicellular tissue during freezing: A network thermodynamic modeling analysis. *Cryoletters* 11: 151–162.

Diller, K. R., and M. E. Lynch. 1984. An irreversible thermodynamic analysis of cell freezing in the presence of membrane-permeable additives. III. Transient water and additive fluxes. *Cryoletters* 5:131–144.

Doyle, R., and M. Rosenberg, eds. 1990. *Microbial cell surface hydrophobicity.* Washington, DC: Amer. Soc. Microbiol.

Dumas, J. P., M. Krichi, M. Strub, and Y. Zraouli. 1994. Models for the heat transfers during the transformations inside an emulsion. *Int. J. Heat Mass Transfer* 37:737–746; 747–752.

Dwyer, H., H. Nirschl, P. Kerschl, and V. Denk. 1994. *Heat, mass, and momentum transfer about arbitrary groups of particles.* 25th Symp. on Combustion, 389–395. Pittsburgh, PA: The Combustion Institute.

Eckert, E., and E. Pfender. 1978. Heat and mass transfer in porous media with phase change. In *Heat Transfer 1978,* Proc. 6th Intl. Heat Transfer Conf., Toronto, vol. 6:1–12. Washington, DC: Hemisphere Publ. Corp.

Eggers, J. 1995. Theory of drop formation. *Phys. Fluids* 7:941–953.

Eggers, J., and T. Dupont. 1994. Drop formation in one-dimensional approximation of the Navier-Stokes equation. *J. Fluid Mech.* 262:205–221.

Engelson, S., C. du Penhoat, and S. Perez. 1995. Molecular relaxation of sucrose in aqueous solution. *J. Phys. Chem.* 99:13334–13351.

Epstein, M., and H. Fauske. 1993. Kinetic and heat transfer controlled solidification of highly supercooled droplets. *Int. J. Heat Mass Transfer* 36:2987–2995.

Erukhimovitch, B., and J. Baram. 1994. Analysis of solidification in the spray atomization process. *Materials Sci. Eng.* A181/A182:1195–1201.

Fahy, G. M. 1980. Analysis of "solution effects" injury. Equations for calculating phase diagram information for the ternary systems NaCL-Dimethylsulfoxide-Water and NaCl-Glycerol-Water. *Biophys. J.* 32:837–850.

Fahy, G. M. 1981. Simplified calculation of cell water content during freezing and thawing in nonideal solutions of cryoprotective agents and its possible application to the study of "solution effects" injury. *Cryobiology* 18:473–482.

Farrant, J., C. A. Walter, H. Lee, and L. E. McGann. 1977. Use of two-step cooling procedures to examine factors influencing cell survival following freezing and thawing. *Cryobiology* 14:273–286.

Figlas, D., H. Arias, A. Fernandez, and D. Alperin. 1997. Dramatic saccharide-mediated protection of chaotropic-induced deactivation. *Arch. Biochem. Biophys.* 340:154–158.

Franks, F. 1982. The properties of aqueous solutions at sub-zero temperatures. In *Water—A comprehensive treatise,* vol. 7, edited by F. Franks, 215–338. New York: Plenum Press.

Franks, F. 1985. *Biophysics and biochemistry at low temperatures.* Cambridge: Cambridge University Press.

Franks, F. 1991. Freeze drying: From empiricism to predictability—The significance of glass transitions. *Develop. Biol. Standard.* 74:9–19.

Franks, F., and R. H. Hatley. 1992. Protein stability under conditions of deep chill. In *Adv. low-temp. biology* 1, edited by P. Steponkus, 141–179. London: JAI Press Ltd.

Franks, F., S. F. Mathias, P. Galfre, S. D. Webster, and D. Brown. 1983. Ice nucleation and freezing in undercooled cells. *Cryobiology* 20:298–309.

Franks, F. et al. 1992. Materials science and the production of shelf-stable biologicals. *Pharm. Technol.* 16(3):32–50.

Fransen, G. J., P. J. Salemink, and D. J. Crommelin. 1986. Critical parameters in freezing of liposomes. *Int. J. Pharm.* 33:27–35.

Freshney, R. I. 1994. *Culture of animal cells.* 3d ed. New York: Wiley-Liss.

Fujikawa, S., and K. Miura. 1986. Plasma membrane ultrastructural changes caused by mechanical stress in the formation of extracellular ice as a primary cause of slow freezing injury in fruit-bodies of basidiomycetes (*Lyophyllum ulmarium [Fr.] Kuhner*). *Cryobiology* 23:371–382.

Fukusako, S. 1990. Thermophysical properties of ice, snow, and sea ice. *Int. J. Thermophys.* 11:353–372.

Fukusako, S., and N. Seki. 1987. Fundamental aspects of analytical and numerical methods on freezing and melting heat transfer problems. *Ann. Rev. Numer. Fluid Mech. Heat Transfer* 1:351–402.

Furlong, C. E. 1987. Osmotic-shock-sensitive transport systems. In *Escherichia coli and Salmonella typhimurium. Cellular and molecular biology,* edited by F. C. Neidhardt et al., 770–796. Washington, DC: Am. Soc. Microbiol.

Gaffney, S., E. Haslam, T. Lilley, and T. Ward. 1988. Homotactic and heterotactic interactions in aqueous solutions containing some saccharides. *J. Chem. Soc. Faraday Trans.* 84:2545–2552.

Gangasharan, J., and S. N. Murthy. 1995. Nature of the relaxation processes in the supercooled liquid and glassy states of some carbohydrates. *J. Phys. Chem.* 99:12349–12354.

Gao, D., S. Lin, P. Watson, and J. Critser. 1995. Fracture phenomena in an isotonic salt solution during freezing and their elimination using glycerol. *Cryobiology* 32:270–284.

Gavish, M., R. Popovitz-Biro, M. Lahav, and L. Leiserowitz. 1990. Ice nucleation by alcohols arranged in monolayers at the surface of water drops. *Science* 250:437–441.

Gazeau, C., M. Jondet, and J. Dereuddre. 1989. Catharanthus cell suspensions observed during freezing and thawing with a cryomicroscope. *Cryoletters* 10:105–118.

Geankoplis, C. J. 1983. *Transport processes. Momentum, heat, and mass.* Boston: Allyn & Bacon.

Glicksman, M. 1989. Fundamentals of dendritic growth. In *Crystal growth in science and technology,* edited by H. Arend and J. Hulliger, 167–183. New York: Plenum Press.

Gorbis, Z. R., and W. A. Kalenderyan. 1975. *Heat exchangers with moving dispersed heat transfer agents* (in Russian). Moscow: Energy.

Graumann, P., and M. Marahiel. 1996. Some like it cold: Response of microorganisms to cold shock. *Arch. Microbiol.* 166:293–300.

Greaves, R. I., and J. D. Davies. 1965. Separate effects of freezing, thawing, and drying living cells. *Ann. NY Acad. Sci.* 125:548–558.

Green, J. L., and C. A. Angell. 1989. Phase relations and vitrification in saccharide-water solutions and the trehalose anomaly. *J. Phys. Chem.* 93:2880–2882.

Griffiths, J. B., C. S. Cox, D. J. Beadle, C. J. Hunt, and D. S. Reid. 1979. Changes in cell size during the cooling, warming, and past-thawing periods of the freeze-thaw cycle. *Cryobiology* 16:141–151.

Grigera, R. 1988. Conformation of polyols in water. *J. Chem. Soc. Faraday Trans.* 84:2603–2608.

Grigull, U., and H. Sandner. 1984. *Heat conduction.* Washington, DC: Hemisphere Publ. Corp.

Grout, B. W., and G. J. Morris. 1987. Freezing and cellular organization. In *The effects of low temperatures on biological systems,* edited by B. W. Grout and G. J. Morris, 147–173. London: E. Arnold.

Grout, B. W., G. J. Morris, and M. McLellan. 1990. Cryopreservation and the maintenance of cell lines. *TIBTECH* 8:293–297.

Gryasnov, S., and R. Letsinger. 1993. Chemical ligation of digonucleotides in the presence and absence of templates. *J. Am. Chem. Soc.* 115:3808–3809.

Guo, H., and M. Karplus. 1994. Solvent influence on the stability of the peptide hydrogen bond. *J. Phys. Chem.* 98:7104–7105.

Haines, T. 1994. Water transport across biological membranes. *FEBS Letters* 346:115–122.

Hammond, S., P. Lambert, and A. Rycroft. 1984. *The bacterial cell surface.* London: Croom Helm.

Hancock, R. E. 1984. Alterations in outer membrane permeability. *Ann. Rev. Microbiol.* 38:237–264.

Hancock, R. E. 1987. Role of porins in outer membrane permeability. *J. Bacteriol.* 169:929–933.

Hartmann, U., B. Nunner, C. Koerber, and G. Rau. 1991. Where should the cooling rate be determined in an extended freezing sample? *Cryobiology* 28:115–130.

Hayashi, H., H. Tanaka, and K. Nakanishi. 1995. Molecular dynamics simulations of flexible molecules. 1. Aqueous solution of ethylene glycol. *J. Chem. Soc. Faraday Trans.* 91:31–39.

Hayes, A. R., and A. Stein. 1989. A cryomicroscope. *Cryoletters* 10:257–268.

Hayes, L. J., and K. R. Diller. 1988. Computational methods for analysis of freezing bulk systems. In *Low-temperature biotechnology,* edited by J. J. McGrath and K. R. Diller, 253–272. New York: ASME.

Hayes, L. J., K. R. Diller, and H. J. Chang. 1986. A robust numerical method for latent heat release during phase change. In *Numerical methods in heat transfer,* edited by J. L. Chen and K. Vafai, 63–69. New York: ASME.

Hayes, L. J., K. R. Diller, H. J. Chang, and H. S. Lee. 1988. Prediction of local cooling rates and cell survival during the freezing of a cylindrical specimen. *Cryobiology* 25:67–82.

Hayes, L. J., K. R. Diller, H. S. Lee, and C. R. Baxter. 1984. On the definition of an average cooling rate during cell freezing. *Cryoletters* 5:97–110.

Heldman, D. 1983. Factors influencing food freezing rates. *Food Technology,* April:103–109.

Henry, S. A. 1983. Membrane lipids of yeast: Biochemical and genetic studies. In *The molecular biology of the yeast* Saccharomyces. *Metabolism and gene expression,* edited by J. N. Strathern, E. W. Jones, and J. R. Broach, 101–158. Cold Spring Harbor: Cold Spring Harbor Laboratory.

Her, L.-M., R. Jefferis, L. Gattlin, B. Braxton, and S. Nail. 1994. Measurement of glass transition temperatures in freeze concentrated solutions of non-electrolytes by electrical thermal analysis. *Pharm. Res.* 11:1023–1029.

Heschel, I., C. Lueckge, M. Roedder, C. Garberding, and G. Rau. 1996. Possible applications of directional solidification techniques in cryobiology. *Adv. Cryogen. Eng.* 41:13–19.

Hilbing, J., and S. Heister. 1996. Droplet size control in liquid jet breakup. *Phys. Fluids* 8:1574–1581.

Hunt, J., and S.-Z. Lu. 1993. Numerical modeling of cellular and dendritic array growth: Spacing and structure predictions. *Materials Sci. Eng.* A173:79–83.

Huppert, H., and G. Worster. 1985. Dynamic solidification of a binary melt. *Nature* 314:703–707.

HyClone. 1992. Freezing and thawing serum and other biological materials. *Art to Science* 11(2):1–7.

Ingraham, J. 1987. Effect of temperature, pH, water activity, and pressure on growth. In Escherichia coli *and* Salmonella typhimurium. *Cellular and molecular biology,* edited by F. C. Neidhardt et al., 1543–1554. Washington, DC: Am. Soc. Microbiol.

Ishiguro, H., and B. Rubinsky. 1994. Mechanical interactions between ice crystals and red blood cells during directional solidification. *Cryobiology* 31:483–500.

Janda, S., and A. Kotyk. 1985. Effects of suspension density on microbial metabolic processes. *Folia Microbiol.* 30:465–473.

Jap, B. K. 1989. Molecular design of PhoE porin and its functional consequences. *J. Mol. Biol.* 205:407–419.

Jap, B. K., and P. J. Walian. 1990. Biophysics of the structure and function of porins. *Quart. Rev. Biophys.* 23:367–403.

Jap, B. K., and P. J. Walian. 1996. Structure and functional mechanism of porins. *Physiol. Rev.* 76:1073–1088.

Jeffrey, G., and W. Saenger. 1991. *Hydrogen bonding in biological structures.* Berlin: Springer.

Jerome, B., and J. Commandeur. 1997. Dynamics of glasses below the glass transition. *Nature* 386:589–592.

Jia, Z., C. DeLuca, H. Chao, and P. Davies. 1996. Structural basis for the binding of a globular antifreeze protein in ice. *Nature* 384:285–288.

Johansson, G., M. Joelsson, and B. Olde. 1990. Partition of synaptic membranes in aqueous two-phase systems at sub-zero temperatures by using antifreeze solvent. *Biochim. Biophys. Acta* 1029:295–302.

Jones, R. P., and P. F. Greenfield. 1987. Ethanol and the fluidity of the yeast plasma membrane. *Yeast* 3:223–232.

Jorgensen, W., and D. Severance. 1993. Limited effects of polarization for $Cl^-(H_2O)n$ and $Na^+(H_2O)n$ clusters. *J. Chem. Phys.* 99:4233–4235.

Karlsson, J. O., E. G. Cravalho, and M. Toner. 1993. Intracellular ice formation: Causes and consequences. *Cryoletters* 14:323–334.

Karshikoff, A., V. Spassov, S. W. Cowan, R. Ladenstein, and T. Schirmer. 1994. Electrostatic properties of two porin channels from *Escherichia coli*. *J. Mol. Biol.* 240:372–384.

Kasai, S., and M. Mito. 1993. Large-scale cryopreservation of isolated dog hepatocytes. *Cryobiology* 30:1–11.

Kawai, H., M. Sakurai, Y. Inoue, R. Chujo, and S. Kobayashi. 1992. Hydration of oligosaccharides: Anomalous hydration ability of trehalose. *Cryobiology* 29:599–606.

Kececioglu, I., and B. Rubinsky. 1989. A continuum model for the propagation of discrete phase-change fronts in porous media in the presence of coupled heat flow, fluid flow, and species transport processes. *Int. J. Heat Mass Transfer* 32:1111–1130.

Keeney, P. 1982. Development of frozen emulsions. *Food Technology,* November:65–72.

Keller, J., and M. Miksis. 1983. Surface tension driven flows. *SIAM J. Appl. Math* 43:268–277.

Kempler, G., and B. Ray. 1978. Nature of freezing damage on the lipopolysaccharide molecule of *Escherichia coli* B. *Cryobiology* 15:578–584.

Kinney, K., S. Xu, and L. Bartell. 1996. Molecular dynamics study of the freezing of clusters of chalcogen hexafluorides. *J. Phys. Chem.* 100:6935–6941.

Kleinstreuer, C., J. Comer, and H. Chiang. 1993. Fluid dynamics and heat transfer with phase change of multiple spherical droplets in a laminar axisymmetric gas stream. *Int. J. Heat and Fluid Flow* 14:292–300.

Koch, A. L. 1984. Shrinkage of growing *Escherichia coli* cells by osmotic challenge. *J. Bacteriol.* 159:919–924.

Kochs, M., P. Schwindke, and C. Koerber. 1989. A microscope stage for the dynamic observational freezing and freeze-drying in solutions and cell suspensions. *Cryoletters* 10:401–420.

Koebe, H., A. Werner, V. Lange, and F. Schildberg. 1993. Temperature gradients in freezing chambers of rate-controlled cooling machines. *Cryobiology* 30:349–352.

Koerber, C. 1988. Phenomena at the advancing ice-liquid interface: Solutes, particles, and biological cells. *Quart. Rev. Biophys.* 21:229–298.

Koerber, C., G. Rau, M. D. Cosman, and E. G. Cravalho. 1985. Interaction of particles and a moving ice-liquid interface. *J. Cryst. Growth* 63:649–662.

Komatsu, N., Y. Sato, and M. Osumi. 1987. Biochemical and electron-microscopic evidence for membrane injury in yeast cells quickly frozen with liquid nitrogen. *J. Ferment. Technol.* 65:127–131.

Konrad, J.-M. 1991. The physics of freezing soils and an engineering frost heave approach. In *Freezing and melting heat transfer in engineering,* edited by K. Cheng and N. Seki, 581–612. New York: Hemisphere Publ. Corp.

Koo, K.-K., R. Ananth, and W. Gill. 1991. Tip splitting in dendritic growth of ice crystals. *Physical Rev.* A44:3782–3790.

Kotyk, A. 1987. Effects of yeast suspension density on the accumulation ratio of transported solutes. *Yeast* 3:263–270.

Kresin, M., and G. Rau. 1992. The glass transition of hydroxyethyl starch. *Cryoletters* 13:371–378.

Kristiansen, J. 1992. Leakage of a trapped fluorescent marker from liposomes: Effects of eutectic crystallization of NaCl and internal freezing. *Cryobiology* 29:575–584.

Kruuv, J., and D. J. Glofcheski. 1990. Survival of mammalian cells following multiple freeze-thaw cycles. II. Independence of cryoprotection using glutamine with dimethyl sulfoxide, hydroxyethyl starch, propylene glycol, and glycerol. *Cryoletters* 11:215–226.

Kruuv, J., D. J. Glofcheski, and J. R. Lepock. 1988. Survival of mammalian cells following multiple freeze-thaw cycles: Implications and data analysis. *Cryoletters* 9:11–20.

Kruuv, J., J. R. Lepock, and A. D. Keith. 1978. The effect of fluidity of membrane lipids on freeze-thaw survival of yeast. *Cryobiology* 15:73–79.

Kurz, W., and D. Fisher. 1989. *Fundamentals of solidification.* Aedermannsdorf: Trans Tech.

Lacroix, M., and V. Voller. 1990. Finite difference solutions of solidification phase change problems: Transformed versus fixed grids. *Numer. Heat Transfer, Part B* 17:25–41.

Langer, J. 1984. Dynamics of dendritic pattern formation. *Materials Sci. Eng.* 65:37–44.

Le Gal, F., P. Gasqui, and J.-P. Renard. 1995. Evaluation of intracellular cryoprotectant concentration before freezing of goat mature oocyte. *Cryoletters* 16:3–12.

Leroux, B., H. Bizet, J. Brady, and V. Ivan. 1997. Water structuring around complex solutes: Theoretical modeling of α-D-glucopyranose. *Chem. Phys.* 216:348–363.

Leslie, S., E. Israeli, B. Lighthart, and J. Crowe. 1995. Trehalose and sucrose protect both membranes and proteins in intact bacteria during drying. *App. Environ. Microbiol.* 61:3592–3597.

Levin, R. L. 1988. Osmotic behavior of cells during freezing and thawing. In *Low-temperature biotechnology. Emerging applications and engineering contributions,* edited by J.J. McGrath and K. R. Diller, 177–188. New York: ASME.

Levin, R. L., E. G. Cravalho, and C. E. Huggins. 1977. Water transport in a cluster of closely packed erythrocytes at sub-zero temperatures. *Cryobiology* 14:549–558.

Levin, R. L., M. Ushiyama, and E. G. Cravalho. 1979. Water permeability of yeast cells at sub-zero temperatures. *J. Membrane Biol.* 46:91–124.

Levine, H., and L. Slade. 1988a. Principles of "cryostabilization" technology from structure/property relationships of carbohydrate/water systems: A review. *Cryoletters* 9:21–63.

Levine, H., and L. Slade. 1988b. Thermomechanical properties of small-carbohydrate-water glasses and "rubbers." *J. Chem. Soc. Faraday Trans.* 84:2619–2633.

Li, J. 1996. Inelastic neutron scattering studies of hydrogen bonding in ices. *J. Chem. Phys.* 105:6733–6755.

Lin, S., and D. Gao. 1990. A study on freezing process of a ternary system used for preservation of biological cells. In *Proc. 9th Intl' Heat Transfer Conf.,* vol. 3, edited by G. Hetsroni, 79–84. Heat Transfer 1990.

Lipp, G., M. Rodder, C. Koerber, and G. Rau. 1992. Interaction of particles and ice-water interfaces during bulk freezing. *Cryoletters* 13:229–238.

Lipp, G., S. Galow, C. Koerber, and G. Rau. 1994. Encapsulation of human erythrocytes by growing ice crystals. *Cryobiology* 31:305–312.

Liu, K., M. Brown, C. Carter, R. Saykally, J. Gregory, and D. Clary. 1996. Characterization of a cage form of the water hexamer. *Nature* 381:501–503.

Liu, Q., and J. Brady. 1996. Anisotropic solvent structuring in aqueous sugar solutions. *J. Am. Chem. Soc.* 118:12276–12286.

Liu, R., and L. Orgel. 1997. Efficient oligomerization of negatively charged B-amino acids at –20° C. *J. Am. Chem. Soc.* 119:4791–4792.

Lunardini, V. J. 1991. Conduction with freezing and thawing. In *Freezing and melting heat transfer in engineering,* edited by K.C. Cheng and N. Seki, 65–130. New York: Hemisphere Publ. Corp.

Luo, H., and H. Svendsen. 1996. Theoretical model for drop and bubble breakup in turbulent dispersions. *AIChE J.* 42:1225–1233.

Luyet, B. J. 1966. Anatomy of the freezing process in physical systems. In *Cryobiology,* edited by H.T. Meryman, 115–138. London: Academic Press.

MacFarlane, D. R., M. Forsyth, and C. A. Barton. 1992. Vitrification and devitrification in cryopreservation. In *Adv. low-temp. biology* 1, edited by P. Steponkus, 221–278. London: JAI Press Ltd.

MacLeod, R. A., and P. H. Calcott. 1976. Cold shock and freezing damage to microbes. *Symp. Soc. Gen. Microbiol.* 26:81–109.

Madura, J., M. Taylor, A. Wierzbicki, and J. Harrington. 1996. The dynamics and binding of a type III antifreeze protein in water and on ice. *J. Molec. Struct.* (Theochem) 388:65–77.

Malik, K. 1987. Preservation of biotechnologically important microorganisms in culture collections. *Progress Biotechno.* 4:145–186.

Mansour, N., and T. Lundgren. 1990. Satellite formation in capillary jet breakup. *Phys. Fluids* A2:1141–1144.

Marcus, Y. 1985. *Ion solvation.* Chichester: John Wiley.

Marcus, Y. 1987. Thermodynamics of ion hydration and its interpretation in terms of a common model. *Pure & Appl. Chem* 59:1093–1101.

Marcus, Y. 1993. Thermodynamics of solvation of ions. Part 6: The standard partial volume of agneous ions at 298.15 K. *J. Chem. Soc. Faraday Trans.* 89:713–718.

Mashimo, S. 1993. High order and local structure of water determined by microwave dielectric study. *J. Chem. Phys.* 99:9874–9881.

Mashimo, S. 1994. Structure of water in pure liquid and biosystem. *J. Non-Crystall. Solids* 172–174:1117–1120.

Mashimo, S., N. Miura, and T. Umehara. 1992. The structure of water determined by microwave dielectric study on water mixtures with glucose, polysaccharides, and L-ascorbic acid. *J. Chem. Phys.* 97:6759–6765.

Masters, K. 1979. *Spray drying handbook.* 3d ed. London: Halsted Press.

Materer, N., U. Starke, A. Barbieri, M. Van Hove, G. Somorjai, G. Kroes, and C. Minot. 1995. Molecular surface structure of a low-temperature ice Ih (0001) crystal. *J. Phys. Chem.* 99:6267–6269.

Materer, N., U. Starke, A. Barbieri, M. Van Hove, G. Somorjai, G. Kroes, and C. Minot. 1997. Molecular surface structure of ice (0001): Dynamical low-energy electron diffraction, total-energy calculations and molecular dynamics simulations. Submitted for publication.

Matsutani, K., Y. Fukada, K. Murata, A. Kimura, I. Nakamura, and N. Yajima. 1990. Physical and biochemical properties of freeze-tolerant mutants of a yeast *Saccharomyces cerevisiae. J. Ferment. Bioeng.* 70:275–276.

Mazur, P. 1963. Kinetics of water loss from cells at sub-zero temperatures and the likelihood of intracellular freezing. *J. Gen. Physiol.* 47:347–369.

Mazur, P. 1966. Physical and chemical basis of injury in single-celled microorganisms subjected to freezing and thawing. In *Cryobiology,* edited by H. Meryman, 213–315. London: Academic Press.

Mazur, P. 1984. Freezing of living cells: Mechanism and implications. *Am. J. Physiol.* 247:C125–C142.

Mazur, P., and N. Rigopoulos. 1983. Contributions of unfrozen fraction and of salt concentration to the survival of slowly frozen human erythrocytes: Influence of warming rate. *Cryobiology* 20:274–289.

Mazur, P., W. F. Rall, and N. Rigopoulos. 1981. Relative contributions of the fraction of unfrozen water and of salt concentration to the survival of slowly frozen human erythrocytes. *Biophys. J.* 36:653–675.

McGann, L. E. 1979. Optimal temperature ranges for control of cooling rate. *Cryobiology* 16:211–216.

McGann, L. E., H. Yang, and M. Walterson. 1988. Manifestations of cell damage after freezing and thawing. *Cryobiology* 25:178–185.

McGrath, J. J. 1981. Thermodynamic modeling of membrane damage. In *Effects of low temperatures on biological membranes,* edited by G. J. Morris and A. Clarke, 335–377. London: Academic Press.

McGrath, J. J. 1987. Cold shock: Thermoelastic stress in chilled biological membranes. In *Network thermodynamics, heat and mass transfer in biotechnology,* edited by K. R. Diller, 57–66. New York: ASME.

McGrath, J. J. 1988. Membrane transport properties. In *Low-temperature biotechnology. Emerging applications and engineering contributions,* edited by J. J. McGrath and K. R. Diller, 273–330. New York: ASME.

Mehl, P. 1995. Effect of pH on the ice crystallization in aqueous solutions of 1,2-propanediol. Consequence for vitrification solutions. *Cryoletters* 16:31–40.

Meryman, H. T. 1966. Review of biological freezing. In *Cryobiology,* edited by H.T. Meryman, 1–114. London: Academic Press.

Meryman, H. T. 1971. Cryoprotective agents. *Cryobiology* 8:173–183.

Meryman, H. T. 1974. Freezing injury and its prevention in living cells. *Ann. Rev. Biophys. Bioeng.* 3:341–363.

Miyajima, K., K. Machida, T. Taga, H. Komatsu, and M. Nakagaki. 1988. Correlation between the hydrophobic nature of monosaccharides and cholates, and their hydrophobic indices. *J. Chem. Soc. Faraday Trans.* 88:2537–2544.

Mizuno, Y., and G. Wakahama. 1983. Structure and orientation of frozen droplets on ice surfaces. *J. Phys. Chem.* 87:4161–4167.

Morel, J.-P., C. Lhermet, and N. Morel-Desrosiers. 1988. Interactions between cations and sugars. *J. Chem. Soc. Faraday Trans.* 84:2567–2571.

Morris, G. J., G. H. Coulson, and K. J. Clarke. 1988. Freezing injury in *Saccharomyces cerevisiae:* The effect of growth conditions. *Cryobiology* 25:471–482.

Morris, G. J., L. Winters, G. E. Coulson, and K. J. Clarke. 1986. Effect of osmotic stress on the ultrastructure and viability of the yeast *Saccharomyces cerevisiae. J. Gen. Microbiol.* 132:2023–2034.

Mozes, N., A. Leonard, and P. Rouxhet. 1988. On the relations between the elemental surface composition of yeast and bacteria and their charge and hydrophobicity. *Biochim. Biophys. Acta* 945:324–334.

Mozes, N., and P. Rouxhet. 1990. Microbial hydrophobicity and fermentation technology. In *Microbial cell surface hydrophobicity,* edited by R. Doyle and M. Rosenberg, 75–105. Washington, DC: Am. Soc. Microbiol.

Muldrew, K., and L. E. McGann. 1990. Mechanism of intracellular ice formation. *Biophys. J.* 57:525–532.

Mullis, A. 1995. A numerical model for the calculation of the growth velocity of nonisothermal parbolic dendrites. *J. Appl. Phys.* 78:4137–4143.

Mullis, A. 1996. A free boundary model for shape preserving dendritic growth at high undercooling. *J. Appl. Phys.* 80:4129–4136.

Nakae, T. 1986. Outer membrane permeability of bacteria. *CRC Crit. Rev. Microbiol.* 13:1–62.

Nakamura, A., and R. Okeda. 1976. The coagulation of particles in suspension by freezing-thawing. II. Mechanism of coagulation. *Colloid Polymer Sci.* 254:497–506.

Nei, T. 1981. Mechanism of freezing injury to erythrocytes: Effect of initial cell concentration on the post-thaw hemolysis. *Cryobiology* 18:229–237.

Nei, T., T. Araki, and T. Matsusaka. 1968. Freezing injury to aerated and non-aerated cultures of *Escherichia coli*. In *Freezing and drying of microorganisms,* edited by T. Nei, 3–16. Tokyo: Univ. Tokyo Press.

Neidhardt, F. C. 1987. Escherichia coli *and* Salmonella typhimurium. *Cellular and molecular biology.* Washington, DC: Am. Soc. Microbiol.

Neidhardt, F. C., J. Ingraham, and M. Schaechter. 1990. *Physiology of the bacterial cell.* Sunderland, MA: Sinauer Associates, Inc.

Nicolajsen, H., and A. Nvidt. 1994. Phase behavior of the system trehalose-NaCl-water. *Cryobiology* 31:199–205.

Nikaido, H. 1979. Nonspecific transport through the outer membrane. In *Bacterial outer membranes,* edited by M. Inouye, 361–408. New York: J. Wiley.

Nikaido, H., and M. Vaara. 1985. Molecular basis of bacterial outer membrane permeability. *Microbiol. Rev.* 49:1–32.

Nikaido, H., and M. Vaara. 1987. Outer membrane. In Escherichia coli *and* Salmonella typhimurium. *Cellular and molecular biology,* edited by F.C. Neidhardt et al. 7–22. Washington, DC: Am. Soc. Microbiol.

Nikaido, H., K. Nikaido, and S. Harayama. 1991. Identification and characterization of porins in *Pseudomonas aeruginosa. J. Biol. Chem.* 266:770–779.

Obuchi, K., S. C. Kaul, H. Iwahashi, M. Ishimura, and Y. Komatsu. 1990. Membrane fluidity estimated by in vivo NMR corresponded to freezing-thawing viability of yeast. *Cryoletters* 11:287–294.

Ohrai, T., and H. Yamamoto. 1991. Frost heaving in artificial ground freezing. In *Freezing and melting heat transfer in engineering,* edited by K. Cheng and N. Seki, 547–580. New York: Hemisphere Publ. Corp.

Ohtomo, M., and G. Wakahama. 1983. Growth rate of recrystallation in ice. *J. Phys. Chem.* 87:4139–4142.

Panchangula, M., P. Sojka, and P. Santangelo. 1996. On the three-dimensional instability of a swirling, annular, inviscid liquid sheet subject to unequal gas velocities. *Phys. Fluids* 8:3300–3312.

Papageorgiou, D. 1995a. On the breakup of viscous liquid threads. *Phys. Fluids* 7:1529–1544.

Papageorgiou, D. 1995b. Analytical description of the breakup of liquid jets. *J. Fluid Mech.* 301:109–132.

Pearson, B. M., P. J. Jackman, K. A. Painting, and G. J. Morris. 1990. Stability of genetically manipulated yeasts under different cryopreservation regimes. *Cryoletters* 11:205–210.

Pegg, D. E. 1976. Long-term preservation of cells and tissues: A review. *J. Clin. Path.* 29:271–285.

Pegg, D. E. 1981. The effect of cell concentration on the recovery of human erythrocytes after freezing and thawing in the presence of glycerol. *Cryobiology* 18:221–228.

Perera, L., and M. Berkowitz. 1993. Erratum: Many body effects in molecular dynamics simulations of $Na^+(H_2O)n$ and $Cl^-(H_2O)n$ clusters. *J. Chem. Phys.* 99:4236–4237.

Pitt, R. E. 1992. Thermodynamics and intracellular ice formation. In *Adv. low-temp. biology* 1, edited by P. Steponkus, 63–99. London: JAI Press Ltd.

Plum, G., C. Koerber, and G. Rau. 1988. The effect of cooling rate on the release of encapsulated marker from single bilayer dipalmitoyphosphatidylcholine liposomes after a freeze-thaw process. *Cryoletters* 9:316–327.

Quinn, P. J. 1985. A lipid-phase separation model of low-temperature damage to biological membranes. *Cryobiology* 22:128–146.

Raetz, C. 1990. Biochemistry of endotoxins. *Ann. Rev. Biochem.* 59:129–170.

Rall, W. F., P. Mazur, and J. J. McGrath. 1983. Depression of the ice-nucleation temperature of rapidly cooled mouse embryos by glycerol and dimethyl sulfoxide. *Biophys. J.* 41:1–12.

Ray, B., M. L. Speck, and W. J. Dobrogosz. 1976. Cell wall lipopolysaccharide damage in *Escherichia coli* due to freezing. *Cryobiology* 13:153–160.

Recommendations for the processing and handling of frozen foods. 1986. Paris: International Institute of Refrigeration.

Reid, D. 1983. Fundamental physicochemical aspects of freezing. *Food Technology,* April:110–115.

Ren, H., Y. Wei, C. Hua, and J. Zhang. 1994. Theoretical prediction of vitrification and devitrification tendencies for cryoprotective solutions. *Cryobiology* 31:47–56.

Rips, I. 1992. Ion solvation in clusters. *J. Chem. Phys.* 97:536–546.

Robarts, A. W., and U. B. Sleyter. 1985. *Low-temperature methods in biological electron microscopy.* Amsterdam: Elsevier.

Roos, Y., and M. Karel. 1991. Nonequilibrium ice formation in carbohydrate solutions. *Cryoletters* 12:367–376.

Rubinsky, B., and T. Eto. 1990. Heat transfer during freezing of biological materials. *Ann. Rev. Heat Transfer* 3:1–38.

Saito, S., and I. Ohmine. 1994. Dynamics and relaxation of an intermediate size water cluster $(H_2O)108$. *J. Chem. Phys.* 101:6063–6075.

Saitoh, T. 1991. Numerical and analytical aspects in freezing and melting. In *Freezing and melting heat transfer in engineering,* edited by K.C. Cheng and N. Seki, 131–175. New York: Hemisphere Publ. Corp.

Sajbidor, J., E. Breierova, and A. Kockova-Kratochvilova. 1989. The relationship between freezing resistance and fatty acid composition of yeasts. *FEMS Microbiol. Letters* 58:195–198.

Samarskii, A. et al. 1993. Numerical simulation of convection/diffusion phase change problems: A review. *Int. J. Heat Mass Transfer* 36:4095–4106.

Schaff, J., and J. Roberts, 1996. Toward an understanding of the surface chemical properties of ice: Differences between the amorphous and crystalline surfaces. *J. Phys. Chem.* 100:14151–14160.

Schekman, R., and P. Novick. 1983. The secretory process and yeast cell-surface assembly. In *The molecular biology of the yeast* Saccharomyces. *Metabolism and gene expression,* edited by J. N. Strathern, E. W. Jones, and J. R. Broach, 361–398. Cold Spring Harbor: Cold Spring Harbor Laboratory.

Scheiwe, M. W., and G. Rau. 1981. Biokaltetechnik: Verfahren der Gefrier-Konservierung in der Medizin. *Chem. -Ing.-Tech.* 53:787–797.

Schievenbusch, A., G. Zimmerman, and M. Mathes. 1993. Comparison of different analysis techniques to determine the cellular and dendritic spacing. *Mat. Sci. Eng.* A173:85–88.

Schu, P., and M. Reith. 1995. Evaluation of different preparation parameters for the production and cryopreservation of seed cultures with recombinant *Saccharomyces cerevisiae. Cryobiology* 32:379–388.

Schultz, S. G. 1980. *Basic principles of membrane transport.* Cambridge: Cambridge University Press.

Schwartz, G. J., and K. R. Diller. 1983a. Osmotic response of individual cells during freezing. I. Experimental volume measurements. *Cryobiology* 20:61–77.

Schwartz, G. J., and K. R. Diller. 1983b. Osmotic response of individual cells during freezing. II. Membrane permeability analysis. *Cryobiology* 20:542–552.

Sears, P., and C.-H. Wong. 1996. Intervention of carbohydrate recognition by proteins and nucleic acids. *Proc. Natl. Acad. Sci. USA* 93:12086–12093.

Sen, K., J. Hellman, and H. Nikaido. 1988. Porin channels in intact cells of *Escherichia coli* are not affected by Donnan potential across the outer membrane. *J. Biol. Chem.* 263:1182–1187.

Shalaev, Y., and A. Kanev. 1994. Study of the solid-liquid state diagram of the water-glycine-sucrose system. *Cryobiology* 31:374–382.

Shalaev, Y., F. Franks, and P. Echlin. 1996. Crystalline and amorphous phases in the ternary system water-sucrose-sodium chloride. *J. Phys. Chem.* 100:1144–1152.

Sharp, R. J. 1984. The preservation of genetically unstable microorganisms and the cryopreservation of fermentation seed cultures. *Adv. Biotech. Process.* 3:81–109.

Shavit, U., and H. Chigier. 1996. Development and evaluation of a new turbulence generator for atomization research. *Exper. in Fluids* 20:291–301.

Shepard, M. L., C. S. Goldston, and F. H. Cocks. 1976. The H_2O-NaCl-glycerol phase diagram and its application in cryobiology. *Cryobiology* 13:9–23.

Shier, T. 1988. Studies on the mechanism of mammalian cell killing by a freeze-thaw cycle: Conditions that prevent cell killing using nucleated freezing. *Cryobiology* 25:110–120.

Silvares, O. M., E. G. Cravalho, W. M. Toscano, and C. E. Huggins. 1975. The thermodynamics of water transport from biological cells during freezing. *J. Heat Transfer ASME,* November:582–588.

Silver, B. 1985. *The physical chemistry of membranes.* Boston: Allen & Unwin.

Singhri, R., C. Schorr, C. O'Hara, L. Xie, and D. Wang. 1996. Clarification of animal cell culture process fluids using depth microfiltration. *Biopharm* 9, April:35–41.

Spelt, J. K., D. R. Absolom, W. Zingy, C. van Oss, and A. Neumann. 1982. Determination of the surface tension in biological cells using the freezing front technique. *Cell Biophys.* 4:117–131.

Sputtek, A., A. Brohm, I. Classen, U. Hartman, C. Korber, M. W. Scheiwe, and G. Rau. 1987. Cryopreservation of human platelets with hydroxyethyl starch in a one-step procedure. *Cryoletters* 8:216–229.

Sputtek, A., G. Singbartl, R. Langer, W. Schleizner, H. Henrich, and P. Kuhnl. 1995. Cryopreservation of red blood cells with the nonpenetrating cryoprotectant hydroxyethyl starch. *Cryoletters* 16:283–388.

Steponkus, P. L., and M. F. Dowgert. 1981. Gas bubble formation during intracellular ice formation. *Cryoletters* 2:42–47.

Stuart, S., and B. Berne. 1996. Effects of polarizability on the hydration of the chloride ion. *J. Phys. Chem.* 100:11934–11943.

Sundari, S., B. Raman, and D. Balasubramanian. 1991. Hydrophobic surfaces in oligosaccharides: Linear dextrins and amphiphilic chains. *Biochim. Biophys. Acta* 1065:35–41.

Sutanto, E., T. Davis, and L. Scriven. 1992. Adaptive finite element analysis of axisymmetric freezing. *Int. J. Heat Mass Transfer* 35:3301–3312.

Suzuki, T., and F. Franks. 1993. Solid-liquid phase transitions and amorphous states in ternary sucrose-glycine-water systems. *J. Chem. Soc. Faraday Trans.* 89:3283–3288.

Takahashi, T., and R. Williams. 1983. Thermal shock hemolysis in human red cells. I. The effects of temperature, time, and osmotic stress. *Cryobiology* 20:507–520.

Takanaka, N., A. Ueda, T. Daimon, H. Bandow, T. Dohmaru, and Y. Maeda. 1996. Acceleration mechanism of chemical reaction by freezing: The reaction of nitrous acid with dissolved oxygen. *J. Phys. Chem.* 100:13874–13884.

Talsma, H., M. J. van Steenbergen, and D. J. Crommelin. 1992. The cryopreservation of liposomes. 2. Effect of particle size on crystallization behavior and marker retention. *Cryobiology* 29:80–86.

Taylor, M. J. 1987. Physicochemical principles in low-temperature biology. In *The effects of low temperatures on biological systems,* edited by B. W. Grout and G. J. Morris, 3–71. London: E. Arnold.

Thom, F. 1988. Deformation of the cell membrane at low temperatures. II. Elastical deformability of the erythrocyte membrane. *Cryoletters* 9:308–315.

Thom, F., and G. Matthes. 1988. Deformation of the cell membrane at low temperatures. I. A cryomicroscopial technique. *Cryoletters* 9:300–307.

Thwaites, J., and N. Mendelson. 1991. Mechanical behavior of bacterial cell walls. *Adv. Microbial Physiol.* 32:173–222.

Tjadhadi, M., H. Stone, and J. Ottino. 1992. Satellite and subsatellite formation in capillary breakup. *J. Fluid Mech.* 243:297–317.

Todt, J. C., W. J. Rocque, and E. J. McGroarty. 1992. Effects of pH on bacterial porin function. *Biochemistry* 31:10471–10478.

Toner, M. 1993. Nucleation of ice crystals inside biological cells. *Adv. Low-Temp. Biology* 2:1–52, London: JAI Press Ltd.

Toner, M., E. Cravalho, M. Karel, and D. Armant. 1991. Cryomicroscopic analysis of intracellular ice formation during freezing of mouse oocytes without cryoadditives. *Cryobiology* 28:55–71.

Vaara, M., W. Z. Plachy, and H. Nikaido. 1990. Partitioning of hydrophobic probes into lipopolysaccharide bilayers. *Biochim. Biophys. Acta* 1024:152–158.

Vicks, B., and D. J. Nelson. 1993. The boundary element method applied to freezing and melting problems. *Numer. Heat Transfer, Part B* 24:263–277.

Viswanath, R., and Y. Jaluria. 1993. A comparison of different solution methodologies for melting and solidification problems in enclosures. *Numer. Heat Transfer, Part B* 24:77–105.

Voller, V. R. 1985. Implicit finite-difference solutions of the enthalpy formulation of Stefan problems. *IMA J. Numer. Anal.* 5:210–214.

Voller, V. R. 1987. A heat balance integral method based on an enthalpy formulation. *Int. J. Heat Mass Transfer* 30:604–607.

Voller, V. R. 1997. An overview of numerical methods for solving phase-change problems. *Adv. Numer. Heat Transfer* 1:341–375; Taylor & Francis, Hauts, UK. edited by W. Minkowycz and E. Sparrow.

Voller, V. R., and C. R. Swaminathan. 1993. Treatment of discontinuous thermal conductivity in control-volume solutions of phase-change problems. *Numer. Heat Transfer, Part B* 24:161–180.

Voller, V. R., and M. Cross. 1981. Accurate solutions of moving boundary problems using the enthalpy method. *Int. J. Heat Mass Transfer* 24:545–556.

Walcerz, D. B. 1995. Cryosim: A user-friendly program for simulating cryopreservation protocols. *Cryobiology* 32:35–51.

Wales, D., and I. Ohmine. 1993. Structure, dynamics, and thermodynamics of model $(H_2O)8$ and $(H_2O)20$ water clusters. *J. Chem. Phys.* 98:7245–7256.

Wang, C., W. Cheb, V. Tran, and R. Douillard. 1996. Analysis of interfacial water structure and dynamics in α-maltose solution by molecular dynamics simulation. *Chem. Phys. Letters* 252:268–274.

Wellman, A., and G. Stewart. 1973. Storage of brewing yeasts by liquid nitrogen refrigeration. *Appl. Microbiol.* 26:577–583.

Whitfield, C., and M. Valavano. 1993. Biosynthesis and expression of cell-surface polysaccharides in gram-negative bacteria. *Adv. Microbial Physiol.* 35:135–235.

Wilcox, W. R. 1980. Force exerted on a single spherical particle by a freezing interface: Theory. *J. Colloid Interface Sci.* 77:213–218.

Wisniewski, R. 1992. Unpublished results.

Wisniewski, R. 1996. Unpublished results.

Wisniewski, R. 1998a. Developing Large-Scale Cryopreservation Systems for Biopharmaceutical Products. *BioPharm* 11(6):50–60.

Wisniewski, R. 1998b. Large-scale cryopreservation of cells, cell components, and biological solutions. *BioPharm* 11g(9): 42–61.

Wisniewski, R., and V. Wu. 1992. Large-scale freezing and thawing of biopharmaceutical drug product. *Proceedings of the International Congress: Advanced technologies for manufacturing of aseptic and terminally sterilized pharmaceuticals and biopharmaceuticals.* Basel, Switzerland, February 17–19.

Wisniewski, R., and V. Wu. 1996. Large-scale freezing and thawing of biopharmaceutical products. In *Biotechnology and biopharmaceutical manufacturing, processing, and preservation,* edited by K. Avis and V. Wu, 7–59. Buffalo Grove, IL: Interpharm Press, Inc.

Wolber, P. 1993. Bacterial ice nucleation. *Adv. Microbial Physiol.* 34:203–237.

Wolfinbarger, L., V. Sutherland, L. Braendle, and G. Sutherland. 1996. Engineering aspects of cryobiology. *Adv. Cryogen. Eng.* 41:1–12.

Yang, C.-H., S. Rastogi, and D. Poulikakos. 1993. Solidification of a binary mixture saturating an inclined bed of packed spheres. *Int. J. Heat and Fluid Flow* 14:268–278.

Yao, L., and J. Prusa. 1989. Melting and freezing. *Adv. Heat Transfer* 19:1–95.

Yoo, J. 1991. Effect of fluid flow induced by a rotating disk on the freezing of fluid. *Int. J. Heat and Fluid Flow* 12:257–262.

Yoo, H., and S. Ro. 1991. Melting process with solid-liquid density change and natural convection in a rectangular cavity. *Int. J. Heat and Fluid Flow* 12:365–374.

Zabrodsky, S., and H. Martin. 1988. Fluid-to-particle heat transfer in fluidized beds. In *Heat exchanger design handbook,* vol. 2, edited by E. Schlunder, 2.5.5.1–2.5.5.7. New York: Hemisphere Publ. Corp.

Zhang, X., and O. Basaran. 1996. Dynamics of drop formulation from a capillary in the presence of an electric field. *J. Fluid Mech.* 326:239–263.

Zhao, L., H. Y. Zhu, T. C. Hua, Y. B. Zhou, and J. Xu. 1988. Low-temperature preservation for producing strains of streptomycin and gentamycin. *Cryoletters* 9:144–151.

Zlotnik, H., M. Fernandez, B. Bowers, and E. Cabib. 1984. *Saccharomyces cerevisiae* mannoproteins form an external cell wall layer that determines wall porosity. *J. Bacteriol.* 159:1018–1026.

3

LARGE-SCALE CRYOPRESERVATION: PROCESS DEVELOPMENT FOR FREEZING AND THAWING OF LARGE VOLUMES OF CELL SUSPENSIONS, PROTEIN SOLUTIONS, AND BIOLOGICAL PRODUCTS

Richard Wisniewski

NASA Ames Research Center, Mountain View, CA

INTRODUCTION

Current biotechnology, biopharmaceutical, and biological industries are moving away from the concept of dedicated manufacturing plants toward the use of multiproduct facilities operating on a campaign basis. The product can be manufactured during a relatively short campaign, frozen, and stored to be gradually thawed and further processed according to the downstream processing and fill and finish plant capacities, or according to market demand. Manufacturing steps may be performed at multiple locations (in certain cases, the bulk may be produced in one plant, the purification conducted in another, and the fill and finish of purified bulk at the third location). Contract manufacturing may require such multisite manufacturing processing as well. The preservation by freezing and later reconstitution by thawing may be required at various stages of the manufacturing process.

An important factor associated with product freezing and storage in the frozen state can be extension of the product life. An important gain may be accomplished by freezing purified and formulated bulk and storing it in a frozen state prior to the filling and finishing operation. In such a case, the formulation does not include any poorly defined components that might affect the product in any way during freezing, storing, or thawing steps. The frozen bulk could be stored for 1–2 years and then thawed prior to filling and finishing with the shelf life of the final product being determined by the date of the filling and finishing steps. A similar approach to raw fermentation broths, concentrated cell paste, or extracted raw product prior to the purification may require experimental knowledge of biological material behavior during freezing, storage, and thawing. In all cases, there is a need for large-scale freezing and thawing processes that can be conducted in a well-controlled way with no detrimental effects to the processed product.

Biological solutions and cell suspensions can be frozen with the purpose of storage in a frozen state prior to lyophilization. Traditionally, freezing in vials, flat trays, bottles, bags, and flasks is performed. Freezing in vials can be done on cooled shelves (typically prior to lyophilization) or by exposure to cold gas (cabinet freezers or cold nitrogen gas freezers), to cold liquid (cooled ethanol, glycol, or brine), or to liquefied gas (liquid nitrogen). A combination of exposure to cold gas and liquefied gas is also used (a two-step freezing). Cylindrical bottles can be frozen in liquid baths or in cabinet or blast freezers. Freezing in liquid baths can be accomplished in a quiescent or agitated bath fluid. Fluid agitation can be accomplished by mechanical mixer or by recirculation. Fluid agitation improves heat transfer and shortens the freezing time. Placement of bottles in the fluid bath could be vertical, horizontal, or slanted. Vertical position ensures symmetrical conditions of freezing (a cylinder freezing pattern) except of the bottle bottom and the top gas pocket. Tall, small-diameter bottles can provide a better freezing pattern and shorter freezing times than short, large-diameter bottles. Horizontal or slanted positions give an asymmetrical freezing pattern due to the presence of a gas pocket and increased surface area. Freezing can also be done in a bottle that rotates in a liquid bath. The heat transfer coefficient on a fluid side is high due to the bottle rotation, but the system is mechanically complex, and too fast rotation may cause exclusion of solutes from the freezing mass and their concentration in the liquid phase. Freezing of bottles in the cabinet or blast freezers is hindered by a low heat transfer coefficient on the gas side, and freezing times are long. Gas recirculation in the blast freezer increases the heat transfer coefficient and shortens the freezing time. Gas velocity distribution and overall pattern need to be carefully adjusted to ensure similar heat transfer conditions for all bottles placed in the freezer. Lyophilization requires a relatively thin layer of frozen material to ensure efficient freeze-drying. Typically the freezing is accomplished on cooled shelves in the lyophilizer. After lyophilization, the vials are sealed by stoppers, and the trays are removed with the top surface exposed. An increase in surface contact area can be accomplished by placing the liquid product in a flat flask or container. Such a configuration can assure a thin layer of liquid (advantageous for lyophilization), fewer containers, and product protection (no exposure during unloading). After lyophilization, a sealing step of the container side is applied. Figure 3.1 shows some of the typical approaches to freezing in industry.

Cell suspensions are traditionally frozen in plastic bags or bottles using the cabinet and blast freezers or liquid baths. The freezing process is long, capacities are relatively low, and operations are labor intensive and costly due to the number of bags or bottles involved in the process. Subsequent thawing is also slow and labor intensive. Scale-up of such a process is difficult but may be done by adding more bags or bottles and building larger freezers.

Anticipated conditions for the freezing and thawing of large volumes of cell suspensions are more challenging than the conditions for freezing and thawing of protein solutions (Wisniewski and Wu 1992, 1996; Wisniewski 1998 a, b). In the case of freezing/thawing of protein solutions, the major interest has been in how to preserve the biological activity of the molecule. In the case of killed cells or cell fragments containing active product, the interest is similar (i.e., how to preserve product activity). In the case of preservation of viable cells, the goal is to preserve cell viability (i.e., preserving the activity of a vast number of cell biomolecules and organelles with varying

Figure 3.1. (a) Freezing in vials. The pattern of internal dendritic ice crystal formation depends on the vial material and on the quality of contact with the vial-cooled shelf. Glass thermal conductivity is lower than that of Teflon and the frozen product; therefore, the heat conduction through vial walls affects the dendritic pattern. Teflon thermal conductivity is comparable to that of frozen product. Therefore, the heat conduction (and the heat flux) through the walls is similar to the one through the product, and the dendritic pattern should not be significantly affected by the walls. (b) Freezing in bottles. The cylindrical bottles can be frozen in a liquid bath or gas environment. Fluid agitation and a gas flow at high velocity increase the heat transfer coefficient and shorten the freezing time. Rotation of the bottle allows a circumferential freezing pattern with the liquid product remaining in the center until final freezing. (c) Freezing in flasks on a cooled shelf. Done prior to lyophilization: ensures thin product layer and larger surface than for vials.

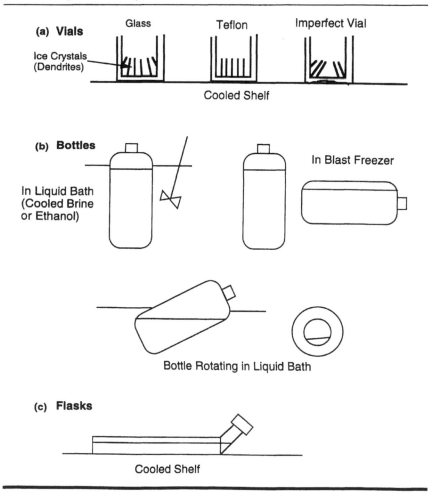

characteristics and other complex cell components). These requirements can make large-scale cell freezing a much more complex and
demanding process than freezing bulk protein, killed cells, or cell
fragments, with many more parameters to consider. Cryopreservation of suspensions of killed cells or cell fragments with the active
product present may involve an issue of product degradation (e.g.,
when certain proteases are present). Rapid cooling and freezing
under well-controlled conditions may reduce the level of product
degradation. The issue becomes critical when large volumes of
product are processed by freezing and thawing since the traditional
methods of freezing and thawing may expose product to other
solutes during extended periods of time.

This chapter also describes 3 case studies of novel processes
for cryopreservation for biological products in large scale that
have been developed by the author.

Process and System Development

The major objective of a freeze-thaw biopharmaceutical process is
to freeze large volumes of biological material in the form of aqueous solutions or suspensions of killed cells, cell fragments, or
viable cells or microorganisms. The process must be acceptable
from the viewpoint of required product characteristics, biological
containment, current Good Manufacturing Practices (cGMPs), and
safety. Material handling before and after freezing should minimize personnel exposure to the product and protect the product
from contamination. An optimal process design will be one that
physically contains the product during all operations related to
freezing. Frozen material should be easily handled and subjected
to a subsequent thawing step. A cleaning method for the system
should be automated and, if possible, eliminate any personnel
exposure to the processed material.

The freezing process development requires knowledge of fundamentals of properly designed experiments, information from
which can be used in large-scale process design. In addition to the
phase-change problem (solidification and melting), the ice nucleation and freezing of supercooled solutions under various conditions must be well understood. Saito et al. (1992) summarized the
fundamental factors affecting the freezing of supercooled water.
Convection in the liquid (agitation or liquid jet), vibration, and shock
have not helped initiate freezing of supercooled water. Collision or

rubbing of solids with physical contact have facilitated freezing. Bartell and Huang (1994) reported on the supercooling of submicroscopic droplets of water to about 200 K. Large clusters of such supercooled water showed no evidence of structural changes until they reached about 200 K. At that temperature, they began to freeze to cubic ice. This work provides insight into freezing under supercooling conditions. However, in practice, the supercooled water may not reach temperature below – 40° to – 42° C.

PROCESS AND SYSTEM DESIGN OPTIONS
Evaluation of the Existing Processes and Systems

When large volumes of cell suspensions are involved, traditional, small-scale methods of freezing in cabinet-type freezers usually will not be feasible. The freezers capable of freezing large quantities of biological materials may be classified as follows:

- Forced convection (air or gas) freezers: blast or fluidized bed
- Direct contact freezers: plate, band, and drum
- Evaporation/sublimation freezers: using liquid nitrogen, liquid/solid carbon dioxide, or liquid fluorocarbons
- Liquid immersion freezers: using brine, glycol, ethanol, or other low-temperature fluids

For freezing biologicals (tissues, dense cell, or cell fragment suspensions) and foodstuffs, the following reported approximate freezing rates can be considered, depending on the freezing conditions (Recommendations 1986):

- Bulk freezing in a batch air blast freezer room   1 mm/hr (0.00028 mm/sec)
- Quick freezing in a tunnel air blast freezer   3–15 mm/hr (0.00083–0.00417 mm/sec)
- Plate freezing   15–35 mm/hr (0.00417–0.00972 mm/sec)
- Quick freezing in a continuous air blast freezer   10–20 mm/hr (0.00277–0.0055 mm/sec)
- Freezing in liquefied gases   30–100 mm/hr (0.0083–0.0278 mm/sec)

In the case of a small liquid droplet freezing in bulk liquefied nitrogen, the freezing rate may reach values higher than those listed here, which have been collected in industrial installations for freezing large slabs of material.

Cabinet Freezers

Freezing large quantities of aqueous products in traditional cabinet or chest freezers typically requires a large number of such freezers due to the limited freezer heat removal capacity (freezers are designed for handling small samples). The freezer cannot be filled with large containers of aqueous solution since the cooling and freezing step will take a very long time, and the freezer temperature setpoints cannot be reached within a reasonable time (freezer may warm up above the setpoint after putting a heavy load into it and may take a long time before returning to the original setpoint). Therefore, the freezer can be only partially filled with product containers, with the product load depending on the heat removal capacity of the freezer. If there is no vigorous and well-distributed air recirculation within the freezer, the heat transfer coefficients between the container surface and the air can be low and the freezing times long. For example, for almost stagnant air $U = 5 \dfrac{W}{m^2 \cdot K}$, a 1-L Teflon bottle with a 9-cm diameter may require more than 10 hours to freeze at the temperature of –40° C. An estimated heat flux from the bottle, considering the U value, will be about 11 W, and from the freezing time, the average heat flux will be about 9 W (this heat flux will be higher than average at the beginning and decline during freezing). The same bottle placed in the freezer with a forced-air recirculation $U = 18 \dfrac{W}{m^2 \cdot K}$ may take less than 4 hours to freeze. Considering the U value, an estimated heat flux from the bottle can be around 42 W, and from the freezing time, an average heat flux will be about 29 W. The cabinet freezer with a significant forced air recirculation approaches the design of the blast freezer, based on a rapid air recirculation with a goal to maximize gas-container heat transfer and accelerate freezing rates.

Blast Freezers

When biological products are frozen in large walk-in freezers with or without air recirculation, the freezing rate can be very slow and not uniform. In a room designed for storing frozen products, the refrigeration/air cooling system may not be able to handle the duties of freezing after a load enters the cold room; the temperature may increase initially and then slowly decline as the heat load decreases during the freezing process.

The blast freezer is a large cabinet or room with vigorous forced-air recirculation. The internal air flow patterns are arranged to provide uniform cooling and freezing conditions within the freezer volume. The air cooling coils are of extended design (finned and in multiple rows), and the compressor size may be of several tens of HP. Usually the liquid is poured into containers of low thickness such as plastic bags or round or flat bottles. Bags placed in an ordinary freezer with stagnant air may require a long time to freeze completely. For example, 1.5-L bags may need several hours to approach the freezer temperature of –85° C, although complete freezing may be accomplished earlier (when all eutectic and glassy substances solidify). When freezing is performed in a blast freezer, freezing fronts move toward the center from the flat side surfaces exposed to outside cooling (using a forced convection of circulating air). The heat transfer coefficient on the air side may be estimated as 15–24 $\dfrac{W}{m^2 \cdot K}$ (Geankoplis 1983).

Temperature distribution within the blast freezer shall be tested for uniformity. Even in small chambers there could be temperature differences from point to point (Koebe et al., 1993).

Flat bags of small thickness and large outside surface provide good conditions for relatively rapid and uniform freezing of sample volumes. The containers are placed on racks and shelves. The configuration of shelves with flat containers may be compared to a stack of parallel plates being cooled. A recent analysis of such configurations can be found in the work by Morega et al. (1995). It is important to have similar air flow conditions (velocity, turbulence, and temperature) around all bags placed in the blast freezer to achieve similar freezing conditions (the design characteristics of a freezer are very important for consistent product quality). Maldistribution of air flow and temperature may cause nonuniformity of freezing. Use of thin, flat, horizontally positioned bags may not only reduce the cell settling effects and the freezing time but also minimize natural convection effects in the liquid phase. The filling of bags may also affect the freezing process in the blast freezer.

Figure 3.2. Freezing in bags in a blast freezer with the heat transfer coefficient $U = 20 \left[\dfrac{W}{m^2 \cdot K} \right]$ at the temperatures 203 K, 223 K, and 243 K. The freezing time depends on bag thickness and freezer temperature.

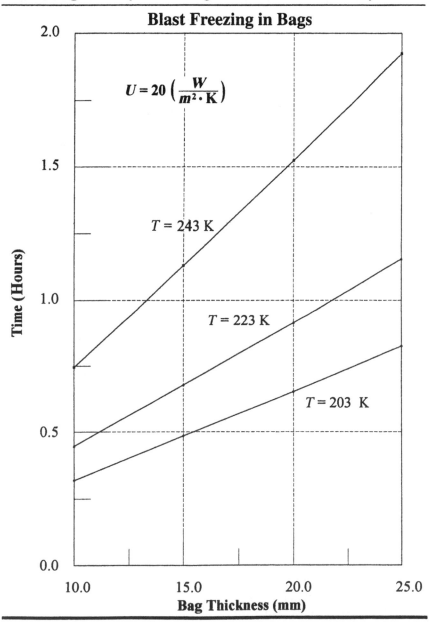

Any air/gas pockets at the upper side of a bag may significantly hinder heat transfer at the top surface due to increased thermal resistance. Figure 3.2 shows freezing times of bags with aqueous biological products at various bag thicknesses (10–25 mm) in the blast freezers at 3 temperatures: –30° C (243 K), –50° C (223 K), and –70° C (203 K). For a typical bag, one may expect a heat transfer coefficient of about $20 \dfrac{W}{m^2 \cdot K}$.

The design of supporting structures may influence the freezing process (e.g., there could be additional heat conduction through the supporting structure, which is cooled by a stream of cold air). The shape and configuration of extended surfaces could be optimized to maximize the heat flux. However, air blast freezing with additional heat conduction through the supporting shelves may cause certain asymmetry in the freezing if the upper surface is only cooled by an air stream via convection.

Dispensing product into small containers and freezing it in a blast freezer or in a freezer with cooled shelves may create logistic problems of handling large volumes of material and large numbers of containers. Economics and operational feasibility must be evaluated at this point. Product in bottles can also be frozen in blast freezers. However, bottle freezing may be slower than freezing in thin plastic bags of the same volume due to the difference in the external heat transfer surface in relation to the freezing distance. A critical characteristics of blast freezer designs include the uniformity of both temperature and gas velocity distributions. Typically the blast freezer application requires intensive validation work, particularly if containters vary with products.

The blast freezing concept has been executed in a variety of designs, such as tunnel, belt, spiral, and fluidized bed freezers. Tunnel freezers are large insulated boxes or tunnels with refrigeration coils and air recirculating fans. The product is placed in stacks or trays and moves along the freezer by means of a mechanical chain drive(s). It is important that product layers are uniformly spaced, allowing forced flow of air. The freezers are usually equipped with a system of baffles permitting adjustment of air flow patterns. Higher air velocities improve and accelerate freezing. If the fan motors are located within the cabinet, they generate heat, which must be removed by refrigeration coils. The tunnel freezer can accommodate a variety of product containers. However, only one type of container should be processed at a time to ensure complete and uniform freezing of all samples. A belt freezer may include a horizontal or vertical spiral belt enclosed in an insulated box with air recirculating fans and refrigeration coils. Belts are made of stainless steel mesh. The freezers operate in a continuous mode. Fluidized bed freezers, although used in the food industry for freezing berries, may not be adapted to handle containers with biological products (other than capsules or very small plastic vials).

Direct Contact Freezers

High cooling rates may be obtained using direct contact between cooled metallic surfaces and thin layers of biological sample. Preferably, the metal should have high thermal conductivity, although conductivity across the wall of the container and the ice formed within the product may be the major barrier for heat transfer under good contact conditions (e.g., when no significant air gaps exist). The thermal conductivity of stainless steel is about 15–16 $\dfrac{W}{m \cdot K}$; for titanium it is about 23 $\dfrac{W}{m \cdot K}$; for copper it is near 400 $\dfrac{W}{m \cdot K}$; and for aluminum about 230 $\dfrac{W}{m \cdot K}$.

Very rapid freezing, which can produce a layer of vitrified water next to the surface (Mueller 1988b), is obtained by spraying the solution on a large block of well-conducting material (such as copper), which is cooled earlier by the cryogenic coolant. The heat transfer is by conduction, and the temperature drop rate is much faster than in the case of droplet immersion into cryogenic coolant, since there is no boiling of coolant nor formation of a gas layer between the cooled surface and the liquid bulk of the cryogenic agent. Such a method is used for very small samples only—for example, for specimen preparation in electron microscopy. Spraying liquids on a cooled, moving, smooth metal belt does not reproduce those conditions due to the limited cooled mass of a belt in contact with a freezing solution. This method of belt freezing is similar to the smaller-scale drum freezing in which a thin product layer is applied on a cooled drum surface. Control over freezing may be improved by maintaining temperature zones along the tunnel the belt is moving through, but limitations to this exist since the layer of freezing liquid is thin and solidifies rapidly on the belt. Fast solidification is also required to prevent spilling, but this is less critical for viscous or pasty materials.

A large-scale processing option can be the use of shelves that are internally cooled by a recirculating heat transfer fluid (e.g., most of the heat transfer is performed by conduction in a mode similar to freezing in a lyophilizer). The heat transfer occurs via direct contact between the cooled metal plate and the product container. The quality of contact may significantly affect heat conduction (Irvine and Taborek 1988; Veziroglu et al. 1976) and the resulting freezing rate and uniformity. Cooling can be severely hindered by the presence of air gaps due to the very low thermal conductivity of air (air thermal conductivity is approximately 30 times lower than glass). Hwang et al. (1994) analyzed the effects of wall conduction and interface thermal resistance on the phase change problem in solidification.

Freezing of large liquid volumes subdivided into large numbers of vials is practiced in the lyophilization of pharmaceutical products (see Figure 3.1). The freezing is affected by the temperature distribution in lyophilizer shelves, the quality of contact shelf-vial, and uniformity of vial walls. The freezing in vials also may experience "corner effects," that is, cooling occurs from the corner along the bottom and vertical wall when the good contact shelf-vial is near the circumference of the vial bottom. In such a situation, the first dendritic ice crystals may occur at the corner and not be perpendicular to the vial bottom since heat flow direction depends on the shelf-vial contact area and heat conduction along the vertical wall. The uniformity of contact shelf-vial bottom is of utmost importance since, after formation of the first ice, the wall effect is usually not significant because the glass thermal conductivity is approximately $2 \div 3$ times lower than that of ice. Containers made of Teflon (PTFE) have a wall thermal conductivity similar to that of ice.

Bulk quantities of biological products can be dispensed into flat trays and placed on cooled shelves. Here, the quality of contact between the shelf and the tray is also of utmost importance, and any air gaps may hinder heat transfer. Metal trays are better than plastic- or glass-made due to their much higher thermal conductivity.

A product in soft packages or containers is frequently pressed between two cooled plates to improve contact conduction heat transfer. For rapid freezing, the thickness of containers or packages cannot be too much and, in practice, is limited to about 50 mm (2 in) or less. The bags, similar to blood/serum bags, can be used to freeze liquid biopharmaceutical products in the plate contact freezer. The plate freezer may have heat transfer coefficients between the slab of frozen material and the plate within the range of $100 \div 250 \dfrac{W}{m^2 \cdot {}^{\circ}C}$. In the multiplate freezer, the plates are usually horizontal, although vertical configurations have been used as well. The plates close and open by means of a hydraulic cylinder in a fashion similar to vial-stoppering shelves used in lyophilizers. Typical freezers may have 10–20 plates. Automated loading/unloading systems have been in use to reduce manual labor. The heat transfer fluid is usually cooled in a heat exchanger where the refrigerant expansion takes place as a step in the mechanical refrigeration cycle. The refrigerant also can be expanded directly in the product-contacting plates. Use of plate freezers may significantly shorten the freezing process when compared with a blast freezer. Calculated approximate freezing times (using a quasi-stationary approach) of a slab with a thickness of 20 mm will be 1 hr 7 min for an external heat transfer coefficient of $20 \dfrac{W}{m^2 \cdot {}^{\circ}C}$ (freezing in the blast freezer) and 11 min 10 sec for an external heat transfer coefficient of $200 \dfrac{W}{m^2 \cdot {}^{\circ}C}$ (freezing in the contact freezer), both at the cooling agent temperature of –50° C. At the cooling agent temperature of –30° C, these times will be 1 hr 45 min and 18 min 37 sec. The quasi-steady approximations give results close to the above under an assumption that the slab center temperature reaches a temperature of 10° C above the cooling agent temperature (here –40° and –20° C).

Scraped heat exchangers and drums may be considered, but their available heat transfer surfaces are small. In addition, freezing on drums produces flaky material that occupies large volumes, and problems were reported regarding subsequent lumping of flakes during storage (Weibel 1987). Such flakes could be, however, compacted into granules or pressed into briquettes. Cooled stainless steel band systems may produce a similar shape of frozen material if the product is sprayed as a thin layer. The band freezer may also handle products in thin, flat containers. The freezing then can be accomplished in a continuous fashion using conduction heat transfer. The conductive cooling through a smooth belt surface makes it advantageous to use a double belt cooler with a light pressing applied to the moving containers for a better heat conduction. Good contact and two-sided cooling shorten an overall solidification time. For a single-band freezer, the product thickness can typically be below 15 mm and for a double-band freezer below 35–40 mm. Contact freezers with product containers may need a brief warming step just before container removal if there is any sticking of container walls to the cooling surfaces.

Contact cooling and freezing can be very beneficial, particularly when a large temperature difference is available (e.g., when the plate or belt temperature may not only be constant at low set-point but may be continuously lowered during freezing to very low levels). This method of operation may maintain the heat flux approximately constant (e.g., the temperature difference: cooling wall/freezing would increase with an increase in thickness of the layer of frozen material [maintaining approximately steady heat flux of heat transfer by conduction]). Maintaining steady conditions of solid-liquid interface movement (heat flux per unit area and interface velocity) creates similar conditions of ice crystal growth at the interface. Specific thermal conductivity change of the frozen mass with temperature can be considered for more accurate estimates. The value of thermal conductivity of ice at 0° C is 2.20–2.25 $\dfrac{W}{m \cdot °C}$. The conductivity increases with temperature decrease (Dharma-Wardana 1983). The thermal conductivity of ice in the range of −173 to 0° C can be estimated according to the formula (Fukusako 1990):

$$k_i = 1.16(1.91 - 8.66 \cdot 10^{-3} \cdot t + 2.97 \cdot 10^{-5} \cdot t^2)$$

expressed in $\dfrac{W}{m \cdot °C}$, where t is the temperature in degrees C.

Custom-designed freezing trays, with cooling baffles, and use of a refrigeration system with a recirculating cooling agent (i.e., a silicone fluid) can also be considered. The cost of such systems is relatively high due to slow freezing times with many trays required. An additional factor is that the freezing rate may not be well controlled if rapid freezing is required.

This concept may involve a deep layer of cell suspension with vertical baffles or heat transfer surfaces to cool the product. Discharge could be performed by turning the shelves upside down and circulating a warm heat transfer agent for a short time to allow the frozen blocks to slip out. A tapered shape for the product-containing cups may facilitate freezing and product discharge. Such a design could also be used to perform a thawing operation.

Problems with system cleaning can be anticipated (due to a complex geometry) in the case of freezing/thawing of viscous and sticky suspensions. Cell settling during the freezing process is anticipated in such a design.

Another option for consideration can be a large chamber with many flat trays, with the product being pumped onto the trays and distributed in thin layers. The cooling agent could then be recirculated through the internal channels of shelves, causing freezing at a relatively rapid pace (this system is similar in concept to that of a bulk lyophilizer). Anticipated problems can include material handling after freezing (removal of the frozen blocks) and thawing, emptying, and cleaning the trays and chamber with arrays of shelves. A variant of this design may also serve as a storage freezer (e.g., after freezing, the product does not leave the chamber and remains on the shelves until it is thawed and pumped out). Selected zones of shelf arrays may be sequentially thawed. Factors to be considered can be a varying viscosity of cell suspensions and anticipated problems with distribution of material on the trays by means of pumping only. Highly viscous materials may not be processed in such a system. An efficient cleaning-in-place system must be devised for this equipment. Capital expense to build a chamber(s) for large product volumes can be high.

A proprietary concept of containers for the cell or protein solution freezing in a blast freezer employing external (for heat transfer increase on the gas side), extended surfaces has been developed by the author. This concept brings the freezing conditions using the blast freezer closer to the conditions of freezing in the contact freezer. The process is described in Case Study 3.3. Figure 3.3 shows (a) freezing of product in bags placed on a cooled shelf, (b) placed on supporting structure in the blast freezer, (c) compressed between two plates with extended heat transfer surfaces on the gas side, and (d) bags compressed between the cooling plates. Bags placed on a cooled shelf experience directional heat flow with parallel dendritic crystal growth. The crystal growth pattern may be disturbed if the contact between the bag and the shelf is not uniform. The bag placed on a support in the blast freezer may freeze in an asymmetric pattern due to thermal conduction from the supporting structure and the presence of a gas pocket at the upper bag wall. Hanging the bags vertically may improve freezing conditions and create a more uniform freezing pattern with solidifying fronts moving from both flat walls and meeting in the center in a symmetrical fashion. High heat

Figure 3.3. Freezing of the product in bags: (a) bag placed on a cooled shelf; (b) bag placed on a support in the blast freezer; (c) bag compressed between two plates with extended heat transfer surfaces on the gas side (Case Study 3.3); and (d) bag compressed between plates cooled by recirculating heat transfer fluid.

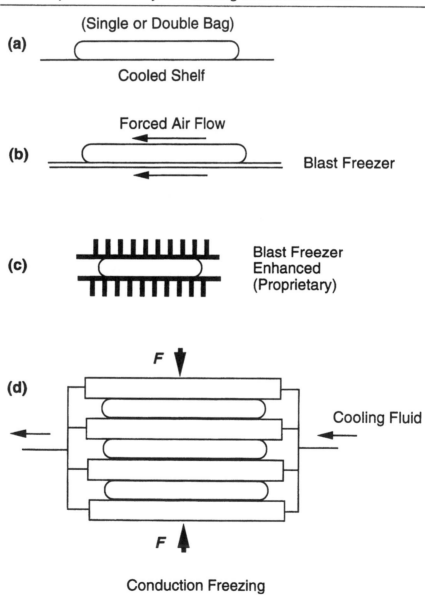

transfer coefficients and short freezing times can be accomplished when the bags are compressed between the cooled plates using a recirculating cooling fluid. If there are gas pockets in the bags, a vertical configuration of bags should be applied to ensure rapid freezing with uniform freezing pattern.

Figure 3.4 shows enhanced freezing in the blast or plate freezer: The effective heat transfer coefficients *(U)* are 50 and 70 $\dfrac{W}{m^2 \cdot K}$ for the blast freezer with bags compressed between the plates with extended heat transfer surfaces and with an external forced convection (see Case Study 3.3) and 150 $\dfrac{W}{m^2 \cdot K}$ for the bags compressed between the cold plates cooled by a flow of liquid through the internal channels. At higher heat transfer coefficients, the cooling temperature and the bag thickness have a smaller influence on the freezing time than they have at a low heat transfer coefficient. This suggests that for the typical blast freezer operation with relatively low heat transfer coefficients to obtain the short freezing times, the bag thickness should be small and the cooling temperature low. Heat transfer intensification between a medium and a product (an increase in heat transfer coefficient) provides more operational flexibility in choosing the bag or container thickness and in the cooling agent temperature.

Direct Cooling with Liquefied Gases

The heat of evaporation of liquid nitrogen is 85.6 BTU/lb or 199.1 kJ/kg, and the heat of sublimation of solid carbon dioxide is 246.3 BTU/lb or 572.9 kJ/kg (AIRCO 1982). Using these numbers, the theoretical amounts of liquid nitrogen and solid carbon dioxide can be estimated per mass unit of frozen and cooled product (the latent heat of phase change water-ice is about 333 kJ/kg). In practice, the amount of used liquid nitrogen or solid carbon dioxide will be higher due to heat gains from the environment. In the case of delivery of carbon dioxide in a liquid form with transition into a solid form prior to contact with product, one may expect no more than about 113 BTU/lb or 262.8 kJ/kg of useful latent heat for product cooling by solid carbon dioxide sublimation since only part of the expanding liquid is turning into solid, and part leaves the nozzle in a gaseous form. The cooling capacity of this gas can be utilized if the expansion nozzle is located within the mass of product.

Figure 3.4. Freezing of biologicals in bags of various thicknesses. The freezing times depend on the bag thickness, the external heat transfer coefficient, and the temperature of the cooling media. *U* **is the heat transfer coefficient in** $\dfrac{W}{m^2 \cdot K}$ **.**

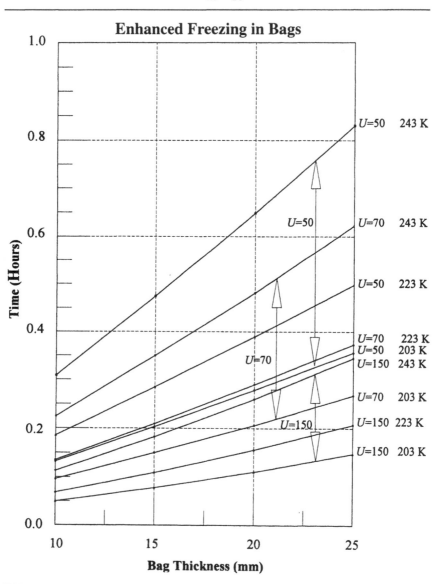

Freezers with Liquid Nitrogen Spraying/Evaporation

Another option that can be considered is the use of a freezing tunnel, as found in the food industry, with liquid nitrogen sprayed by rows of nozzles over the product located on a moving conveyor belt in an insulated tunnel. The product packages/containers are placed on the belt. The nitrogen boils and evaporates on the surface of the containers, and vapors leave the system, passing through the product precooling zone. The design also can be in the form of a cabinet or a spiral belt/slide type. For thorough freezing, the consumption of liquid nitrogen would exceed 1.7 kg per 1 kg of product.

Spraying liquid nitrogen on the surface of materials to be frozen involves boiling nitrogen droplets on the surface of cooled material, and a Leidenfrost type of droplet evaporation occurs. Since the Leidenfrost temperature for liquid nitrogen is about 90 K (–183° C), the Leidenfrost boiling may be anticipated during freezing operations by spraying this cryogen onto surfaces of freezing materials. The size of the droplet should match the container design and its movement. Droplets of liquid nitrogen with a diameter of 1.9 mm on a surface with the temperature of 20° C require approximately 8.2 seconds to completely evaporate (Chandra and Aziz 1994).

Immersion of Product Droplets in Liquid Nitrogen

The concept of cooling/freezing with bulk liquefied gases is also frequently used. The idea is to create almost instant cooling and freezing by immersing product droplets into a pool of liquefied gas. Liquid carbon dioxide may also be used but elevated pressures are associated with the liquid form. Liquid nitrogen has been typically used for such applications (Schmidt and Akers 1997). Liquid products can be dripped into a pool of liquid nitrogen and collected in a wire mesh basket that can be periodically removed, or a continuous removal mechanism could be applied in the form of a conveyor to unload the frozen granules. Weibel (1987) described a system used at Hoffman-LaRoche. The system (Biofreezer) has been designed based on a principle of freezing the biomass drops in a liquid nitrogen pool to form small beads. Harvest of the beads has been automated. The system has operated under hygenic and biosafe conditions. It has been used in the production of Roferon-A (alpha interferon). Weibel also reviewed investigated freezing techniques used prior to the Biofreezer design. These included freezing on plates or small blocks and drum methods. The Biofreezer design allowed cleaning-in-place and steam sterilization. During freezing by pouring product droplets into a liquid nitrogen pool, the cooling and freezing rates are very high but very difficult to control. If the ability to control the cooling and freezing rates is part of the freezing process scope, such a pool operation may not be suitable.

A certain level of control over the freezing rate can be accomplished using a liquid nitrogen cascade system. An insulated box may contain a liquid nitrogen pool at the bottom, which is pumped by an internal pump to the top of the cascade of inclined trays. The liquid nitrogen flows downward from tray to tray until it reaches the bottom pool. The product can be coarsely sprayed via a nozzle on the top tray into the flowing liquid nitrogen stream. The droplet surface rapidly solidifies, and granules freeze internally during their travel over the tray surfaces. Final freezing may occur in the nitrogen pool. In such a design, there is limited control over the frozen particles' residence time on the trays. This can be controlled to a certain degree by the number of trays and their slope. Longer trays (for example, of a spiral compact design) may ensure residence times sufficient to freeze the granule completely during its travel on the tray. The frozen granules are collected at the bottom of the tray using a mechanical conveyor. For example, the system built by IQF (Mississauga, Canada) has a spiral tray with claimed granule residence time of approximately 14–17 seconds. Such a time is usually sufficient for freezing granules of 2–3 mm in diameter. Depending upon the droplet forming systems and characteristics of the liquid product (density, viscosity, and surface tension), the droplet sizes may reach larger diameters, and the needed residence time and resulting length of the trays should be increased.

The freezing rate at the beginning of the process may be controlled by a ratio of recirculated liquid nitrogen and liquid feed. Slower initial freezing rates may be accomplished by delivering a limited amount of liquid nitrogen onto the top tray (to cause freezing of the outside shell and to form a particle). Such particles may then roll over the cascade of trays being exposed to cold nitrogen gas, with limited addition of liquid nitrogen along the way. Under such conditions, the freezing rate may be slower than by immersion into liquid nitrogen, but the overall length of the trays may be impractically high. Such a process approach requires the formation of large uniform droplets, and a liquid dispensing system may perform this task better than spraying liquid through nozzles where there is always a certain droplet size distribution pattern and many fine droplets may be formed. The droplet formation may be based on the principle of breakup of the free-flowing, vertical liquid thread. This type of droplet generation is described in Chapter 2 (page 128).

Part of the recirculated liquid nitrogen stream may be directed onto lower trays at one or more points. Tray design should ensure a controlled pattern of particle motion to avoid clump formation. Such a design may work well on a continuous basis but may pose operational problems if many consecutive batches need to be divided by system cleaning/sanitization steps. If this is the case, liquid nitrogen needs to be drained and the system warmed up. After reaching ambient temperature, standard CIP and sanitization cycles can be applied. Subsequently, the system is filled with liquid nitrogen and cooled down to operating conditions.

A paper by Ryan et al. (1995) described a cryogenic pelletizer of the spiral cascade type used for preserving biologicals. This system was based on the IQF design and could produce frozen granules in a continuous fashion. The system has limited production capacity and can be sensitive to liquid characteristics that may affect droplet formation (e.g., its diameter). Estimated time of droplet freezing in nitrogen boiling at droplet surface using simple formulas can be about 8.7–10.4 seconds for the sphere 5 mm in diameter, and slightly longer using numerical methods of calculation. A system with droplets falling into a nitrogen pool can be insensitive to droplet size, and its performance does not depend upon the liquid nitrogen circulation rate, its distribution quality, or length of the trays. The freezing is completed regardless of the droplet size, and the frozen drops are removed from the liquid nitrogen pool by a mechanical conveyor.

In the case of falling liquid droplets into liquid nitrogen, there is a very rapid solidification of granule surface and subsequent fast freezing of granule interior. Such a freezing pattern produces a large amount of fine ice crystals. High levels of undercooling and rapidly moving temperature fields occur. The solid-liquid interface also moves rapidly through the spherical droplet. The rapid solidification of droplets may produce the planar-cellular and fine dendritic structures (Jones 1984; Kurz and Fisher 1989). Such a very rapid freezing can produce glassy states between ice crystals with the water content W_g' much higher than obtained during slow freezing. High water content corresponds to low T_g' temperature. As a result, the required storage temperature for rapidly frozen product is much lower than for product frozen under optimized conditions. A thorough review on rapid solidification processing was published by Jacobson and McKittrick (1994). Prediction of the formation of microstructures during rapid solidification may be difficult using the phase diagram alone (Kurz and Gilgien 1994), and the phase equilibria and kinetics have to be considered together. The aqueous droplet freezing times in liquid nitrogen, depending on the heat transfer coefficient and droplet diameter, can be found in the author's work in this volume, page 126.

The cells may experience intracellular ice formation as described earlier. Formation of a large number of fine ice crystals will generate a very large crystal surface within the cells, cryoprotectants, and other solutes distributed in thin layers in intercrystalline spaces. This ice crystal pattern is different than in controlled dendritic growth where the ice crystals are fewer and grow in a regular pattern and the spaces between dendrites are larger (e.g., the cells may be embedded in eutectics, glassy, vitrified solutes, and cryoprotectants). For very small droplets a significant supercooling of liquid phase may occur, followed by a very rapid solidification of the whole droplet volume.

Heat Transfer in Liquid Nitrogen Boiling

Cooling of droplets using liquid nitrogen is associated with liquid nitrogen boiling and convection heat transfer between the liquid nitrogen and the frozen granule. The liquid nitrogen boiling at solid surfaces has been thoroughly investigated in the field of cryogenics. Early works on heat transfer phenomena in nitrogen boiling were summarized by Frost (1975). In quenching a sphere, the heat flux may vary, depending on nitrogen boiling regime, from a minimum of about 2000 $\frac{BTU}{ft^2 \cdot hr}$, $\left(6.3 \ \frac{kW}{m^2} \right)$ to a maximum of 47,000 $\frac{BTU}{ft^2 \cdot hr \cdot °F}$, $\left(148 \ \frac{kW}{m^2} \right)$ (Merte and Clark 1964). The nucleate boiling of liquid nitrogen on a flat disk gave a maximum heat flux of about 20 $\frac{W}{cm^2}$, $\left(200 \ \frac{kW}{m^2} \right)$ (Kosky and Lyon 1968). The heat transfer during boiling depends on a superheat. Heat transfer coefficients at low-temperature differences between solid surface and liquid (a superheat) of 1–3° F (0.5–1.6° C) are in the range of 200–250 $\dfrac{BTU}{hr \cdot F \cdot ft^2}$,

$\left(1,135 - 1,418 \dfrac{W}{m^2 \cdot K} \right)$ and increase to more than 2000 $\dfrac{BTU}{hr \cdot F \cdot ft^2}$,

$\left(11,350 \dfrac{W}{m^2 \cdot K} \right)$ at a superheat near 10° F, (5.5° C), (Lyon 1964). In

liquid nitrogen boiling, the heat flux increases, with an increase in a superheat from near zero° C, and the maximum heat flux occurs at a

superheat of about 5–10° C, reaching about 100 $\dfrac{kW}{m^2}$ (Peyayopanakul

and Westwater 1978). This corresponds to a heat transfer coefficient

of about 13,330 $\dfrac{W}{m^2 \cdot °C}$. The heat flux declines with a further

increase in a superheat and reaches a minimum of near 5 $\dfrac{kW}{m^2}$ at

about 20° C superheat. The corresponding heat transfer coefficient

is then about 250 $\dfrac{W}{m^2 \cdot °C}$. Further increase in a superheat produces

a slight increase in heat flux, reaching about 20 $\dfrac{kW}{m^2}$ at 100–120° C

superheat (heat transfer coefficient of about 182 $\dfrac{W}{m^2 \cdot °C}$). Transient

nitrogen boiling during quenching of metal disks demonstrated that the process changes when disk thickness decreases below a certain value (e.g., when the cooled sample is smaller than a threshold limit). The maximum heat flux occurred at a higher temperature difference for thin samples (Peyayopanakul and Westwater 1978). Hartmann and Scheiwe (1984) determined the minimum heat flux for a film of boiling liquid nitrogen at a surface of vertical plates. The range of superheat was between 20° and 180° C. The minimum heat flux of

about 6–7 $\dfrac{kW}{m^2}$ occurred at 20° C and reached about 20 $\dfrac{kW}{m^2}$ at 180° C

superheat (corresponding values of heat transfer coefficient: 325 and

111 $\dfrac{W}{m^2 \cdot °C}$). Work was conducted in connection with quenching of

cell suspensions in liquid nitrogen for cryopreservation purposes. Timmerhaus and Flynn (1989) provided a summary of nitrogen boiling data in the form of a chart showing numerous experimental results and correlations. Recent experiments with liquid nitrogen boiling around heated wire (Duluc et al. 1996) showed the peak heat flux of about 142 $\frac{kW}{m^2}$ occurring at a temperature difference near 12–14 K and reaching about 250 $\frac{kW}{m^2}$ at a temperature difference of 200 K.

Since the liquid nitrogen density at its boiling point is about 0.807 g/mL (*CRC Handbook of Chemistry and Physics* 1995), the droplets of aqueous solution will tend to immerse in the nitrogen pool, causing intensive nitrogen boiling at the entry point. Due to the high superheat, an initial boiling is anticipated to be of a film type and may pass through transitional nucleate boiling while lowering the surface temperature of the freezing granule, finally reaching a state of convective liquid cooling. The process is thus very transitional in nature.

The granule freezing process will have a changing surface heat flux following the granule surface and liquid nitrogen temperature difference and, accordingly, a changing boiling heat transfer coefficient (the surface temperature may change from 0° C to –196° C and the temperature difference from 196° C to about 0° C). According to these data, the heat transfer coefficients for most boiling would change from about 120 to 250 $\frac{W}{m^2 \cdot °C}$. A brief peak at the end of granule surface temperature decline would raise the heat transfer coefficient to more than 10,000 $\frac{W}{m^2 \cdot °C}$ when the temperature difference drops to the range of 5–10° C. This increase in heat flux will also accelerate the end of freezing, since there is only a small amount of water left in the center of the granule, and the temperature difference between the granule surface and center may be high (a high thermal conduction driving force).

The cooling rates of samples immersed in liquid nitrogen may be increased by induction of nucleate boiling on the sample surface after immersion by wrapping the container in cloth or mesh. For example, the reported cooling rates for 48.3-mg samples in sealed aluminum pans wrapped in a cloth were up to 3160° C/min (Spieles et al. 1995). Another method of increasing the heat transfer rate is to maintain the sample movement through a nitrogen pool to affect the sample-liquid interface area and reduce the effects of gas cushion there (Han et al. 1995). The 2-mm-diameter solid samples experienced a cooling rate of about 500° C/min, whereas, for samples of 0.2 mm in diameter, the cooling rates were about 3200–3300° C/min. Methods of augmentation of boiling and evaporation were reviewed by Bergles (1988) and Marvillet (1995), including porous surfaces, meshes, and active techniques (surface rotation, wiping, vibration, etc.). Recent work by Drach et al. (1996) reports on transient heat transfer from surfaces of defined roughness to boiling liquid nitrogen. The rough surfaces exhibited higher critical transient heat fluxes during transition from the nucleate to film boiling.

Liquid Immersion Freezers

Containers with product may be immersed in a cooling agent (cryogen, brine, glycol, etc.) and held there until the contents freeze. This operation can be performed batchwise or in a continuous fashion. The NaCl eutectic solution has been used this way for freezing foodstuffs. Lower temperatures may be reached using the eutectic solution of $CaCl_2$ (a eutectic point at −55° C, 29.8 percent). The external cooling agent is recirculated through a refrigeration system, or direct expansion coils may be a part of the cooling bath. A disadvantage of such a system is that the containers have to be cleaned after their removal from the bath to get rid of traces of cooling medium. In the past dichlorodifluoromethane (Freon R12) was used as a coolant due to its good cooling characteristics (the melting point at 115 K and the boiling point at about 243 K) in the form of a bath or spray. This coolant has been abandoned for environmental reasons.

The cooling liquids should have a low melting point, and their boiling point should be far removed from it. The liquid should have a high heat capacity, high thermal conductivity, high density, low viscosity, and be safe to use. Table 3.1 shows the characteristics of some cooling liquids and cryogens.

In practice, ethanol works well down to about 170 K at which point it becomes very viscous. Liquid propane can be used for low-temperature cooling due to its low melting point. A spherical sample of about 1 cubic millimeter immersed in liquid propane may reach cooling rates up to 400–500° C/sec.

Product is frequently dispensed into bottles, which are placed in large numbers in a cooled bath. The heat transfer coefficient between the bath fluid and bottle surface may vary depending on the type of bath fluid and on the level of agitation. Figure 3.5

Table 3.1. Thermal Properties of Cooling Media.

Liquid	Melting Temperature (K)	Boiling Temperature (K)	Specific Heat $\left(\dfrac{kJ}{kg \cdot K}\right)$	Thermal Conductivity $\left(\dfrac{W}{m \cdot K}\right)$
Ethanol	156	352	1.89	0.206
Isopentane	113	301	1.72	0.182
n-Pentane	143	309	1.89	0.177
Propane	84	231	1.94	0.222
Liquid nitrogen	63	77	2.0	0.153

Figure 3.5. Freezing of biologicals in bottles and cylindrical containers of various diameters. The freezing times depend on the container diameter, external heat transfer coefficient, and temperature of the cooling media. U is the heat transfer coefficient in $\dfrac{W}{m^2 \cdot K}$.

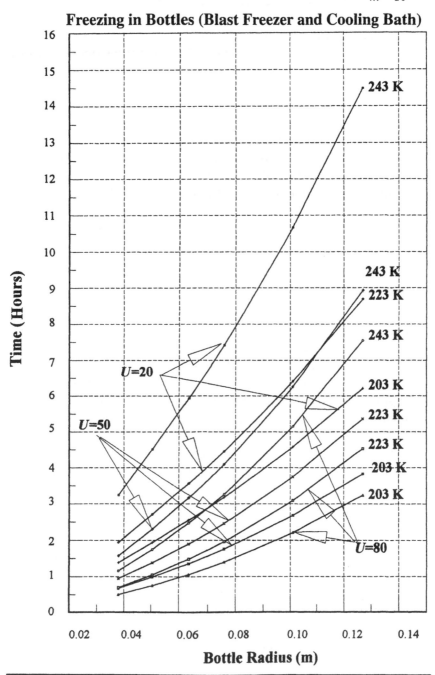

shows that the freezing times in bottles depend on the bottle radius, the heat transfer coefficient, and the coolant temperature. Bottles of a small diameter can be frozen relatively fast with less dependency upon the cooling agent temperature and the heat transfer coefficient. Freezing time of larger bottles depends more on these parameters and can be very long for the low heat transfer coefficients and the high temperature of the cooling agent. The chart can also be applied to bottles frozen in blast freezers (using the lower value of the heat transfer coefficient). The low value of heat transfer coefficient may also apply to the nonagitated baths, whereas the high value can be used with the agitated baths.

Freezing in the state of an emulsion (aqueous phase dispersed) may involve high undercooling (temperature difference between solidification and melting) of the freezing phase if the droplets are small. For water droplets with a volume of a few cubic micrometers, such undercooling may reach about 38° C (Dumas et al. 1990; see also the material in Chapter 2 on principles of cryopreservation).

Sublimation Freezers

In addition to cooling only, carbon dioxide may be used for freezing pasty or liquid materials. Carbon dioxide can only exist under atmospheric pressure in the form of a gas or solid. The solid form may be advantageous for cooling since it can contact product directly and then sublime. Bulk coolers use either carbon dioxide pellets agitated with the product in a vessel or liquid carbon dioxide injected through nozzles into the mass of product or onto the product surface. This method produces a mixture of solid particles and gas after liquid decompression. Such a mixture can be injected through the bottom of a container or sprayed on the product surface. In both cases, product agitation permits better utilization of carbon dioxide and more uniform freezing of product due to the uniform distribution of carbon dioxide particles through the product volume.

In this process concept, the continuous delivery of solidified carbon dioxide into a vigorously agitated cell suspension causes the temperature to drop to the vicinity of 0° C, and then there is no further temperature drop despite ongoing coolant delivery. The apparent viscosity of the suspension begins to increase, followed by an increase in the agitation power. The suspension reaches a creamy consistency before it breaks into granules. After that, the temperature of the agitated granules decreases, and agitation power returns to a lower level. The viscosity increase is mostly due to the formation of ice microcrystals within the cell suspension. The microcrystals form at the points of contact with the particles of solid CO_2 that fall into the agitated liquid after the liquid temperature reaches a plateau. At this level, the liquid does not cool any further, and the coolant heat of sublimation is used for ice crystal formation and to balance the heat of agitation. The agitation prevents any surface freezing of liquid. After the number of ice crystals reaches a certain critical level, viscous paste forms, which then breaks into granules. The system is described in detail in Case Study 3.1.

Other Types of Freezing Devices

Small freezing devices are frequently based on a cold plate principle and are typically used for freezing small specimens. Cooling of the plate may be accomplished by application of a cryogenic fluid, by expansion of classical refrigerant compressed in a mechanical system, by a thermoelectric principle, by the application of a version of Stirling engine, or by the use of a single-pass thermodynamic expansion.

CASE STUDY 3.1: LARGE-SCALE PROCESS AND SYSTEM DEVELOPMENT FOR FREEZING AND GRANULATION OF CELL SUSPENSIONS USING LIQUEFIED GASES

A Brief Description

This practical case study describes a freezing/granulation process and system that can process aqueous solutions, suspensions, and pasty materials. In the study, the primary products considered for processing in the system were bacteria and yeast cell suspensions. The objective was to expose biological material to uniform and controlled thermal and osmotic conditions during freezing and thawing. A primary focus was on the division of a large original volume into multiple smaller volumes to assure optimal conditions during freezing and thawing.

Contrary to the freezing and thawing of large volumes of protein solutions in which a principle of controlled dendritic growth was applied (Wisniewski and Wu 1992, 1996; Wisniewski 1998a), the principle of freezing large volumes of cell suspensions (volumes above 200 L) involves a rapid "volumetric freezing," i.e., creation of conditions under which multiple ice crystals are nucleated within the whole product volume, and an initial freezing out of water in the entire mass takes place under temperature plateau

Figure 3.6. Experiments using agitated product with addition of cryogenic agents (liquid and solid carbon dioxide and liquid nitrogen): (a) manual addition of ground solid carbon dioxide, manual agitation; (b) manual addition of ground carbon dioxide, injection of liquid nitrogen, mechanical agitation; and (c) manual addition of ground carbon dioxide, injection of liquid carbon dioxide, injection of liquid nitrogen, mechanical agitation.

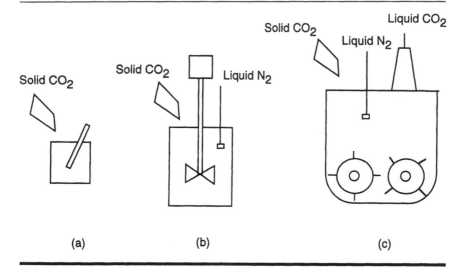

conditions. Figure 3.6 shows the initial experiments using manual agitation and addition of finely ground solid carbon dioxide; mechanical agitation with a manual addition of ground carbon dioxide and injection of liquid nitrogen; and double rotor mechanical agitation with manual addition of carbon dioxide, injection of liquid carbon dioxide, and injection of liquid nitrogen.

Freezing Process Development

Dendritic growth was not an objective in this stage of freezing, although it was considered during the temperature decrease period (beyond the temperature plateau) due to directional cooling taking place. During this second period, the product mass was already divided into small individual volumes (granules) cooled through solidified surface. Addition of external cryoprotecting agents was not considered; only the components of fermentation broth such as glucose, amino acids, peptides and proteins, mineral nutrients, etc., were present in the solution.

Early work involved agitation of a liquid containing the cell suspension (*E. coli* fermentation broth) in a small blender with addition of finely ground solid carbon dioxide. The agitation rate and the rate of the addition of carbon dioxide were changing, and the temperature of the processed mass was recorded. The temperature plateau, mass apparent viscosity increase, and mass breakup into granules were observed under conditions of continuous carbon dioxide addition and agitation. A layer of solid "snow" carbon dioxide was floating on the liquid product surface, and a gaseous pillow formed at the interface due to rapid sublimation effects. When liquid nitrogen was sprayed on the surface of water or aqueous solutions/suspensions, observations were hindered by intense fogging. In both cases, a rapid agitation of liquid was essential for a volumetric uniform cooling.

The first large-scale research runs were performed using an aqueous suspension of baker's yeast (*Saccharomyces cerevisiae*) and solid carbon dioxide ground in a food-grade disintegrator. Large blocks of carbon dioxide were crushed and passed through a disintegrator to produce small particles. Approximately 30 L of yeast suspension were loaded into a stainless steel mixer, and carbon dioxide was gradually added. The run produced good results, with the frozen material forming granules of a size no larger than about 10–15 mm in diameter. The granules were frozen through and, as observed after cutting across, had a porous internal structure. The operation involved the steps of grinding and delivery of carbon dioxide, and due to the nonuniformity of carbon dioxide particle size and problems with containment, a delivery of liquefied carbon dioxide using expansion nozzles was chosen.

Water, aqueous solutions, and cell and particle suspensions may be treated using this method. The size of granules may be controlled by means of agitation and delivery of cryogen. For example, slow agitation of water with spraying of solid carbon dioxide has produced large granules of sizes within the 20–30-mm range, whereas rapid agitation has produced granule sizes below 5 mm.

Low-Temperature Freezing

Cell preservation and storage in small quantities are frequently conducted using liquefied nitrogen at its boiling temperature ($-196°$ C) under atmospheric pressure. Such low-temperature conditions are usually favorable for even long periods of storage (Malik 1987).

The glass transition in proteins occurs at about 200 K (–73° C), and this phenomenon occurs within a certain temperature range (Rupley and Careri 1991; Angell 1995). This temperature range may vary for different proteins. The glass transition step is associated with a significant change in specific heat and involves protein hydrating water. Sartor et al. (1992) reported that for methemoglobin, the glass liquid transition for bound water is at 169 K (–104° C) and such water starts to crystallize into cubic ice near 210 K (–63° C). Later, Sartor et al. (1994) reported results of similar studies involving lysozyme, hemoglobin, and myoglobin. Cell proteins and, particularly, enzymes may undergo such transition within not exactly known ranges of temperature; therefore, freezing and storage at temperatures below 190–200 K may be beneficial to cell preservation. The glass transition temperature for pure water is about 136 K (–137° C) (Johari et al. 1987).

Carbon dioxide temperature during sublimation from the solid state is about –78° C, and lies near the borderline of protein glass transition. To reach lower temperatures, other cryogenic fluids are required. Spraying with liquid nitrogen can bring temperature down locally to its boiling point, but an average temperature of bulk granulated material may still remain above this level. Spraying a combination of carbon dioxide and liquid nitrogen has been shown to reach very low temperatures.

Principles of System Design and Operation

The process specifications may call for treatment of cell suspensions at various concentration levels from diluted to highly concentrated fermentation broths, as received from high efficiency centrifuges or filtration systems. As reviewed earlier, an elevated cell concentration could have pronounced effects on cell injury. This may be an additional difficulty that needs to be addressed while designing a specific process. The principle of maintaining volumetric homogeneity and uniformity of freezing conditions for the whole product mass during the cell freezing process may be adopted as a possible way of solving such problems.

Pilot Plant Concept

A small pilot installation for testing freezing processes using liquefied gases was designed. The liquefied gases selected were carbon dioxide and nitrogen. Gas delivery could be accomplished by top spraying or by direct injection into the mass of product. Injection of liquefied gases under the surface of liquid products posed several challenges. If the liquefied gas is injected under the surface of freezing material, there is an instant local release of a large volume of gas. Such a gas jet may propel fine frozen particles into the gas outlet duct. Gas injecting nozzles could be contaminated by processed material and, therefore, would require a separate cleaning procedure. Since a major concern was to minimize loss of the product, as well as possible emissions from the system to atmosphere, a top spraying system was selected over gas injection under the liquid surface.

During test runs, spraying a combination of carbon dioxide and nitrogen attained product temperatures as low as −135° C. Nozzles used for liquid nitrogen spraying were of the low-pressure type. Even lower temperatures could be reached with the existing setup, but such temperature levels were not in the project scope. At very low temperatures, special attention must be paid to the design of equipment, including factors such as thermal dimensional changes, selection of construction materials (material brittleness at low temperature must be considered), proper thermal insulation, types of sensors and instruments, dynamic seal design, etc. Using two liquefied gases allows a two-step cooling procedure (e.g., as proposed by Farrant et al. 1977), since the holding period after reaching a predetermined temperature can be easily maintained. This method may allow reaching temperature levels close to those of vitrification of water (136 K or −137° C). MacFarlane (1987) and Fahy et al. (1987) reviewed principles of vitrification phenomena with its possible applications in processing biological materials. The vitrification temperature of water is about −137° C, but temperatures at which aqueous solutions reach a glassy (vitrified) state can be higher, depending on the character of the solutes (Levine and Slade 1988). MacFarlane et al. (1992) reviewed the topic of vitrification and devitrification in cryopreservation. Vitrification of water requires a very rapid freezing rate, which may not be accomplished in large volume samples, but vitrified states of aqueous solutions may be obtained using less demanding procedures, such as the process described here.

A parameter of interest in process design is the cooling rate. This variable can be adjusted by a change in the delivery rate of liquefied gas. Several sizes of spraying nozzles were tested for liquefied carbon dioxide before an optimal delivery rate range was established. A factor included in the adjustment of the liquefied gas delivery rate was its effect on granulation of frozen product. An excessive delivery rate of liquified gas may cause formation of large granules and even lumps of material. During the final freezing process, the cooling surface is the total surface area of granules in the bed volume. The cooling agent is present throughout the whole bed volume due to action of the agitators, which drag the agent from the surface into the bed with simultaneous thorough mixing of the bed.

Pilot Plant Design

A pilot plant was designed and built to test and process relatively large volumes of cell suspensions (typical runs were processing around 35–45 L of product). The central part of the system was a two-rotor, jacketed machine made of 316-L stainless steel. Stainless steel not only satisfies the standard requirements of both pharmaceutical and biotechnology industries with regard to corrosion resistance and cleanability but also performs well under stress at low temperatures. Internal surfaces in contact with the product were polished to a root mean square roughness (RMS) of 0.76–0.89 micrometers (150 grit). The machine had an elevated top on which nozzles were mounted for spraying liquefied carbon dioxide and nitrogen. This elevated top also served as a separation zone for particulate material released during spraying and agitation. Frozen product could be unloaded through the discharge ports located in the side wall. The rotors could assist in material discharge by pushing the bed of granules toward these ports. The rotors had variable speed controllers, allowing independent movement of each rotor. An outlet duct for the released gas was connected to a discharge blower that maintained the entire system under slightly negative pressure. The pressure level was adjusted using a damper located in the exhaust duct prior to the blower. Liquefied gases were delivered in commercial stainless steel Dewar cylinders. The quality of gas used was of certified medical grade.

Instrumentation involved a series of thermocouples and thermometers for temperature measurement, a meter to measure current to the motors, and pen recorders. Pressure was monitored in liquefied gas lines prior to the nozzles and, directly, in the gas supply cylinders. Level indicators on the cylinders provided approximate data on the remaining gas supply. The following parameters were recorded:

- Temperature at two points inside the apparatus (opposite ends)
- Outlet gas temperature
- Current requirement to the agitator motors

The recorder charts were marked during the system's operation to indicate process steps and events. Temperatures measured inside the machine were average values between temperatures of frozen particles and gaseous phase in the bed. Such average temperatures can serve well as a main parameter to monitor and control the process, as long as thorough agitation and a relatively uniform granulation pattern can be achieved. Interiors of particles have higher temperatures than those of outside surfaces during the freezing phase and the subsequent cooling. There is a temperature gradient in the external frozen layer of material due to heat transfer by conduction.

The product was manually loaded from the top and then the machine was closed. The rotors were turned on, and spraying of liquefied gas began. Temperatures and agitation power requirements were continuously monitored. During the run, the jacket was turned on to facilitate freezing and product granule formation. Water at a controlled flow rate and temperature was supplied to the jacket. In a subsequent step, the jacket was rapidly emptied using compressed air. The run was continued until the desired product temperature was reached. The product was discharged through the two end ports into storage containers. By switching media in the jacket, the system also could be used to precool the product prior to a freezing phase, using a circulation of refrigerated water through the jacket. This could be done during the period of time when the batch was being delivered into the machine from the harvest or concentration steps.

The original design of rotors did not work satisfactorily for products with varying consistency. This was improved after the rotor design was changed to one with individual, adjustable paddles. Of concern in the pilot system were the seals on rotor shafts. They frequently leaked product before freezing could begin and were difficult to clean. Cleaning and sanitation of the pilot system were performed manually.

Data were collected for research and development work using two liquefied gases, carbon dioxide and nitrogen. A third option was also investigated that included spraying carbon dioxide at the beginning and finishing with nitrogen. This approach was considered for low-temperature freezing applications because carbon dioxide is limited to only achieving cooling temperatures of around −78° C. If further cooling was required, the second part of the process could be continued and completed with liquid nitrogen sprayed over already frozen material. Nitrogen may be used for final freezing and deep cooling, or it can be used early in a process to accelerate the cooling rate. If liquid nitrogen is used as a cooling agent, the design of the machine should anticipate local temperature shock wherever the nitrogen spray hits exposed parts of the machine. Therefore, the spraying nozzles should be located in such a pattern as to direct the spray only onto the freezing product. At the same time, it should provide a large enough covered area to maintain a uniform thermal treatment of freezing material and avoid localized overcooling.) The system was also tested with a discontinuous gas delivery using a two-or-more-step cooling technique with holding periods at predetermined temperatures. Finishing the run with liquid nitrogen provides nitrogen (inert gas) atmosphere for the granulated product that may be desirable for product storage. Both of the liquefied gas delivery systems were built of stainless steel seamless tubing, which was thermally insulated. Such a system was chosen because of the temporary nature of the pilot installation. Otherwise, a double-wall, vacuum insulated type piping design should be used for liquid nitrogen.

The granulation process that occurred here was fundamentally different from the traditional granulation principles (Sherrington and Oliver 1981), in which liquid is added to an agitated bed of dry material. A remote analogy to the process might be drying/granulation of certain pasty materials in agitated dryers. During the process of water removal, a state of the mass breaking into granules may be reached at a certain liquid water level. The heat transfer mechanism is, however, different; water is removed by adding heat through the wall, not by freezing out and forming a new solid phase (i.e., ice crystals).

Pilot System Performance

In the small-scale pilot installation, batches of up to 45 L were processed. The instrumentation was simple but allowed operators to sufficiently monitor the process. The most important parameters were the temperatures at two points in the machine, the temperature in the outlet gas duct, and the power requirement of the rotors. Other parameters were carbon dioxide and nitrogen pressures and levels in supply tanks, the temperatures of water in inlet and outlet of the jacket, and jacket water flow rate. Operators observed the recorded temperatures and the power requirement, and at appropriate times, they manually turned on water flow through the jacket and emptied the jacket with the aid of compressed air. As during the initial experiments, the product shape was in the form of granules with sizes ranging from small particles of around 1 mm in diameter to relatively large ones of around 25 mm. The majority of particles have been, typically, within the range of 2–7 mm. The temperature curve in time looked typically as represented in Figure 3.7.

This figure provides an explanation of the phenomena occurring during freezing using the freeze/granulation process. Initially, the temperature falls due to the liquid product cooling. Then, near 0° C, the temperature reaches a plateau, and the ice crystal formation and phase change from liquid to solid occurs. The temperature plateau occurs in the entire mass of agitated product (thermocouple readouts at different points have confirmed this). Due to a continuous delivery of carbon dioxide particles, which are well mixed with the product, a volumetric cooling takes place. Ice crystals appear in the entire volume outside of the cells. The presence of ice crystals also increases the product apparent viscosity and makes it highly non-Newtonian. Consequently, the freezing mass becomes more "sticky" and more difficult to mix, and the power meter reading increases. Finally, the whole mass breaks into granules, and the power meter reading declines. After granules are formed, the temperature begins to drop, and after an initial short period of a slow decrease, it reaches an almost steady decline rate. The transition period is caused by freezing of the interior of the granules. The subsequent period of an almost steady rate of temperature drop is the cooling of particles with latent heat released during freezing of central zones of granules. After reaching a predetermined bed temperature, the liquefied gas supply is stopped. The bed temperature levels off and reaches an equilibrium. During

Figure 3.7. The temperature profile of agitated liquid biological product with a surface addition of cryogenic fluids. *T* is the temperature, and *t* is the time. A remarkable temperature plateau near 0° C can be observed. After granule formation, the bed of granules is cooled at the cooling rate = a/b, $\left(\dfrac{\Delta T}{\Delta t} \right)$, depending on the delivery rate of the cryogenic fluid.

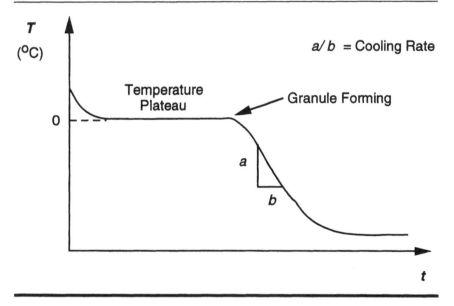

this period, excess solid carbon dioxide particles sublime (the so-called product "degassing" step).

The first design of rotors did not give satisfactory results for some runs when product characteristics varied. A typical problem was the formation of large lumps of material with a frozen skin and a soft, unfrozen center. An improved rotor design provided much better uniformity of granule size and eliminated lump formation. The rotor design should satisfy both liquid and granular material agitation requirements and, in addition, play a critical role in granule formation.

When the coolant delivery was temporarily stopped during the declining bed temperature, the agitated bed temperature began to increase, mostly due to release of latent heat from freezing granule interiors, from heat of agitation, and to a small degree, from some heat gains from the environment. Such stoppage at later stages of freezing showed only a small temperature increase since most of the granules were already thoroughly frozen. The coolant delivery schedule may be matched with the freezing pattern; for example, at the end of the freezing process, the delivery of coolant can be decreased to keep pace with reduced cooling demand to balance the latent heat of phase change from freezing the granule centers and, subsequently, freezing a smaller number of larger granules.

The pilot freezing machine was also tested for a rapid thawing process using its external heating jacket. The results were positive, although the thawing process was not as efficient as the freezing process and slower due to the limited heat transfer area of the jacket alone and limited temperature difference between the thawed material and the jacket (jacket temperature was maintained near the original cell cultivation temperature).

Production System

Design. The final system design for the large-scale operation was based on experience derived from performance of the pilot unit. Attention was paid to principles of containment and system automation, as well as to cleaning-in-place (CIP) of equipment. The freezing/granulating machine was built to the author's specifications as a customized system. One of the most difficult design problems to solve was the rotors' seals. The first option considered was a mechanical seal. The containment concept called for a double seal with a barrier fluid. Seals should be able to withstand shaft vibrations during product phase change from liquid to solid. Seals can be lubricated by a barrier fluid during the cooling phase when the product is still in a liquid form. After the phase change, the product becomes solid, and the barrier fluid may freeze as well. Under the low-temperature conditions, seals may not have any lubrication at all, and a barrier fluid can be a pure gas. Since pasty and viscous cell suspensions may be processed in the system, the seal design should be completely cleanable in place with an easy access for the cleaning solution and a potential for vigorous motion of this solution against all surfaces in contact with processed material. Material selection for seal components should consider the temperature shock that occurs when liquefied gas delivery begins.) The barrier gas was kept under pressure higher than the machine interior to prevent product penetration into the seal chamber. This pressure in the seal chamber also protected the seal edges. The barrier gas leaks into the machine were accepted since they did not influence product characteristics. During a CIP cycle, the cleaning solution can pass through the seal chamber to assure that any product penetration there is washed away. Since the cleaning solution is inside the machine as well, both sides of the seal can be cleaned during the CIP cycle. Rotation of shafts provided additional enhancement to the seal cleaning process. A review of aseptic dynamic seals was summarized by Wisniewski 1988), who also discussed internal seal conditions and their possible influence on biological materials (Wisniewski 1989).

Careful analysis was performed on the design of rotors to optimize the freezing process, to control granulation, and to avoid formation of large material lumps. The optimized parameters were angle of mixing surfaces, their areas, distances between them, and interaction between rotors. Since the frozen microorganisms can be damaged and disrupted by applying mechanical stress on a frozen mass (Hughes et al. 1971; Coakley et al. 1977; Omori et al. 1989, 1990), the rotors were designed to minimize the mechanical stress on agitated material by optimizing the shape of the mixing surfaces and by selecting of their angles of attack and blade profile. A sufficient level of agitation was assured by the selection of surface area and distances between agitating surfaces. Rotor design also played an important role in the mixing of solid carbon dioxide particles with liquid product at the beginning of the process. The agitators forced solid carbon dioxide particles, which normally settled on top of the agitated mass, into the bulk material volume. The rotor optimization process was done in the pilot unit where shafts were equipped with many points of attachment for different kinds of mixing devices. The agitator blades could also be rotated to test different angles of attack. Several blade types were tested before the final design was established for the production unit.

Heat transfer in the agitated mass with a top spray of cryogenic fluid (called cryogen) cannot be easily described due to the influence of many equipment design factors such as spray-covered area, cryogen delivery rate, intensity of agitation, agitator design, level of processed material, etc. Heat transfer during cooling of formed granules is also complex and depends on the aforementioned factors. The relationships found in literature for agitated beds of granular material (heat transfer wall-bed and gas phase-bed) cannot be directly applied, although if the spray is uniform and the bed is well mixed, the gas-agitated bed heat transfer data may be used as a starting point. Estimates of cryogen evaporation and sublimation within the bed also should be introduced.

The bed temperature drop after an initial plateau at constant coolant delivery indicates the diminishing overall latent heat of phase change; that is, there is a cooling of the solidified part of granules simultaneously occurring with the internal freezing. The presence of ice crystals within the freshly formed granule accelerates the thorough freezing process but may interfere with formation of regular dendritic ice crystal structure.

The agitation of liquid phase serves a double purpose: to maintain cells in suspension (uniform concentration, no settling) and to create conditions for volumetric heat transfer. At the beginning and during the temperature plateau, the cryogen is sprayed onto agitated liquid, and adequate agitation is needed to provide uniform volumetric cooling. In this study, four thermocouples located across the agitated volume indicated uniform temperature distribution and similar (in time) temperature profiles for each point. (This can also indicate the quality of agitation in the liquid phase.) During the cooling of the bed of granules, these four thermocouples also indicated a good temperature uniformity.

The agitated material may be exposed to cyclic temperature variations while passing near the surface covered by a carbon dioxide or liquid nitrogen spray. This phenomenon is more pronounced for liquid nitrogen spray—with droplets rapidly evaporating on the surface of an agitated bed of granules—with limited penetration of liquid nitrogen into the bed. In the case of carbon dioxide, its sublimation rates are slower than liquid nitrogen evaporation. When agitation is vigorous enough (Cheremisinoff 1986), the solid carbon dioxide particles can penetrate thoroughly all across the bed of granules. In a mixed bed, the heat transfer from solid particles of carbon dioxide to granules depends on the quality of mixing (e.g., on rapid intake of the bed surface layer where carbon dioxide particles are settling and mixing homogeneously within the whole bed). The goal in this study was to ensure a uniform volumetric sublimation of carbon dioxide in the whole bed. Rotor elements such as paddles may cross the bed surface, as in liquid agitation, to carry the surface layer into the bed bulk.

There are no useful heat transfer correlations to be applied to such a volumetric sublimation cooling of the agitated bed of solidifying granules. Available heat transfer data for agitated beds of granular material pertain to the heat transfer between the wall and agitated bed (Muchowski 1988; Schlunder 1982). Fluidized bed correlations for flowing gas particle heat transfer (Schlunder 1982; Zabrodsky and Martin 1988) may be applied for the first estimate, but they do not exactly represent phenomena occurring in the bed. In liquid nitrogen spraying, the cooling occurs mostly in the top layer of the bed due to intensive nitrogen evaporation. The agitation should ensure rapid exchange of granules between the top zone and the bed bulk.

The agitation intensity depends on the geometry of the freezing apparatus and the geometry and movement pattern of the agitators (Malhotra et al. 1990; Sterbacek and Tausk 1965). The mixing quality depends on rotational speed, shape of mixing elements and their number, ratio of mixing blade width to bed height, as well as vessel wall-mixing element clearance. Local temperature distribution across the bed of agitated material is more uniform for carbon dioxide spraying than for liquid nitrogen spraying. In addition, local temperature differences through the granules exposed to the sprays are much lower for carbon dioxide than for liquid nitrogen. Therefore, the cold shock the cells may be exposed to is much milder in the case of carbon dioxide spraying.

During the process, the speed of agitation was changed depending on the stage of freezing, with one of the goals being to minimize mechanical stress. Since the rotors can turn in either direction with individually controlled speeds, additional control over the mechanical stress acting upon the product can be applied, whether or not mechanical damage to the cells is intended. Maximum agitator tip velocities encountered in the typical agitated industrial blenders/processors for granular solids are about 1.5 m/sec for ribbon blenders and up to 6.5 m/sec for double rotor mixers/processors. Changes in mechanical stress on the cells may be accomplished by changing rotor speed or direction of rotation, turning the rotors at different speeds, etc. The final rotor design performed well on a variety of cellular products (e.g., yeasts, various *E. coli* strains, inclusion bodies, and disrupted cell suspension). The rotors also provided assistance during product discharge by pushing product toward the gates.

The spraying nozzles for liquid carbon dioxide and nitrogen were located at the top of the machine. Since there were concerns about possible carryover of solid carbon dioxide particles into the gas exhaust duct at high gas delivery rates and a loss of some cooling capacity, a system of internal baffles was designed to direct gas/solid jets from the nozzles towards the surface of the product. The nozzles were located at a sufficient distance from the walls of the machine to prevent any solid carbon dioxide or other material buildup on the walls. Diameters of orifices in the nozzles and diameters of the nozzles themselves were optimized to provide the best operating conditions for the freezing system.

Delivery of liquefied gas was monitored and controlled by a microprocessor using automated valves of a high-pressure, low-temperature design. After each run, the gas delivery system was automatically emptied and vented to prevent any entrapment of carbon dioxide in liquid form under high pressure. The gas delivery system was thermally insulated. There were several pressure relief devices along the gas delivery system as well as pressure monitoring for safety reasons. Originally, the system was supplied with medical-grade carbon dioxide and nitrogen from multiple mobile liquefied gas containers. Later, a permanent liquid carbon dioxide storage tank was installed. Since there was still a small carryover of fine carbon dioxide particles into the exhaust duct, a cyclone separator was installed. Fine particles of carbon dioxide were separated from the exit gas in the cyclone and collected in a bin below. Since the process was a batch type, the particles collected in the bin were sublimed after each run by applying heat prior to a cleaning cycle.

A cleaned gas stream left the cyclone and passed through a set of final filters. Gas filtration included prefilters and a final set of HEPA filters to assure proper system containment. Specialized low-temperature filters were developed to meet the author's specifications. The filter media and internal sealers were tested at low temperatures. The frame sealing material was selected after testing for mechanical stress at a temperature of $-80°$ C. Some of the standard filter materials cracked under temperature shock alone, even without applying any mechanical stress. For very low-temperature freezing, (i.e., when the liquid nitrogen spraying is extended and the product temperature reaches below $-80°$ to $-85°$ C level), a duct heater can bring the gas temperature close to $-80°$ C to protect the filters from excessive thermal stress. The concept of final filtration of the gas stream was applied because of the possibility of some carryover of fines beyond the cyclone. The whole system was kept under negative pressure by an exhaust blower located after the filters. The ducting, cyclone, and filter housing were thermally insulated to avoid external moisture condensation/frosting and for safety reasons. The frozen product was discharged through the double gates into containers and then sealed. Product discharge was conducted under negative pressure conditions. The product was weighed continuously during the discharge. The rotors, as in the pilot unit, were equipped with variable speed drives, could turn in both directions, and were used to assist in product discharge.

Controls. The system was controlled by a microprocessor-based system with a CRT display, data storage, and printing capabilities (Micromax® system by Leeds and Northrup). The product delivery, the freezing cycle, and the cleaning cycle were automated with a possibility to run selected steps in a manual mode. The monitored and controlled parameters were as follows:

• Temperature	Temperature at four points inside the processor
	Temperature in the outlet gas duct
• Pressure	Pressure in the processor
	Pressure in the seals
	Pressure differences across the final filters
	Pressures of liquefied carbon dioxide and nitrogen
• Flow rates	Flow rates of liquefied carbon dioxide and nitrogen, and their individual total volumes
	Temperatures of inlet and outlet of heat transfer agent to the jacket
	Flow rate of heat transfer agent in the jacket
• Volume	Volume of the liquid product delivered to the processor
	Volumes of the cleaning/rinsing solutions delivered to the processor

- Power   Power demand by each rotor
Speed and direction of rotation of each rotor
- Weight of product Weight of discharged product
- Air quality   Quality of air in the processing room (oxygen and carbon dioxide level monitoring)

Key measurements served to fine-tune the process during a run. The most important parameters were initial material load volume, material temperature, temperatures inside the processor, power requirement by the rotors, and flow rates of liquefied carbon dioxide and nitrogen.

The control concept was to use all 4 temperature measurement points inside the machine with the goal of obtaining an average temperature. An allowance was built into the control logic to reject one of them if there was any significant deviation in temperature read at any time. (This feature can prevent false readings due to an accidental buildup of product around a temperature sensor or due to a sensor's malfunction.) Pressure was controlled and monitored at several points to ensure that the system worked under negative pressure and operated within a range of safe conditions. In case of a pressure increase in the system, an alarm was initiated. If pressure increased beyond a double-alarm level, the liquefied gas supply was cut off and the entire system shut down. Depending on the values of key parameters, the jacket was automatically turned on and off and emptied using compressed air. Operation of the jacket could also be a part of mechanical stress control upon the freezing product control program. This was possible due to a strong agitating rotor/machine-wall interaction during the mixing process. During the freezing run, the speed of the rotors was changing to optimize cooling of the liquid product, liquid/solid phase transition, and final product freezing and degassing, i.e., removal of solid carbon dioxide via sublimation. The process run parameters were stored, and selected variable trends were printed in a continuous chart (trends included temperatures inside the machine, power demands, and pressure). Monitored variables also could be viewed on the CRT screen in the form of trending charts. Step events and alarm conditions were displayed on a screen and documented using a printer.

In addition to sequenced events, the control system also performed PID (proportional-integral-derivative) control loops (an example is a control loop for the jacket temperature). For safety reasons, the oxygen and carbon dioxide gas monitoring detectors were installed in the processing room and in the area where gas supply containers were located. Should dangerous gas composition occur in the room, an alarm would sound and the carbon dioxide/nitrogen supply would be shut down.

Cleaning-in-Place (CIP).   The CIP cycle of the freezing/granulating system was combined with the rest of the manufacturing system regarding delivery of cleaning and rinsing solutions, although cleaning of the freezing system was done separately. Cleaning solution was pumped through the product supply piping and was collected in the freezing machine until its level was above the shaft seals. The rotors were then turned on, and the bottom drain valves were opened to supply the CIP recirculating pump located below the freezing machine. Heating was applied to the jacket to maintain the cleaning solution at an elevated temperature. After running the rotors at high speed, the CIP pump was started in a discharge mode to dump the most contaminated liquid into a biowaste treatment system. A new batch of fresh cleaning liquid was supplied to the machine, and liquid recirculation was begun. Due to size of the system, there was a sequence of turning on different groups of spray balls located in the machine and in the gas outlet ducting. The area requiring the most cleaning was the machine interior, due to the typically sticky and viscous consistency of the product. Flooded surfaces below the seal level were the most contaminated, and they received the most vigorous cleaning treatment due to the rotor's mixing action. Surfaces at the top of the processor were much less contaminated, and they were cleaned sufficiently by liquid jets from the spray balls. Those jets were arranged to clean sight glasses, gasketed areas, and both sides of the gas flow separation baffles.

The nozzles for spraying liquefied carbon dioxide stayed clean due to the cleaning action of gas/solid jets—with carbon dioxide particles sweeping the nozzle's internal surface. The internal nozzle surfaces were configured to prevent any buildup of solid carbon dioxide, since a nonuniform delivery of carbon dioxide could initiate formation of product lumps. The exhaust duct system received the CIP treatment all along the lines located prior to the exhaust filters. Spray balls were inserted into the ducting and the cyclone separator. Configuration of sprays in the cyclone ensured cleaning of cyclone walls and its internal tube. After recirculation, the cleaning solution was discharged, a fresh portion of liquid was supplied, and the recirculation cycle was repeated. After discharge of the cleaning solution, cycles of final rinses followed in a similar pattern.

Production System Performance

The large system worked on the principle of containment. It could accommodate large batches of liquid product (up to 300 L) and complete the freezing cycle in a short time (between 30 and 45 min per batch, depending on the load, cell concentration, and the cryogenic fluid delivery pattern). The subsequent CIP cycle did not exceed 2 hours. System performance was satisfactory for a variety of products with different characteristics, but the major operation was processing of *E.coli*-based products.

The freezing cycle included the following steps:

- Filling the machine with product
- Cooling liquid product
- Liquid/solid phase transition (ice crystals formation)
- Forming granules
- Freezing interior of granules
- Cooling product to storage temperature
- Product degassing (in the CO_2 cycle only)
- Product discharge

All operations, except the product discharge step, were controlled by a microprocessor-based control system. For safety reasons, an automatic product discharge was manually initiated by operators. The process parameters were collected and stored, and selected data were printed out as a hard copy for manufacturing documentation purposes. Granule size can be kept within a relatively narrow range (typically 1–3 mm), and this range will depend on the type of product. Figure 3.8 shows an example of the temperature profile and the power requirement inside the freezer during one of the test runs. The granules were removed during test runs at the period of temperature decline and cut open to determine

Figure 3.8. (a) The product temperature profile and (b) the agitation power during a freeze-granulation run. The power peak occurs at the breaking of an agitated mass into granules. The product temperature stays at a plateau when the product is in a liquid form and decreases after the granules are formed.

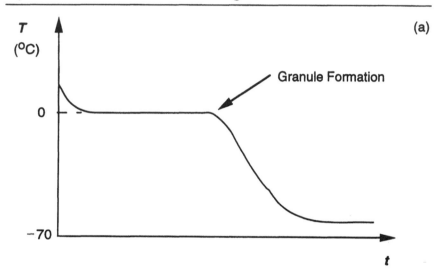

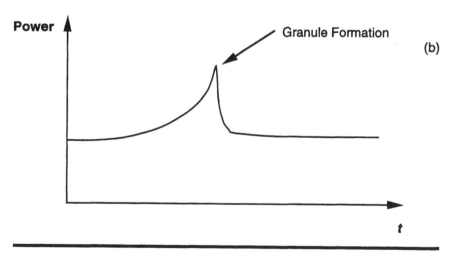

the progress of freezing. Particular attention was paid to the larger granules since they freeze last.

The volume of the agitated bed of granular material is much larger than the original volume of liquid due to the porosity of the bed, (i.e., the bed may exceed the original liquid volume by more than 67 percent). The apparent porosity of agitated granular material depends on agitator design and on intensity of agitation. In the extreme case of highly turbulent agitation, a low-porosity suspended bed may be obtained.

Example of Production Run. In one of the production runs, which used a bacterial (*E. coli*) cell suspension concentrated in a disk stack centrifuge, the following freezing parameters were recorded:

- Cooling liquid from 18° C   8 min
  to near 0° C
- Initial water freezing at a   9 min
  temperature plateau near 0° C
- Cooling from near 0° C to about   15 min
  –60° C in a granular form
  (cooling rate approximately 4° C/min)
- Low-temperature stabilization   5 min

Freezing and Granulation in the Production System. The agitated bulk of liquid or pasty material is first cooled by the sprayed liquid or solidified cryogenic agent. After reaching the freezing temperature level (at or slightly below 0° C), the first ice crystals begin to form in agitated material. Solid carbon dioxide particles or liquid nitrogen droplets generate nuclei of new ice crystals in the agitated mass. Since the number of falling particles or droplets is very large, the number of formed nuclei is high, and the number of crystals quickly increases. The agitated mass soon has a large portion of its water frozen out in the form of small ice crystals. The earlier formed crystals can grow in a dendritic form (in a pattern close to equiaxed) at significant growth rates if the undercooling of surrounding liquid is adequate. Such an undercooling can be achieved by delivering the cryogenic fluids at a sufficient rate. Delivery rate and form (particle or droplet size) of the cryogenic agent is critical, and it cannot be too rapid nor too concentrated since it may form large, irregular, frozen-surface pieces of product with soft cores. A slower delivery rate (also in an on-off mode) may allow the ice crystals to grow instead of seeding more small crystals. The mass temperature remains at a plateau during this process.

Since there is a liquid water removal process going on, the cells are also losing water via osmotic transport. This process can be conducted at a close to constant temperature near the 0° C level (resulting in no supercooling and no danger of formation of ice crystals within the cells). Growth of large crystals, which may create external stresses, is also under control, since the process can run under the condition of seeding of large numbers of smaller crystals. Such overall conditions may prove beneficial for cell freezing. An increase in crystal number causes material apparent viscosity to increase. After a significant portion of water is frozen out as crystals, which leaves the viscosity of the mass high, the agitator rapidly begins to break the whole mass into granules. This step is relatively short compared to the temperature plateau during seeding of the crystals. After formation of the granules, the apparent viscosity of the agitated bed rapidly declines and so does the agitation power. Bed temperature begins to drop if the cryogenic agent is continuously delivered. Control of the rate of temperature change is determined by delivery of the cryogenic agent. Changes in delivery rate may lead to different temperature change slopes (e.g., rates from 1° to 20° C/min could be accomplished). Higher rates may require large volumes of cryogen to be delivered in the processor, and the gas venting system should be designed to handle the large volumes of exhaust gas.

The bed temperature, as measured by thermocouples or resistance temperature devices (RTDs) inserted into the bed, is the temperature between the granule surface and the gaseous phase among the granules. The temperature reading is lower than the real temperature of granule interior. Intermittent, on/off delivery of cryogen allows the measurement of temperature closer to material temperature during the off periods, since the gas assumes the temperature of the solids (due to the low gas heat capacity compared to the solid). At a slower cryogen delivery rate, breaks in cryogen supply may not be required to obtain more realistic readings of material temperature.

During the period of bed temperature decrease, the surfaces of granules are the coldest part and the centers may even remain at near 0° C for a very short time. The remaining water freezes from the surface inward, and frozen surface layers conduct heat from the freezing front (that moves toward the center) toward the granule surface. The small size of granules ensures that such a transitional process is very short, that is, that the freezing-out of the remaining water is rapid across the granule. This step may not be critical even for a few larger granules since a large part of the water has already been frozen out during the bulk temperature plateau freezing. The surface cooling at continuously dropping external temperatures maintains a high heat flux with the temperature difference (external vs. freezing front temperature) compensating for increasing heat resistance caused by an increase in the thickness of the frozen layer. Further delivery of cryogen accomplishes granule cooling to a desired level.

Cross sections of granules of different sizes showed no large ice crystals, no significant dendritic patterns, and a relatively uniform interior. The granule initially consists of a frozen solid surface and a pasty center with small ice crystals. This makes the granule formed in such a process different from the falling liquid drops into liquid nitrogen, where the surface freezes immediately and the center of the granule remains in a liquid or pasty form (no ice crystals present). The center is initially at a temperature above 0° C, but the freezing front moves very rapidly from the surface toward the central point.

The granule interior consists of ice crystals, concentrated solutes and cells or cellular products (aqueous solution/suspension). Temperature decrease after granule formation permits controlled process of solute solidification. The cooling rate affects ice crystal growth and solidification of glassy states and eutectics. The final freezing continues in almost volumetric fashion due to the uniform initial distribution of ice crystals, solutes and cells within the granule volume and small temperature gradient across the granule. The remaining water migrates towards the ice crystal surface or forms new crystal nuclei. Glassy states may form close to the equilibrium, since the cooling rate is not very rapid. The ice crystals also form under conditions close to equilibrium during the temperature plateau period and the subsequent cooling. As a result, they are less susceptible to recrystallization during storage and thawing. The cellular products are better protected and there is no significant granule fusion during storage. The glassy states solidify close to equilibrium, e.g., the water content W_g' in frozen state is low. The frozen glass temperature T_g' is close to equilibrium, i.e., high, and therefore, the storage temperature can be higher. (Also see Table 2.6 in Chapter 2 for typical values of W_g'.)

Since most of the free water has been removed during the temperature plateau period in a homogenous (agitated) environment, all the cells uniformly followed the osmotic equilibrium losing intracellular water. After granule formation the cells may be embedded into a mixture of ice crystals and highly viscous, concentrated solution. The osmotic transport of water out of the cells may continue, but this period is short and dehydrated cells quickly become embedded into firstly, the "rubbery" and later, into glassy states. Due to a significant cell dehydration during the temperature plateau, there is no danger of intracellular ice formation during granule cooling and solidification.

The cooling protocols may incorporate varying temperature profiles, including holding periods at certain temperature levels (the varying cooling rates can be obtained by changing cryogen delivery rate, and the holding periods can be obtained by on-off or reduced cryogen delivery). In such a way, the final freezing of the free water may be manipulated prior to a total solidification. The initial composition and concentration of solutes and the cell type and its concentration shall be considered while developing such protocols. Further information on interaction of cells with solutes and glassy state formation can be found in Chapter 2 of this book by the author and his two recent papers (Wisniewski, 1998 a,b).

The freezing conditions accomplished by the agitated mass volumetric freezing technique at the temperature plateau are different from those of an encapsulated liquid in a sphere similar in size to the granule and a subsequent cooling through the surface. In the case of such a capsule, all of the water has to be frozen from the surface toward the center; principles of directional freezing of solution from the cooled surface toward the liquid apply, and the case of solidification in spherical coordinates can be used. At the same rate of surface temperature change, an overall capsule freezing process will be longer than for granules, since much more latent heat needs to be removed from the capsule interior and carried through its surface into the environment. Directional heat flux over a longer period of time may lead to formation of dendritic structures pointing toward the center. Therefore, in the case of capsules, the external heat transfer coefficient is more important than for granules, since more energy is going to be removed through the surface. The agitated beds with the cryogen sprayed into them have a capability of developing high heat transfer coefficients, since heat transfer is not only between the gas and solid but also by direct particle contact with subliming or evaporating cryogen. Uniformity of thermal conditions within the bed is assured by vigorous agitation and the distribution of cryogens over wide areas.

Delivery of Liquefied Gases

Delivery of liquefied gases to the freezing apparatus was investigated to assure acceptable system performance and further process optimization. The top spraying concept was selected, and a comparison of different spraying systems for liquid carbon dioxide and liquid nitrogen was done by preliminary analysis and field testing of various spraying devices.

Delivery of Liquefied Carbon Dioxide. Liquefied carbon dioxide is typically sprayed through small orifices, and the liquid expands from a pressure of about 20 bars ($20 \cdot 10^5 Pa$) to the pressure level in the process equipment (typically near ambient). The expanding liquid converts into gas and fine solid particles. There is a theoretical limit as to how much of the liquid carbon dioxide can be converted into solid particles. Cooling applications utilize solid particles as cooling agents since the cooling capacity of the gaseous phase is lower than that of the solid. The cooling capacity of the solid phase comes predominantly from the heat of sublimation, with carbon dioxide particles changing into their gaseous state. The sublimation temperature of carbon dioxide is −78.3° C, and this is the limit to how low products may be cooled using this technique.

Spraying orifices for carbon dioxide are typically mounted in tubular nozzles. These nozzles provide a restricted volumetric space for the liquid as it changes into a mixture of gas and solid particles. The nozzles can be made of different shapes—a narrow shape to shoot a jet of solid/gas mixture at high velocity, or wide for reduced stream velocity. Manufacturers of these nozzles may claim their efficiencies to be close to the theoretical values for producing a maximum amount of solid particles, but the real efficiency is usually lower than the theoretical level. More liquefied gas should be expected to pass through the nozzle than expected from theory to accomplish any specific cooling or freezing effect. It is suggested by manufacturers of these nozzles that such factors as the curvature of the surface where the jet hits a wall, nozzle diameter, and nozzle length are important factors in maximizing nozzle efficiency. There are claims that a slight taper along the nozzle may improve the liquid-into-solid conversion factor.

Because of additional applications for carbon dioxide particulate jets, such as in the semiconductor industry for cleaning electronic components, further development in spraying nozzle design can be anticipated. There is a significant volume of gaseous phase present in the solid/gas mixture. This stream can carry away very fine particles of solid carbon dioxide, reducing cooling efficiency. If any fine particles of frozen products are present, the gas jet velocity should be low to prevent any particle carryover effect. In this case, the nozzle outlet should be of a large enough diameter to ensure low gas velocities and a free fall of solid carbon dioxide particles. A jet of solid/gas mixture leaving the nozzle can meet a stream of rising warmer gas produced by sublimation of solid carbon dioxide after contact with the surface of cooled material. However, the kinetic energy of carbon dioxide solid/gas mixture is typically much higher than the energy of the rising gas within a spray pattern, and only the finest solid particles from the periphery of the spray might be carried away with a rising gas before they are able to reach the surface of cooled material.

One practical aspect of nozzle design for spraying carbon dioxide onto a freezing liquid surface is prevention of any internal buildup of solid carbon dioxide inside the nozzle, which would cause large pieces of carbon dioxide to fall into the liquid product. Those large pieces could initiate the formation of large chunks of freezing product. Such lumps freeze at a much slower rate than the relatively homogeneous mass of remaining material that is forming small granules. Such a problem, in some cases, was found to be caused by the asymmetrical location of spraying orifices within the main nozzle.

Liquefied carbon dioxide was stored in containers or large tanks under pressure. Varying its pressure is difficult and possible only within a limited range for thermodynamic reasons (since lowering liquid pressure in the delivery system may cause formation of solid and gas phases). Therefore, the control of spraying rates can be accomplished by applying multiple nozzles and turning them on and off, using cutoff valves. Location of these valves is critical in avoiding nozzle plugging by solid carbon dioxide. The valves should be located as closely as possible to the spraying nozzle and may be combined with it into a single assembly.

Delivery of Liquefied Nitrogen. Spraying of liquid nitrogen is performed at much lower supply pressures than for carbon dioxide. The spraying nozzle design is based on different principles from those for liquid carbon dioxide. There is no solid phase formation, and droplets leaving the nozzle evaporate when making contact with a warmer environment. To produce a long spraying pattern, the droplets should be relatively coarse. This will assure a long travel distance while evaporation takes place in the spray state and droplet diameter decreases. Due to gas entrainment in the spray jet (MacGregor 1991), the gas temperature in the chamber may affect the evaporative losses from the spray. In thermally well-insulated and contained chambers, gas temperatures may approach the boiling temperature of liquid nitrogen, but in practical applications they will be much higher.

Spraying patterns that form a solid cone with a circular or a square cross section are recommended over a hollow cone spray pattern produced by centrifugal atomizers with high tangential velocities. The center of a solid cone spray pattern is surrounded by vaporizing edges. Temperatures within the spray core are low enough to ensure sufficient spray range and delivery rates of liquid nitrogen to the surface of the cooled material. In the case of a hollow cone, the spray is thin, and the surface exposed to the environment is relatively large, causing a more significant evaporation effect. Sometimes flat (fanlike) sprays are used; such patterns are also acceptable if the droplets are coarse and the nozzle capacity is sufficiently high. Droplets that are too fine are not recommended due to the generation of a large evaporative surface, thus shortening the spray range. The presence of gaseous phase in liquid gas supplied to the nozzles may significantly reduce nozzle cooling capacity, due to a reduction in the liquid flow rate.

After the spraying pattern was selected, nozzle locations were established to provide liquid delivery in a uniform pattern without any significant local overcooling. As was mentioned earlier, the liquid spray descending toward the material surface will meet a rising stream of warmer gas. These two streams can move in opposite directions if the gas outlet is located at the top of the apparatus. Coarse droplets are not affected by the gas stream produced at the surface of cooled material due to spray dynamics.

During a start-up, while the liquified gas delivery system is still at ambient temperature, liquid nitrogen evaporates inside the delivery piping, and only the gaseous form leaves the spray nozzle. A two-phase (gas/liquid) flow follows, but once the delivery system cools sufficiently, only a liquid phase leaves the nozzle. If the gaseous phase is present in supply lines, a nozzle's cooling capacity will be drastically lowered due to reduction in the liquid flow rate. An in-line flowmeter will show erroneous results. Control of the cooling rate can be accomplished by throttling nitrogen flow to the nozzles or by applying many nozzles and turning them on and off using cutoff valves.

After liquid nitrogen drops reach the material surface, there is rapid evaporation of liquefied gas. General analysis of film heat transfer of impacting sprays was proposed by Deb and Yao (1989). These authors also compared dense and dilute sprays. They concluded that dense sprays do not show significant dependence on droplet diameter and velocity as the dilute sprays may, and this is due to a strong interaction among droplets. Phenomena occurring during film boiling from liquid nitrogen sprays on a heated plate were studied by Awonorin (1989). He concluded that the parameters affecting heat transfer between the spray and the plate are mean droplet size, mean temperature difference, and local mass velocity of a spray. Droplets hitting the heated plate can bounce, rebounce, and roll along it. The Leidenfrost effect may play a significant role due to the formation of a vapor phase between the droplet and the solid surface. The breakup of droplets impinging the surface under Leidenfrost conditions may depend on impinging velocity and angle, droplet diameter, and the wall temperature (Ko and Chung 1996).

In practical applications with an agitated mass of material, to avoid excessive local overcooling, the spraying pattern should be wide but the spray dense enough to provide sufficient droplet interaction and minimize spray evaporation losses. The agitated mass may not allow nitrogen droplets to bounce and roll, but they may be entrained in the bulk of material and boil and evaporate within the bed below its surface. This can be beneficial due to the better heat transfer and more efficient cooling with utilization of liquid nitrogen. The agitation rate and the pattern of bed movement by the agitator are important process parameters in combination with the gas delivery and may need to be varied during the process run.

Discussion of Case Study 3.1

The system for freezing cell suspensions was built with the goals of providing controlled cooling and freezing rates for large product volumes and controlling the size of frozen samples. The cooling was accomplished by spraying liquid carbon dioxide and liquid nitrogen at controlled rates into a mass of cells to be frozen. The controlled sample size was a frozen granule, which could be produced in large quantities as a bulk material. The process of freezing and granulation involves liquid cooling, holding, ice crystal formation, and water freezing-out periods before granules can form.

The holding period occurs at a temperature plateau near 0° C. During this period, some extracellular water freezes out, and small ice crystals form outside the cells in the undercooled solution. Diller et al. (1985) investigated the influence of controlled ice nucleation on a thermal history during freezing. After ice nucleation was initiated in the undercooled solutions, the temperature increased, and a temperature plateau was observed due to a release of latent heat. After formation of granules, there is an external layer of small crystals mixed with the frozen cells and a solidifying liquid center of the granule, with small ice crystals formed initially in the agitated mass. During the temperature plateau period, some growth of these ice crystals may occur. Further growth of the intergranular ice crystals and cell dehydration depend on the cooling rate. The cooling rate for granules is controlled externally by the rate of delivery (spraying) of liquefied gases. The spraying rate of liquefied gases can vary during the steady temperature (plateau) period and the subsequent period of decreasing temperature and granule cooling. The temperature plateau period can be extended by a reduction in gas delivery rates. During the temperature plateau, the liquid is in an undercooled condition and undercooling is maintained by the supply of cryogenic fluids. Such conditions not only create new nuclei (new ice crystals) in the points of contact of cryogenic fluid droplets (or solid particles, as in the case of carbon dioxide) but also promote growth of the already formed ice crystals. At gentle agitation, the crystals can grow into an equaxial dendritic pattern. More vigorous agitation may cause breakage of formed dendrites and creation of new inoculation points. Breaking of large crystals may occur, in particular, at the end of the temperature plateau period when the physical interactions among the numerous ice crystals occur. The growth conditions of dendritic crystal tip are different from those for cooling the liquid through the wall. The heat flow (mainly from the solidification latent heat of fusion) is from the growing solid tip of dendrite into the undercooled liquid (the dendrite temperature is slightly higher than that of the liquid); whereas, the cooling through the wall, the heat flow is from the tip through the dendrite column toward the wall (the dendrite tip temperature is lower than the temperature of the liquid, and temperature continuously decreases along the dendrite length). The growth rate of the free-floating equiaxial dendrite depends on undercooling of the liquid phase. The overall undercooling has three major components: tip curvature undercooling, solutal undercooling (depending upon the solute distribution between the solid and liquid phases and on the solute concentration distribution in liquid phase next to the dendritic tip), and thermal undercooling. Since in aqueous solutions the solute distribution coefficient is close to zero (i.e., there is solute exclusion from solid phase) and since due to agitation there is no significant gradient of solute in the liquid phase between the bulk liquid and dendritic tip surface, the solutal undercooling component at the crystal surface may diminish. Due to an increase in concentration of solutes in the liquid phase, a bulk liquid undercooling will occur. More information on dendritic crystal growth can be found in Chapter 2, page 55, by this author). It is difficult to create and maintain high liquid undercooling in larger samples of aqueous solutions, and therefore, the rate of crystal growth (solid-liquid interface velocity) is limited. In this case, the rapid solidification (freezing-out) of water is possible because of the existence of a very large number of formed and growing ice crystals and a continuous undercooling of liquid.

Predicted freezing times for the granules, using a computer model, were confirmed by experiments. Larger granules were removed from the machine at different times, cut in half, and their interiors examined for the degree of freezing-through. The results yielded good agreement with the computer model predictions. The model used rates of temperature change derived from early experimental runs, and analysis was performed on the granules of larger size since they freeze late. The principle of that model was a stepwise calculation of freezing front movement toward the center of the sphere with accompanying release of latent heat at the solid-liquid interface. An additional difficulty in system performance analysis resulted from the fact that the sprays of liquefied gases were localized due to physical constraints of the system design.

Initial concern about possible particle carryover causing premature plugging of the final filters has not materialized. Since particulate material trapped on the filters was composed of fines of carbon dioxide, as soon as the process ended and the filters warmed up, these solids disappeared due to sublimation, and the filter capacity was restored. The final filters performed well, and there was no filter deterioration due to a repeating temperature shock. Performance of the CIP system has been found satisfactory even for freezing very viscous and sticky products. Figure 3.9 shows system schematics, Figure 3.10a shows a view of the freezing-granulation production apparatus, and Figure 3.10b shows a schematic view of the manufacturing system.

The system performs well with the low (diluted fermentation broth) and high (centrifuged products at 15–22 percent dry weight) concentrations of cell suspensions. Viscosity of the entering liquid depends on cell concentration, cell type, and type of fermentation broth. The highly concentrated suspensions typically produce smaller granules with diameters of 1–2 mm. Suspensions with lower concentrations of cells produce slightly larger granules. Figure 3.11 shows a view of the pilot/small production scale freeze-granulator able to process up to 55 L of product.

During the granule-forming period, the power peak of agitation may reach more than 200 percent of the agitation power required for the liquid. The agitation power begins to increase as soon as the temperature plateau is established. This occurs because the product contains an increasing amount of ice crystals. During runs with concentrated cell paste received from a centrifuge (a continuously operating disk stack type machine), freezing-out of water during the temperature plateau could reach beyond 40 percent of the total water content. The amount of free extracellular water decreases, followed by a change in intracellular water level due to osmotic equilibration. Due to a relatively slow dewatering process (a temperature plateau fre-

Figure 3.9. The schematics of the manufacturing system in the very low temperature version. The product can be loaded directly from a bioreactor or can be concentrated in a centrifuge or in a filtration system prior to the freeze-granulation step. The liquefied cryogenic gases are delivered to the processing apparatus through sets of spraying nozzles. The exhaust gas passes through a particle separator, heater, and a set of filters. The final gas filter is a low-temperature HEPA. Gas is discharged using an exhaust blower. The schematic shows the main system controls. The heating jacket operates during the phase transition in freeze-granulation and during thawing. The CIP supply from a plant system, the CIP pump, and distribution valves are shown. The CIP connections to different system components are not shown for drawing clarity.

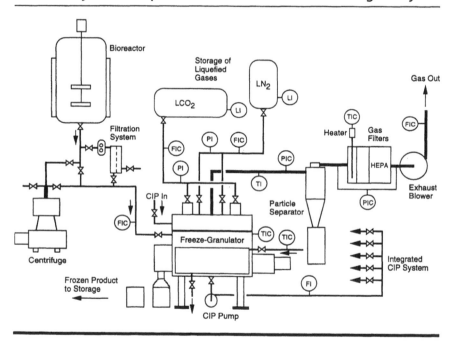

quently lasts 9–10 min but may be extended by slower cryogen delivery), the cells can follow an equilibrium path between intracellular and extracellular water. Since the process occurs at about 0° C, there is no danger of intracellular ice formation. The intracellular water moves across cell walls and is frozen out in the form of ice crystals. The suspension of ice crystals and cells in a concentrating solution reaches a state that corresponds to a high apparent viscosity, and the

---

Figure 3.10a. The freezing-granulation production apparatus. Two carbon dioxide spraying nozzles are located on the top, the product discharge chute is on the right side, and the drive assembly is on the left side.

(a)

---

mass breaks into individual granules. The granule breaking point may occur when the volume of free water reaches a level comparable to or lower than the volume of ice crystals. The slurry with non-Newtonian characteristics primarily determines at which level of free water the granule breaking point occurs. The agitator design also

Figure 3.10b. The general view of the manufacturing system. The freezing apparatus is shown from the product discharge side. The particle separator and the gas filters are shown on the right side.

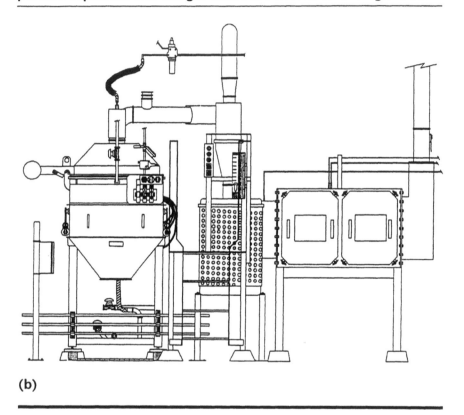

(b)

has some effect upon this point. This process of granule formation cannot be described by classical approaches to granulation (Sherrington and Oliver 1981) or particle breakage (Hill and Ng 1996).

In one of the examples, the estimated amount of free water during granule formation was about 56 kg, the amount of extracellular ice crystals was 66 kg, and the mass of partially dehydrated cells was about 70 kg. After granule formation, the continuous delivery of cryogen causes agitated bed temperature to drop at the rate of approximately 2–3° C/min. During this period, the remaining extracellular water freezes out with associated further cell dehydration. Due to the small granule size and low water content, a rapid movement of the freezing front occurs within granules. For small granules, the Biot number will stay below 0.1, and therefore, one may

Figure 3.11. A pilot/small production scale freeze-granulator. Open top shows the central nozzle for spraying liquefied gas. Product discharge gates are shown under the control panel. The gas exhaust system is not shown. Courtesy: Integrated Biosystems Benicia, CA

expect the granule temperature to follow closely the measured average agitated bed temperature. A condition of low-shear agitation is important for preserving cell integrity due to the presence of a large number of free ice crystals in the agitated mass.

Frozen granules are put into storage in –70° C or –86° C freezers and subsequently thawed according to the needs of the manufacturing plant. The system performs successful freezing of large volumes of biological material on a daily basis. The freeze-granulator has also positively performed with using solutions, such as buffers, growth media, protein and peptide formulations, pharmaceutical suspensions. In an extreme case, the test using pure water has been successfully conducted—the granules consisted compacted ice crystals. The process and its implementation have been a subject of the pending patents.

CASE STUDY 3.2: CRYOPRESERVATION OF BIOLOGICAL LIQUID PRODUCTS IN FREEZE-THAW VESSELS WITH EXTENDED HEAT TRANSFER SURFACES (FINS) OF NOVEL DESIGN

Advanced Design of Large-Scale Cryopreservation Vessels with Extended Heat Transfer Surfaces

Phase change heat transfer can be significantly enhanced using extended heat transfer surfaces such as rods or fins (Kalhori and Ramadhyani 1985; Wisniewski and Wu 1992, 1996; Wisniewski, 1998a). The extended heat transfer surfaces act as efficient heat conducting devices. The author has developed and designed systems for freezing protein solutions using large volume containers with extended internal cooling surfaces (Wisniewski and Wu 1992, 1996; Wisniewski, 1998a). The large volume of liquid is subdivided into compartments by means of finned heat transfer pipes forming internal heat exchangers. The finned heat transfer surfaces can perform very efficiently during the solidification processes under the condition that the fin material thermal conduction coefficient is much higher than the one of frozen product. Fin efficiency also depends upon its thickness. In general, thicker fins perform better during solidification (a fin also can be longer if its thickness increases). After formation of frozen material on the fin surface, the fin becomes a cold wall, and further buildup of the solidifying material continues.

The subcooling of the solid phase during thawing tends to inhibit the melting heat transfer for flat or cylindrical heat transfer surfaces. Heat transfer involving extended surfaces is much less affected by such subcooling. This is because of the heat conduction through the external surface elements—the solid mass temperature is rapidly brought to the vicinity of the fusion/melting temperature. In both cases—the solidification and melting—significant natural convection effects take place, affecting the heat transfer and the shape of the solid-liquid interface. Stainless-steel-made fins require a certain thickness to provide good fin thermal efficiency due to relatively low thermal conduction coefficient of stainless steel. The heat transfer and solidification principles in such a system are based on a controlled dendritic freezing occurring in small compartments formed by extended heat transfer surfaces. In the case of protein solutions, a satisfactory distribution of solutes within the frozen volume has been accomplished (see the publications by Wisniewski and Wu 1992, 1996, for details). The original containers with internal finned heat exchangers were, however, not the optimal designs due to certain manufacturing restrictions (for example, the bottom of the container was beyond the reach of finned heat transfer surfaces). The system manufacturing restrictions have also limited the number of fins. As a result, the freezing times were relatively long, and container filling volume flexibility was restricted. More recently, the author has developed a series of designs with different fin configurations delivering much better control over the freezing process, much shorter freezing times, and complete filling volume flexibility (patents pending). The size of these new designs permits one to treat batches up to 250–350 L in a single container. These designs also facilitate validation and have scale-down modules to test and optimize the freezing and thawing conditions in small volumes (in the range of several hundred mL). These modules can reproduce freezing and thawing conditions as they occur in the full size containers and can be used for product stability studies. Larger designs are feasible, but they become more complicated and expensive due to the complex patterns of the internal heat transfer surfaces. The author has designed vessels of large volume (up to 500 L) with the multiple active (with recirculating heat transfer fluid) and passive (interacting fins) heat transfer surfaces.

Figure 3.12 shows a novel design of the freezing/thawing container with an external jacket and an internal finned heat

Figure 3.12. Design of a container for large-scale cryopreservation using the extended heat transfer surfaces (fins). The vessel has two cooling surfaces: an external jacket (heat transfer through the vessel wall) and a central bayonet cylinder with fins (heat transfer through the cylinder wall and the fins).

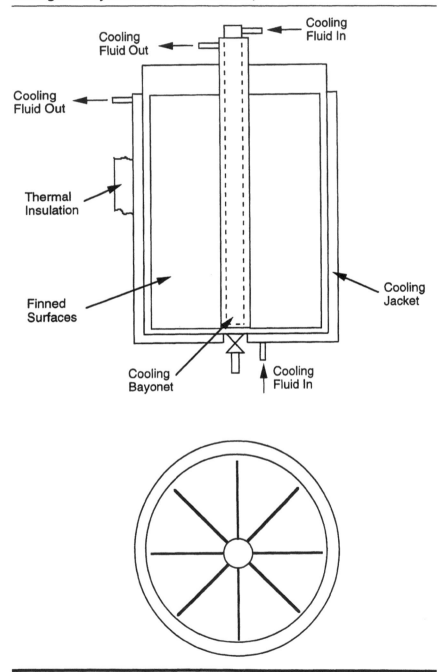

exchanger. The internal heat exchanger is designed using a bayonet flow pattern (a tube inside a tube). Fins extend from the central tube wall toward the jacketed vessel walls. The freezing pattern is symmetrical and repeats in each zone between two fins. The points where the freezing fronts (solid-liquid interfaces) meet are identical in each zone. Such a pattern facilitates validation of the freezing and thawing steps. A temperature sensor with multiple temperature sensing points permits monitoring of the progress of solidification and melting processes. The multiple measuring points allow adequate control of freezing and thawing processes for various vessel loads (e.g., smaller than a full load). The fins reach close to the vessel's bottom to make sure that there is uniform freezing there and that no liquid product is entrapped in the lower part with a solidified top. The solidification using a heat exchanger with fins submerged into a large liquid volume shows that with increasing fin length its thermal efficiency decreases. This deficiency can be reduced by use of fin configurations providing conditions for fin interactions among themselves and with cooled surfaces. With interacting fins, the length of a thermally efficient fin can be increased. For the configuration shown in Figure 3.12, the effective fin length is almost doubled due to the fin interaction with the cooled vessel wall. This permits using a vessel of larger diameter with longer fins. The external container walls are thermally insulated.

The extended heat transfer configurations and the interactive fin designs are a subject of pending patents. Figure 3.12 represents a schematic view of the container. The actual aseptic design is drainable, and the corners are made with certain radii. The product-contacting surfaces are mechanically polished and electropolished to satisfy the cGMP requirements. The bottom product valve is made of stainless steel and is of aseptic type.

Figure 3.13 shows examples of interacting fins with patterns of solidifying layers of material. The principle of interacting fins was applied to the central heat exchanger with radially located fins. The fins approach the cooled wall. As the layer of frozen material forms on the wall and touches the fin tip (a thermal bridge forms), the fin begins to conduct heat from the liquid phase toward the cooled wall. The fin effectiveness increases as the frozen material layer thickness at the cooled wall increases. The fin surface cools down below the freezing temperature, and a solidified layer forms

Figure 3.13. Patterns of interacting extended heat transfer surfaces (fins). T_c is the temperature of a cooling fluid flowing inside the finned pipes.

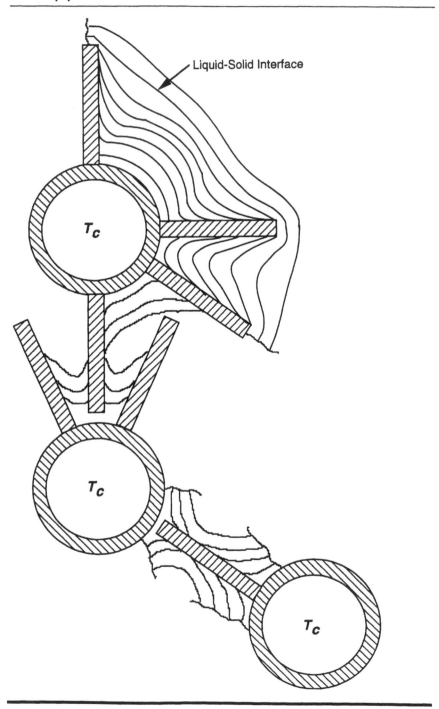

along the fin. Figure 3.14 demonstrates the beginning of freezing on the fin with the thermal bridge made of frozen material present. Buildups of solidified material are seen at both ends of the fin.

Temperature distributions along the fin and across the center of vessel compartment (a liquid cavity) are shown. The temperature in the liquid phase remains close to 0° C (with a slight undercooling, depending on liquid phase composition) and decreases within the solidified layer toward the cooled walls. Temperature in the fin stays close to 0° C in the area of liquid-fin contact and decreases along the fin in the areas covered with solidified material. Such a heat transfer pattern (fin surface temperature in the liquid zone being close to the solidification temperature) causes a rapid coating of fin surface with the solidified material.

Figure 3.15 shows a later stage of freezing (e.g., after layers of frozen material have formed on the cooled walls and fins). The fin temperature is now below the solidification temperature along the whole fin. At both ends of the fin, the temperature decreases. The central part of the fin acts as a cold wall, interacting with the adjacent solidified layers and conducting the latent heat of solidification released at the liquid-solid interface. The temperature distribution across the compartment cavity shows the liquid temperature close to the solidification temperature and decreasing temperature across the solidified layers. A similar pattern of temperature distribution occurs in cross section: from the liquid phase throughout the solidified material to the fin.

The geometry of the compartment enclosed within the two cooled walls and the fins forms a shape that is advantageous for rapid, controlled freezing. Figure 3.16 shows freezing configurations of the finned compartment, the cylinder, and the slab with flat walls. The dimensions of these configurations are similar. Freezing patterns in the finned compartment also depend on the angle between the fins (here shown as a 45-degree angle). The cylindrical configuration (with the radius R_c) shows a significant reduction in freezing time when compared to the slab with flat walls of a comparable distance between the cooled walls ($s = 2R_c$). For example, the slab-shaped chamber with the thickness $s = 0.15$ m/and the cylinder-shaped chamber of 0.075 m radius will require 3 hrs 26 min and 1 hr 44 min respectively (at the cooling fluid temperature of –50° C and a turbulent flow of cooling fluid). The almost triangular finned compartment (as shown) is approaching the cylindrical

Figure 3.14. An early stage of freezing in a vessel with the central finned heat exchanger. The fin tip interaction with the cooled wall through the thermal bridge of frozen material has produced solidified material buildups (marked separately) along the fin surface. T_c is the cooling fluid temperature. Temperature distributions along the fin and across the compartment cavity are shown. (Drawing not to scale.)

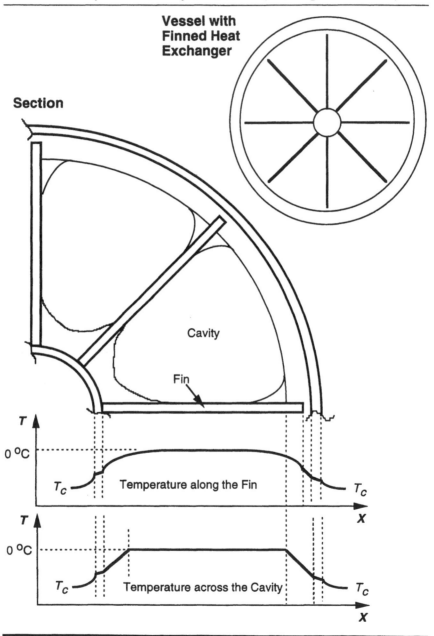

Figure 3.15. A late stage of freezing in a vessel with the central finned heat exchanger. The fin temperature is below the solidification temperature level (marked as close to 0° C) and the fin acts as a cooled wall. T_c is the cooling fluid temperature.

Section of the Vessel with Internal Finned Heat Exchanger

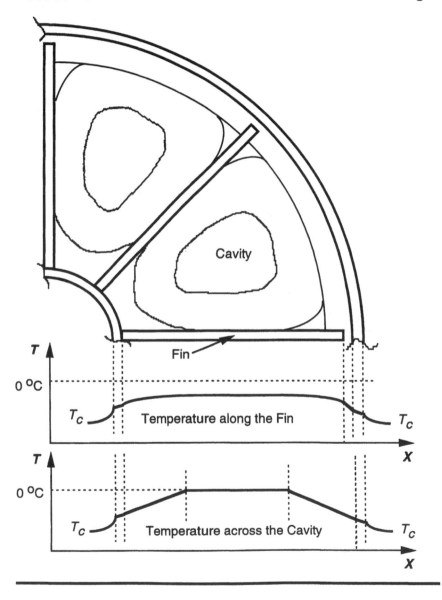

Figure 3.16. Freezing in the chambers of various configurations of comparable dimensions. The freezing rates in the cylindrical chamber are shorter than for the slab chamber with flat walls. The finned compartment freezing pattern may approach the cylinder freezing but also depends on the angle between fins.

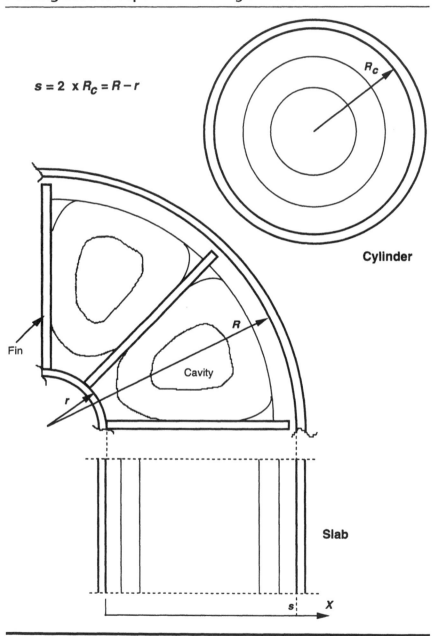

$$s = 2 \times R_c = R - r$$

Cylinder

Fin

Cavity

Slab

configuration (with fins acting as the cooling surfaces). At small angles between fins, a significant corner effect takes place at the central tube (i.e., the freezing accelerates there).

The finned configuration with active cooling surfaces in the form of a central pipe and external jacketed wall and the passive cooling surfaces (fins) approach the cylindrical freezing configuration that provides rapid freezing rates. Freezing aqueous solution in a small-capacity (research and pilot scale, maximum product volume 20 L) vessel can take less than 2 hours since liquid is cooled to near 0° C. The average velocity of a solidification front may exceed 40 mm/hour. Overall processing times include cooling the vessel and product from initial temperature to near 0° C and subsequent freezing. The finned configuration of internal heat transfer surfaces can provide rapid freezing rates in large-volume vessels.

Figure 3.17 shows an external view of the research/pilot scale vessel with 20-L product capacity. The external jacket and the top of the bayonet heat exchanger are thermally insulated. The refrigeration system with cooling fluid recirculation capability is shown on the right (the connecting cooling fluid hoses are not shown). It has a built-in temperature controller that can provide cooling operation with a fixed and changing (ramping) temperature setpoint. The cooling fluid may reach –85° C when the cascade refrigeration system is applied. The controller can be interfaced with an external computer system. Figure 3.18 shows the details of the internal heat exchanger with fins (patent pending). The central tube is welded to the top cover. All internal surfaces are polished by abrasives and electropolished. There are 2 (inlet/outlet) connections of the cooling fluid shown together with quick coupling connectors.

Since research and stability freezing and thawing runs may not permit use of expensive product in large volumes, a research cartridge has been designed that simulates phenomena occurring in the full-scale vessels. The cartridge comprises a section of the large vessel with two fins, part of the external jacket, and part of the central bayonet and is a horizontal section (slice) of the large vessel. The design ensures that the geometry and heat transfer conditions are the same as in the large vessel. The heat transfer along the fins and building of the thermal bridge between the external wall and fin tip are closely preserved. The cartridge product volume is only several hundred mL, and the freezing process progresses at the

Figure 3.17. The research/pilot scale 20-L vessel with the refrigeration system. Courtesy: Integrated Biosystems, Benicia, CA.

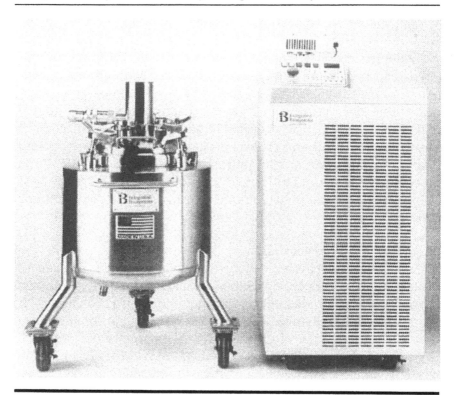

same rate as in the process vessel. The research cartridge can be connected in parallel to the production vessel with the cooling fluid recirculated through the vessel and cartridge. Such simultaneous runs demonstrate that the freezing progress is well reproduced since it may differ by only around a few percent between the vessel and cartridge. The cartridge has been used for research and development work as well as for stability studies. Other design patterns have been developed for larger vessel diameters to ensure optimal performance of extended heat transfer surfaces (e.g., maintaining the effective thermal length of fins).

The system as designed for protein solutions may perform differently for cells due to a possible cell settling with cell concentration increase at the container's bottom. Any agitation in the liquid phase introduced to help maintain cells in suspension may

Figure 3.18a. The internal heat exchanger with fins for 125-L capacity vessel. Courtesy: Integrated Biosystems, Benicia, CA.

(a)

Figure 3.18b. The internal heat exchanger with fins for 20-L capacity vessel.
Courtesy: Integrated Biosystems, Benicia, CA.

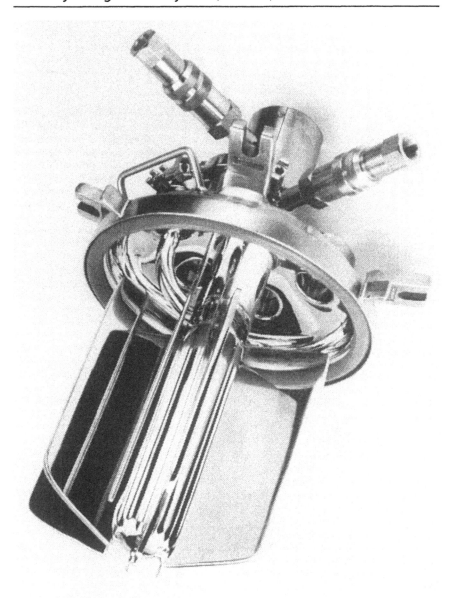

(b)

cause undesirable solute and cell rejection from a freezing mass and cryoconcentration effects (Deshpande et al. 1982). The cell settling depends on the cell size, its density, and viscosity and density of the liquid. These characteristics depend also on the cell cultivation/fermentation conditions. For example, the settling rates of the yeast *Saccharomyces cerevisiae* may depend on nutrient composition—the cells grown on saccharose and glucose settle faster than the cells grown on sucrose with molasses or on potato dextrose broth (Wisniewski 1996, unpublished results). The finned surfaces for cell suspension freezing should be designed differently than those for solutions, (i.e., long vertical fins are replaced with a shorter type). Vertical distances should be reduced (i.e., shallow freezing modules configured).

The central finned bayonet design has a limitation regarding the internal vessel diameter, above which the freezing rates may become too low (slow freezing front movement) due to the combination of a thickness of the frozen material and available lowest temperature of the cooling fluid, and the resulting freezing times may become long. Fin efficiency depends on fin material, fin length and its thickness. Fin efficiency decreases after length becomes larger than the optimal value. This decrease can be compensated by increase in fin thickness. However, thick fins are heavy and it is difficult to obtain thick plates with high surface quality, e.g., there are practical limitations regarding fin thickness. As a result, the fin cannot be too long, since its performance may become far from optimal. Limitation in fin length also limits the vessel diameter for the single bayonet design. Increase in vessel volume can be accomplished by increasing vessel height.

The finned bayonet design cannot be used in large diameter vessels due to long distances between the heat transfer surfaces, since the fin length would be extended beyond the optimal value if design shown on Figure 3.12 is used. Therefore, for larger designs, additional finned heat transfer surfaces are inserted into the vessel. The design preserves the concept of thermal bridges between the active heat transfer surfaces and fins.

Designs of larger diameter vessels may require different fin configurations if the single bayonet may not be sufficient. For the large-diameter, large-capacity vessels, the author has developed designs with multiple cooled surfaces and interacting heat transfer fins (patents pending).

The tall vessels made of stainless steel may not only be beneficial regarding the freezing and thawing patterns but may also reduce a possibility of product degradation due to the wall-air-liquid interface effects during handling of liquid product and thawing. For example, the free liquid surface (air-liquid interface) in the 150-L vessel with 508 mm (20 in) internal diameter may be approximately 5 times smaller than in the case of product filled in 1-L plastic bottles (for a vertical bottle position; for a horizontal bottle position, this ratio can be even higher).

The small freeze-thaw vessels at 5–20 L volume using the principle of fin-assisted heat transfer and thermal bridge formation between the fin tip and active heat transfer surface have been also developed for products made in small quantities.

System Design and Performance

In this process, cooling is accomplished through the wall and through extended fin surfaces. The cooling fluid recirculation is maintained by a recirculating pump, and the temperature level (including a setpoint variable in time) is controlled by controls associated with the refrigeration system. The ramp/soak feature of the temperature setpoint control for the cooling fluid is important for maintaining constant conditions for an optimum dendritic ice crystal growth (as described in Chapter 2, page 55) in this volume by the author). Due to the refrigeration cycle thermodynamic principles, there is a certain level of the cooling fluid lowest temperature, depending on whether a single compressor or a cascade system (two compressors) is used. Therefore, it is beneficial to design the process to freeze to a temperature level no lower than about –40° C if a single compressor design is desirable. Fine-tuning of the refrigeration system may provide an additional few degrees of extra temperature margin for a single compressor system. For example, the tap or chilled water cooled condensers/refrigerant condensing coolers may provide cooler liquid refrigerant to the expansion valve than do the air or tower water cooled condensers. The design of an expansion valve is critical due to a wide load variation. The load is the highest during the solidification stage (because of removal of the latent heat of fusion) and substantially declines after solidification is finished and the solid mass is cooled. The recirculating pump should be of very high hydraulic efficiency with a thermally insulated and removed motor (to minimize heat gains by conduction). The piping runs should be as short as possible to minimize holdup of expensive heat transfer fluids and to reduce the size of the holding reservoir.

Duplication of research and development conditions (e.g., freezing to −70° C or −86° C) can be difficult on an industrial scale. Such low temperatures require expensive cascade refrigeration systems. In addition, the viscosity of recirculated heat transfer fluids increases with temperature decrease, and the heat transfer efficiency deteriorates (lowering of the heat transfer coefficients due to the laminar flow in the jackets and internal heat exchangers). The importance of the heat transfer coefficient on the cooling agent side regarding the freezing rate at certain temperature levels was discussed earlier. An associated increase in pressure drop may adversely affect the recirculating pump performance. Special low viscosity, inert heat transfer fluids for very low temperatures (for temperatures of −70° C to −85° C) can be very expensive. Storage of the containers is easier at temperatures near or higher than −40° C than at temperatures below −70° C; the walk-in cold storage rooms cannot be used at this low temperature for personnel safety, and the container handling should be mechanized and automated. The author has designed a frozen product storage facility where the previously described containers can be chained in clusters with the recirculating cooling fluid flowing through each jacket. Such clusters are located in a thermally insulated enclosure (box) to reduce heat gains. The refrigeration system needed for the cooling fluid recirculation can be small since it has to balance only the heat gains from the environment and recirculating pump. The insulated box configuration and piping design permit withdrawing of individual containers and easy piping re-connection. The temperature and flow rates of recirculating fluid are monitored and recorded as is the temperature inside the insulated enclosure. Temperatures in individual containers are also recorded for documentation purposes.

Studies of freezing using the large-scale vessels and research cartridges demonstrated that rapid, controlled freezing rates are possible using this approach. Developed dendritic crystal growth has been observed. The dendritic crystal patterns first form at the directly cooled surfaces and later on the fins (i.e., after the frozen layer covers the fins and the fin temperature drops to below 0° C). The dendritic patterns grow toward the center of the cavity. The system performance has been significantly better than the author's earlier designs (as described in Wisniewski and Wu 1992, 1996); that is, the freezing process has been more uniform across the whole product volume, the freezing time has been shortened several times, and it has been performed under controlled cooling conditions to promote optimal dendritic crystal growth. The dendritic crystal growth is conducted at such front velocities that the water trapping in glassy states is minimized (W_g' is close to equilibrium) and the glassy state temperature T_g' is close to T_g' for equilibrium. The action of extended heat transfer surfaces (fins) has been improved by applying the concept of the thermal bridge being built of frozen product between the cooled wall and the fin tip. The fin configuration provides enhanced heat transfer conditions not only during freezing but also during thawing due to the difference between the thermal conductivity of metal and liquid product. Natural convection patterns in the liquid phase develop near the active heat transfer surfaces as well as near the fins. The new designs have increased the fin surface area multiple times over the early design. The heat transfer surfaces' thermal performance has been enhanced by introduction of proprietary turbulizers on the heat transfer fluid side; the turbulizers increase the heat transfer coefficient on the fluid side when the flow of the cooling fluid might become laminar due to the viscosity increase at very low temperatures.

CASE STUDY 3.3: CRYOPRESERVATION OF BIOLOGICAL LIQUID PRODUCTS IN BAGS IN CONTACT WITH PLATES WITH EXTENDED HEAT TRANSFER SURFACES (FINS)

Freezing of Biological Products in Bags Compressed between Plates with Extended Heat Transfer Surfaces

Extended surfaces may be used to increase the structure(shelf)-air heat transfer coefficient. Thereby, the heat transfer (and the freezing rate) in a cabinet or blast freezer can be enhanced by compression of a bag between two plates with extended heat transfer surfaces (fins) on the air side. The horizontal bag position is less advantageous than the vertical if there is a gas pocket in the bag. Also, the vertical fin position facilitates natural convection of air between the fins. Figure 3.19 shows the configuration of a bag compressed between two finned plates in a cabinet freezer.

The heat transfer by natural convection is limited by fin configuration (i.e., there is an optimum fin spacing, fin thickness, fin length, and plate length to provide maximum heat transfer). Extending the

Figure 3.19. Freezing in a cabinet freezer using a bag with product compressed between two plates with extended surfaces (fins) in a position facilitating natural air convection. The air convection currents between fins are shown as dashed arrows. The gas pocket is at the bag's top; it does not deter heat transfer by conduction between plate and bag.

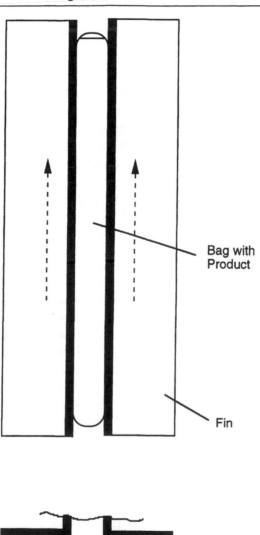

Bag with Product

Fin

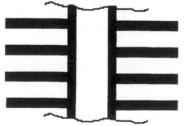

length of the plate may reduce benefits due to the temperature distribution and the resistance to air recirculation. Heat transfer coefficients on the fin side are limited to 3–15 $\dfrac{W}{m^2 \cdot K}$ and in practice may be within a single-digit range. They correspond to much higher effective heat transfer coefficients calculated for an area of the flat plate. This determines the heat transfer coefficients on the flat plate surface (on the bag side) and, subsequently, the freezing rates. The correlations found in the literature (for example, see Burmeister 1993) should be verified for fin and plate geometry (plate height, fin spacing, fin thickness, and fin height) since they may be given for certain dimensional characteristics only. The fin geometry and height of the plate can be optimized for maximizing natural convection heat transfer coefficients (Bar-Cohen 1979).

The use of forced convection heat transfer (air circulation by fan) may significantly improve heat transfer between air and the extended surface (finned plate). In the simplest case, a recirculating fan can be introduced into a cabinet freezer. However, a typical cabinet freezer may have limited heat removal capacity and be incapable of efficiently handling multiple assemblies of bags with finned plates. In practice, a blast freezer will provide much better conditions for freezing such assemblies since the recirculating fans are a part of the design and the refrigeration capacity is sized for large loads and relatively high freezing rates. Heat transfer coefficients can be increased by more than 2–4 times when compared to freezing in a blast freezer using flat heat transfer surfaces. The author's experience with freezing of aqueous products in bags using the rod shelves and the shelves composed of two finned plates in freezers with a relatively slow air recirculation showed that the bags placed on the rod shelves required approximately 3.5–5-fold longer times than the bags compressed between the finned plates. An increase in recirculating air velocity can further reduce the freezing time in the finned plate assemblies.

For GMP purposes the desirable plate material should be stainless steel or other corrosion-resistant nontoxic alloy, but no reasonably priced stainless steel plates with fins are readily available and one may need to settle with aluminum or its alloys or copper, which can be extruded. The author has used the extruded aluminum shapes (raw and with protective coating). The cGMP production application may require a coating on the aluminum/copper shapes to prevent corrosion caused by cleaning agents. The coating could be by epoxy painting, Teflon spray coating, dip coating, etc. The coating should be thin to minimize heat conduction resistance on the bag side.

Finned plate heat transfer depends on the plate/fin material, fin thickness and height, and the distance between fins. One of the parameters to be considered in finned plate selection is the extended heat transfer surface area per unit of the plate flat surface (given in $\frac{cm^2}{cm^2}$ or $\frac{inch^2}{inch^2}$). Available finned plates could have this value in excess of 12. The height of extruded fins may exceed 75 mm (3 in). The extruded plates may have rectangular or trapezoidal fin profile. Most of the analysis regarding longitudinal fins has been done using rectangular fins (Kern and Kraus 1972). The optimization of trapezoidal fins was more recently analyzed by Razelos and Satyaprakash (1993). The extruded fin shapes may have dimensional limitations in fin height and plate width, caused by material properties and available extruding tools. A bonded fin assembly could be used if very high, closely spaced fins are required. The height of the fins may exceed 100 mm (4 in) with fin spacing as close as 6 mm (0.238 in). The fins are bonded to the base plate with bonding techniques providing high fin-plate thermal conductivity. The assemblies can be used for both natural and forced convection conditions of heat transfer. For natural convection, the effective heat transfer coefficients for tall fins (100 mm) can be about 1.6–1.7 times higher than for the short fins (50 mm). The heat transfer for forced convection may be remarkably higher if the air is forced between fins in a controlled flow pattern. The bonded fin-plate assembly with fins 100 mm high at air velocities exceeding 2.3 m/sec may provide effective heat transfer coefficients approximately 4–6 times higher than for natural convection. The bonded fin assemblies can also be made of stainless steel.

The bag wall thermal resistance can be significant, depending on material, wall thickness, and wall structure (smooth or microcorrugated). It can be comparable to the thermal resistance of the frozen product (if a smooth surface made of Teflon) or can be high in the case of thick-walled bags made of low thermal conductivity polymers with surfaces covered with microcorrugations. The contact between the plate and the bag wall also is important. Both surfaces should be smooth, and a certain compression force needs to be applied between the plates. Microcorrugations on the bag surface can reduce the active plate-bag contact surface area and introduce miniature air pockets in the contact zone. The heat transfer flux across such a surface will be significantly reduced (Fletcher 1988), and as a result, the freezing rate will also be reduced. Any air pockets inside the bag may significantly impair the heat transfer plate-bag wall-product. To minimize this resistance, the surfaces of both the plate and the bag should be smooth. Compression pressure applied to the plates may improve the plate-bag contact and decrease the contact resistance. Caution is recommended when using double bags since there might be a gas pocket between the bag walls (this can be critical when the bags are not manufactured as double-wall type but one bag is put into another prior to freezer loading).

The airflow pattern should be along the fins to ensure high thermal efficiency of the fin. There is a temperature gradient along the fins for long plates, which may cause some temperature gradient along the bag/container surface in contact with the finned plate. Assemblies of finned plates in series should be avoided, and if possible, the plates should be placed in parallel configuration. Such configuration may impose requirements on the blast freezer design or space utilization of the existing freezer. The parallel configuration requires uniform air velocity distribution over large face areas prior to the plate assembly. The configuration of fans and air distribution baffles can be considered a part of a new freezer design, but may pose problems in existing freezers—ducts and baffles/vanes airflow pattern modifications may reduce useful volume of the freezer. Extensive validation work may be required to assure uniformity of temperature and air velocity distributions across freezer volume. Repeatability of those distributions must be ensured in production operations.

A test using a thermoanemometer may be conducted to determine the air velocity field. The air velocity field in the empty freezer would not guarantee a uniform heat transfer condition in the loaded freezer unless certain bags/finned plates configuration is ensured. The fans of an axial design may not overcome higher airflow resistances (this should be considered when testing the freezer for a full load of bags, i.e., how densely the freezer is packed), and a bypass airflow may develop around the bags. A rack design for placing the bag/plate assembly may play a role in air distribution as well. Temperature distribution tests (in air passages, in bags, on plates) during freezing may be performed using arrays of small size temperature sensors. For rapid freezing, the bags should be thin, well compressed between the plates, and with no air pockets between the wall and liquid (vertical bag position is recommended).

The heat transfer coefficients for product freezing at the bag wall/flat plate surface may reach levels above $60 \div 70 \dfrac{W}{m^2 \cdot K}$ and depend on fins configuration, airflow pattern around the fins, and air velocity. Experimental data obtained were in good agreement with the heat transfer calculations. Rapid freezing can be accomplished using this approach for bags with a thickness no greater than 20–25 mm, low freezer temperatures, and high air velocities. The average velocities of the freezing front movement were within the range of 30–40 mm/hr. The freezing front velocities exceeding 40 mm/hr have been recorded at temperatures in the blast freezer below –60° C. At air velocities higher than 2 m/sec and temperature below –60° C, the 12.7-mm- (0.5 in)-thick bag with aqueous product may need about 10 minutes to freeze (after liquid product reached proximity of 0° C).

The fins can be in a horizontal or vertical position depending on the airflow pattern within the freezer; the airflow must be along the fins. The bag should be placed in a vertical position to eliminate the detrimental effect of air bubbles on heat transfer. Figure 3.20 shows the assembly of a bag with product compressed between two plates with extended heat transfer surfaces (fins). The fins are in a configuration for horizontal airflow.

An example:
The freezing of an aqueous solution in plastic bags compressed between aluminum plates with extended heat transfer surfaces in the freezer with the air temperature –12° C and low air velocities showed that the heat transfer coefficient on the bag side (flat side of the finned plate) was about $60 \dfrac{W}{m^2 \cdot K}$. The height of fins was 1.25 in (31.75 mm), and the distance between fins was 5/16 in (7.93 mm). The heat transfer surface factor, i.e., the extended heat transfer surface to flat plate heat transfer surface was close to 7.4. The thickness of the filled bags was about 18 mm. The freezing was in a horizontal position with precautions taken to remove air from the bag after filling. However, some air bubbles were still present in the top bag surface. Thorough freezing occurred after 1 hr 5 min, including initial liquid cooling from 18° C. The plate and bag surfaces were both smooth. The air bubbles at the upper bag surface caused an asymmetry in heat transfer from the lower and upper plates into the freezing material due to the insulating character of air. Placement of the bag-plates assembly, with the bag in vertical position, reduced

Figure 3.20. Freezing in a blast freezer using assemblies of bags with product compressed between plates with extended heat transfer surfaces (fins). The fins are configured for a horizontal airflow pattern. The bag is in a vertical position with an air pocket at the top. The air pocket does not deter the heat transfer by conduction between the plate and bag.

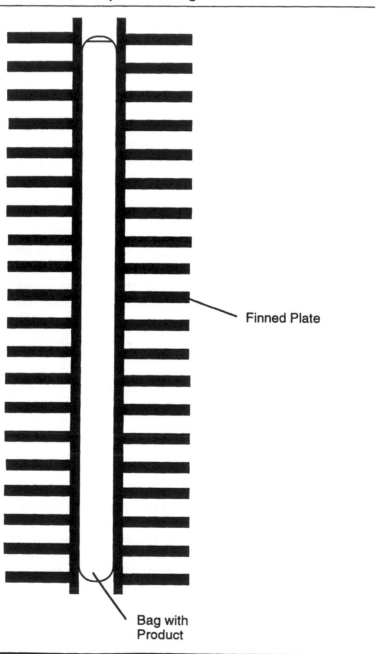

Finned Plate

Bag with Product

the freezing time to below 1 hr since the air bubbles were no longer present in the bag sides in contact with the cold plates. In the latter case, the average velocity of the freezing front was close to 20 mm/hr. Air bubbles may remain attached to the bag wall even after moving the bag to a vertical position, but, in this study, their effect was not significant on the overall freezing rate.

One may also use plates with multiple pin fins. Such a plate is insensitive to the direction of airflow along the plate since the pins are exposed individually to the airflow. The pressure drop for airflow along such a surface can be much higher than for the longitudinal fins, and low velocity or stagnant zones may develop between the fins. Such plates are, therefore, used for small cooled areas, frequently with the impinging airflow (Ledezma et al. 1996). The plates with pin fins may be difficult to clean on the extended heat transfer side.

Freezing between two parallel cooled plates with a small inter-plate distance (as modeled by slab freezing) can provide conditions for well-defined columnar dendritic or cellular crystal growth. The heat flow can be strictly symmetrical and directional. The crystal dendrites may grow from both cooled walls and meet in the slab center. The dendrite growth rate depends on undercooling (e.g., greater undercooling produces faster solid-liquid interface velocity). The temperature of the dendrite decreases from its tip along the column toward the wall. The heat flux depends on the dendrite length, its thickness, and the temperature difference between the dendrite tip and its base, and is based on the thermal conductivity of ice. The solid buildup at the dendritic tip and at the dendritic column sides releases latent heat of fusion, which is being removed by conduction along the dendrite and solidified interdendritic mass of solutes toward the cold wall. The undercooling at the growing dendritic tip typically consists of three major components: the undercooling of the dendritic tip curvature (radius), solutal undercooling (depending on solute distribution between the solid and liquid phases and on solute concentration distribution in the liquid phase next to dendritic tip), and thermal undercooling. Under conditions of freezing of aqueous solutions between the cooled plates, an overall undercooling comprises mostly the solutal and curvature undercooling. The solutal undercooling depends on the solute concentration gradient at the dendritic tip in the liquid phase and on solute distribution between the solid and liquid phases. The factor of the solute distribution between the solid and liquid phases is not important since very little solute may be incorporated into ice. Depending on the freezing front velocity, the solutal undercooling may change (at low velocities, the concentration gradient of solute in liquid at the dendritic tip may flatten), and the thermal undercooling at the dendritic tip is not significant. In practice, the major undercooling is the solutal undercooling with some of the curvature undercooling. In reality, the overall undercooling is small, and it limits the practically possible-to-achieve front velocities (typically to less than 40–45 mm/hr). More information on the dendritic crystal growth can be found in Chapter 2, page 55, by the author. Higher velocities of solid-liquid interface and resulting shorter freezing times may be obtained by lowering the temperature in the freezer. Very short freezing times can be achieved when the slab thickness is small (about 10 mm or less).

Bag handling is the critical part of the operation. Filling the bags and loading them into freezers requires close attention (possibility of bag damage by mishandling, need for accurate bag labeling since each batch is divided into multiple bags, etc.). The author has designed mechanical systems for rapidly loading/unloading the product-filled bags into the plate assembly and their simultaneous compression to ensure not only sufficient bag-shelf contact but also uniform distances between the plates. A patent on the freezing/solidification process using the enhanced heat transfer surfaces in gaseous environments is pending.

A simplified version of a shelf design for a blast freezer with a horizontal airflow is shown in Figure 3.21. Such a shelf may perform in a fashion similar to the liquid-cooled shelf of the contact freezer. The heat transfer is by conduction from the product in a container (from left to right: vial, tray, and bag) to the flat shelf, which is cooled by cold air flowing along the fins. The difference between the blast freezer shelf and the contact freezer shelf is that in the former the product containers are additionally cooled by the flowing air. The flow of cold air above the shelf may be blocked by a system of baffles to provide more controlled freezing conditions similar to the contact plate freezer. As in the contact plate freezer, the freezing process is asymmetrical. Figure 3.21 shows schematically the blast freezer shelf made of finned plate.

As in the case of the finned plate assemblies described earlier, this shelf design may provide nonuniform freezing conditions from shelf to shelf if the shelves are located in series. Parallel shelf configuration assures better uniformity but imposes more difficult conditions for airflow distribution inside the freezer. The parallel shelf

Figure 3.21. The blast freezer shelf with the heat transfer during freezing being predominantly conduction from product containers to the flat shelf and the shelf being mainly cooled by forced air flow convection on the finned surface. The examples of the product containers located on the shelf are, from left to right: vial, tray, and bag. The heat transfer is asymmetrical (mostly by conduction between shelf and product container and some by convection between air and product container and product surface).

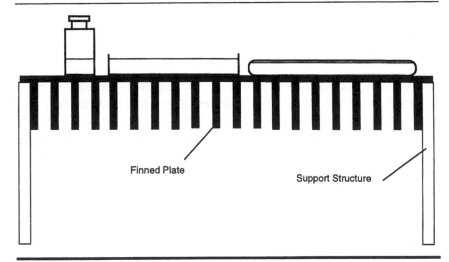

configuration requires wide and short freezer design. To maintain sufficiently high air velocities in a wide design, high air recirculation rates are needed. Air distribution within the wide and short passages is more difficult than in the narrow and long. Although the shelves made of finned plates can be used as a modification of an existing blast freezer design, an optimal solution is to design the freezer and shelves together as a complete system.

It has been the author's experience that with good contact between the plate and container/product, the extended heat transfer surfaces can significantly increase the overall gas-product heat transfer coefficient and shorten the solidification time. Very low temperatures can be obtained if the air is cooled by a cascade refrigeration system, or the recirculating gas is cooled by boiling/evaporation of liquid nitrogen. To reduce heat gains, the fan motors can be located outside and long shafts with low-temperature bearings used. Handling of the product frozen to very low temperatures may introduce logistics problems (rapid frozen product removal and transfer to the storage facility is required since there might be a substantial heat gain from the environment after the finned plates are exposed to the air at ambient temperature). The control of freezing rates can be performed well by programmed changes in air temperature. Due to small bag thickness and large heat transfer surfaces, the lag between the change in air temperature and the freezing product response is relatively short. The response lag also depends on the mass of finned plates, which can be substantial.

THAWING

Introduction

The thawing process is the reverse of freezing—that is, energy must be delivered to balance the requirements of warming the frozen material to its melting temperature (in general, for aqueous solutions to a vicinity of 0° C, although the eutectics and glassy states may soften and melt at temperatures far below 0° C), to provide latent heat of phase change from solid to liquid, and to heat the liquid to a final temperature. The glassy states of aqueous solutions may contain crystal nuclei that do not produce growing crystals due to the viscosity of the glassy state below the glass transition temperature T_g (see Table 2.6 in Chapter 2 page 107). During warming, these nuclei may produce growing crystals at temperatures near T_g. This phenomenon is associated with devitrification (MacFarlane et al. 1992). At a temperature near 220 K (–53° C), there is a release of charges trapped within the ice crystals (Apekis et al. 1983).

The energy needed to thaw is typically delivered through the wall of the container where the thawed mass resides or through the walls of a heat exchanger if the already thawed liquid is recirculated.

The general melting process within a container with warmed walls can be divided into three heat transfer phases:

- An initial phase when the melted liquid layer is very thin and the heat transfer occurs by conduction
- A transition phase when the natural convection develops in the liquid phase
- A quasi-steady phase when the natural convection in the liquid is developed and the heat transfer rate across the liquid is approximately steady

During the last phase, the equivalent Nusselt number (*Nu*) for heat transfer will depend on the size and shape of the melted liquid zone, the orientation of heat transfer surfaces in the enclosure, and the Rayleigh (*Ra*) and Prandtl (*Pr*) numbers. The Prandtl number for water near 0° C is high (around 13.69). Since this is much above 1, a simple correlation for estimation of the Nusselt number (*Nu*) may be used (Lim and Bejan 1992):

$$Nu = 0.35 \cdot Ra^{0.25}$$

where

$$Nu = \frac{q'}{(k \cdot \Delta T)}$$

and the Rayleigh number:

$$Ra = \frac{(g \cdot \beta \cdot \Delta T \cdot H^3)}{(\alpha \cdot v)}$$

and *g* is the gravitational acceleration, *k* is the liquid thermal conductivity, *q* is the overall heat transfer rate $\left(\frac{W}{m}\right)$, ΔT is the temperature difference between heated wall and melting point of solid, ß is the volumetric thermal expansion coefficient of liquid phase, *H* is height, α is the liquid thermal diffusivity, and v is liquid kinematic viscosity.

Since the temperature difference ΔT is limited by product stability (often in the case of protein-containing pharmaceuticals it might not exceed about 25° C), the thawing can be intensified by an increase in heating surface areas and configuring them with a geometry that ensures large values of *H* (height of heat transfer surfaces). Use of extended heat transfer surfaces (such as fins) may further enhance thawing heat transfer.

The temperature profile in the time during thawing may be as important as during freezing, although the reason is different: The temperature range of ice recrystallization should be passed rapidly to prevent formation of large ice crystals. Highly organized dendritic structures may involve complicated phenomena during the warming and thawing process. Thermal conduction through the dendritic structures may be different from conduction in the pure ice due to the structural heterogeneity and differences in thermal conductivities between the dendrites and the substances included in the interdendritic spaces. The eutectics and glassy states included between the dendrites may soften and melt before the ice dendrites reach their melting temperature. Since the cells are present in the interdendritic spaces, they may be exposed to liquefied, concentrated substances before the ice melts and the solution reaches its original concentration.

Recirculation of liquid phase may enhance the melting rate due not only to forced convection effects but also to local ablation phenomena that occur when the jet of liquid is directed against the frozen material surface (Wisniewski and Wu 1996). The crystalline dendritic structure, if formed during freezing, may facilitate thawing using the convection/ablation on the liquid phase side. This may be particularly pronounced if the eutectic and glassy substances among dendrites soften and melt earlier than the dendrites, which are made of ice. Such interdendritic melting weakens the structure of solidified mass and makes it more susceptible to convectional and ablative effects caused by the stream of moving liquid.

Attention must be paid to the possibility of ice recrystallization (smaller ice crystals dissolving and larger ones growing) that may occur during the warming of the frozen mass. The recrystallazation occurs in both intracellular and extracellular ice. For cells with a small amount of fine intracellular ice crystals, rapid thawing may lead to cell recovery without damage. The slow warming may cause irreparable damage due to intracellular ice crystal growth. Design of the thawing process must consider biochemical and functional changes in cells during recovery from deep sub-zero temperatures (de Locker and Pennickx 1987).

Thawing of large masses by heat delivery close to the volumetric mode was sought to provide a complementary volumetric thawing process in addition to the previously described volumetric freezing process. Electrical field or microwave energy was considered the primary option, and heat delivery from enhanced heat transfer surfaces to packed/porous bed or agitate granule bulk was the alternative.

In general, thawing may be as difficult to control in large volumes as freezing. Thermal thawing may be conducted using convection heat transfer from air or liquid, steam at a pressure below atmospheric, heat transfer by conduction (heated plates or blocks), heat pipes, or infrared radiation. Electrical thawing may employ resistance or dielectric or microwave heating. The thawing process can be divided into the following steps:

- Heating the solid to a thawing temperature plateau
- Thawing (melting of ice)
- Heating the liquid to a desire temperature

Most of heating energy is required to melt the ice included in the sample. In thermal thawing, the small samples follow this pattern closely, but larger frozen volumes may experience transitional states; the surface temperature may be at a thawing point, while the temperatures inside a frozen mass may be changing from the original low temperature through rising temperature isotherms up to melting temperature at the surface.

The thawing in still air may be very slow, and forced air recirculation and heating are recommended. Thawing in circulating water requires water velocity at the warmed surfaces (product container walls) to be above 1–2 cm/sec to increase thawing rates. This is usually accomplished by agitating the water bath and maintaining proper configuration of thawed samples to ensure sufficient and uniform water movement around all samples. Thawing of small containers in shaker bathlike thawing devices may accomplish a double purpose: moving the heat transfer agent around the containers and mutual moving of the product solid and liquid fractions within each container. Both phenomena accelerate the overall thawing rate. The 1-mL frozen samples in vials can be warmed at an approximate rate of 485° C/min when immersed in the agitated water bath kept at 37° C (Kruuv and Glofcheski 1992). The vacuum heat thawing systems use low-pressure steam condensing on the product packages. Steam is generated under vacuum at temperatures of 5° to 30° C. The process is performed batchwise in a vacuum-rated chamber. High heat fluxes can be obtained due to release of latent heat of condensation.

These thermal methods depend on surface thawing, and the thawing process is limited by the ability to deliver heat to the surface without damaging the product. Thawing is also limited by internal heat conduction inside the frozen material. The thawing rate depends on the maximum temperature the cells can be exposed to while in a liquid phase, which the temperature of heat transfer surfaces must not exceed, since the wall boundary layer in the liquid may reach a temperature close to the wall temperature. Frequently, this temperature level is established at 36° to 38° C. Such a temperature limit determines the heat transfer flux possible to deliver from the heating surfaces to the thawing material. This allowable temperature difference can serve as a basis for sizing the heat transfer surfaces. The direct contact thawing process of the frozen blocks of material (containerless handling) between the heated plates may be enhanced by introducing grooves in the plate surface to drain away the melted liquid (Saito et al. 1992). Such a design should also consider the feasibility of effective plate cleaning. Cell cryoinjury may occur even during rapid thawing of cryoprotectant-containing solutions due to the thermomechanical stress caused by temperature gradients within the frozen mass (Vorotilin et al. 1991).

Agitation of liquid phase may significantly accelerate thawing due to intensification of heat transfer at the heating surfaces, turbulization of liquid, reduction of temperature gradients in the liquid phase and the presence of ablation effects at the liquid-melting solid interface. A similar effect can be accomplished when a movement of the solid phase, relative to the liquid phase and vessel internal components (heating surfaces) could be induced. The moving solid phase turbulizes liquid in a space between the solid mass and heat transfer surfaces. As a result, an intensification of heat transfer occurs and an ablation may appear at the melting solid surface.

The electrical thawing systems have an ability to deliver heat to the whole frozen volume and, therefore, may provide much more uniform thawing conditions (de Alvis and Fryer 1990). Hexagonal ice Ih is paraelectric, and the paraelectric effect within ice is almost isotropic (Jackson and Whitworth 1996). Calculated electrostatic interactions predicted that the energetically favored structure of ice is antiferromagnetic. The electromagnetic field oscillations are conveyed to the molecules, and their oscillations are dissipated as heat. The amount of heat generated this way depends strongly on the electrical characteristics of the product (Constant et al. 1996).

Electrical resistance thawing typically uses low frequencies (50–60 Hz). It may work well with regular sample size (such as flat blocks) but may locally overheat/underheat samples of irregular shape. The effective voltage of alternating current is $0.707\ E_{max}$, and the effective current is $0.707\ I_{max}$. The electrical resistance of ice changes with temperature and is 3.57×10^7, 1×10^9, and 3.85×10^9 ohm/cm at 0°, –10°, and –19° C, respectively. At low voltages, the power absorption by ice alone will not be sufficient to generate heat needed for melting. The presence of solutes in the frozen mass may, however, lower the resistance, and more heat can be delivered. The resistance thawing may depend upon the material and shape of the container and the concentration uniformity of the frozen product. Frozen aqueous solutions have high electrical resistance, and an overall resistance may only be altered by the choice of the material of the container. Occurrence of liquid may change the thawing conditions since natural convection would take place. The frozen granules/capsules may be thawed in a continuous flow by passing between the electrodes (El Moctar et al. 1993, 1996; Kaji et al. 1982). In certain cases, the temperature of solid particles may reach a temperature higher than that of the liquid. Resistance heating is sometimes combined with other thermal heating, for example, by immersing the samples in a heated aqueous solution.

The thawing can be performed by dielectric or microwave methods. Dielectric heating uses frequencies near 0.01–0.04 GHz, e.g., radio wavelengths (radio frequency heating), while the wavelengths used in microwave processing are about 33 cm for 915 MHz and about 12.2 cm for 2.45 GHz. Dielectric heating is used for electrically insulating products containing dipolar molecules (nonpolar molecules cannot be heated his way). The frequencies used for dielectric heating in most countries are: 13.56, 27.12, and 40.68 MHz. In dielectric thawing, the frozen material is placed between parallel electrodes. Frozen product concentration and shape uniformity (parallel sides must be present) are important for obtaining temperature uniformity within the thawed sample. Sample heterogeneity might cause local temperature differences. The microwave thawing quality may depend on the product concentration (water content) and frozen sample shape uniformity. A runaway heating may occur if there are temperature, concentration, and shape nonuniformities within the frozen sample. Cooled air may be recirculated to improve thermal conditions and prevent overheating near the edges and corners of frozen blocks. Lower frequency processing (for example, 915 MHz) may reduce these surface effects (Mudgett 1986). The internal patterns of temperature distribution may depend on sample shape, size, and ionic content. For example, a cylindrical sample may have the highest temperature in the center at low ionic content, but at a high ionic content the temperature near the surface may be much higher than in the center. Such center effects do not occur in slabs since there is no focusing effect present. 2.45-GHz radiation may thaw completely layers of biolog-ical products up to 35–40 mm. Penetration of microwaves decreases with a frequency increase. For example, the half-power depth for water is approximately 2.3 cm at 2.45 GHz and about 20.0 cm at 915 MHz. Processing of large blocks of frozen biological material may produce much better results at 915 MHz than at 2.45 GHz.

The microwave properties of biological material may be char-acterized by the dielectric constant (the material's ability to store electrical energy) and the loss factor (the material's ability to dissi-pate electrical energy). The ratio of dielectric loss to dielectric con-stant represents the material's ability to be penetrated by an elec-trical field and to dissipate electrical energy as heat. The properties may vary with microwave frequency and material temperature. The dielectric constant for water at 20° C and 25° C is, respectively, 80.37 and 78.54, while for ice it is only 3.2 at the frequency of 2.45 GHz. The dielectric loss factor for water at 25° C is about 12.48, while for ice it is 0.0029 (Schiffman 1986). Water is highly absorp-tive to microwaves and easy to heat, whereas ice can be consid-ered highly transparent to microwaves and almost does not heat. However, the frozen material's loss factors increase rapidly between –10° C and approximately 0° to 1° C.

During thawing, a local runaway heating may occur since absorption of microwave radiation increases with temperature increase, particularly when the level of about –5° C is reached. In some thawing applications, the microwave radiation may be too risky as the only means of energy delivery but may be used in com-bination with other methods to first bring the whole product volume rapidly to about –5° C using microwave heating and after that follow with thermal thawing. Microwave properties of the solution without product (cells, proteins) should be investigated prior to selection of microwave thawing conditions (Macklis and Ketter 1978; Macklis et al. 1979). Heterogeneous product mass may product different local heating conditions. Thawing of large specimens may produce mushy, thawed, and frozen zones, with a runaway heating occurring in the mushy and thawed zones. As a remedy, the power might be reduced to permit thermal equilibration. Coleman (1990) proposed a numerical model of microwave heating of frozen substances with the consideration of superheating and mushy zones. Zeng and Faghri (1994) compared experimental and numerical calculation results for microwave thawing of frozen foodstuffs. The two-dimensional mathematical model considered the presence of three states: frozen, mushy, and thawed. As a model material, a 23 percent solid methylcellulose gel was used. Clemens and Saltiel (1996) described a numerical modeling of materials processing in microwave furnaces. The approach was based on solving Maxwell equations. The presence of coupled nonlinear process is significant in materials that have temperature-dependent electromagnetic properties (for example, when heating may produce an exponential rise in temperature after a critical temperature is reached). Electric field intensity changes within the chamber, depending on the frequency of microwaves and the properties of treated material. The double helix of DNA in biological samples may be affected by microwaves due to its potential to vibrate at about 2 GHz (Zhang and Chou 1996).

Choi and Hong (1991) investigated melting process enhancement by ultrasonic waves. The enhancement of thermal thawing was about 1.6–1.8 times at a frequency of 50 kHz and declined with a decrease in ultrasound frequency.

Heating by shortwave infrared radiation (IR) also may be considered in certain cases. The penetration of such radiation into ice is approximately 30 mm (Charm 1981), but in the presence of water, the absorption in water is mostly at the longer waves (above 1.4 µm). The black body ceramic emitters are used in modern IR devices. They provide uniform radiation and have long operations life (about 5 years).

Thawing of Large Volumes

Thawing large volumes by using heating surfaces involves heating the frozen mass from its original temperature to the ice melting temperature near 0° C. While the surface temperature remains near 0° C as melting progresses, the frozen mass temperature changes dynamically under conditions of transient heat conduction. The temperature histories of each point may vary, depending on the shape of heating surfaces and the shape of frozen mass. In the large frozen volume, its parts may warm up slowly and pass through a temperature range of ice recrystallization during extended time periods. An increase in temperature may cause eutectics to melt and glassy states to liquify in the intercrystalline spaces at temperatures below the ice melting. Since the cells are located in the intercrystalline spaces, these phenomena may have a detrimental effect upon them (i.e., the growing ice crystals might mechanically crush the cells). The cells also might be exposed to concentrated solutes.

As a result, the thawing process is as critical as the freezing process in large-size sample processing. Thawing protocols should be established, considering the sample shape and thermal history of particular points located within the thawed volume.

Thawing of Product in Large Containers

Thawing of products in large containers introduces difficulties similar to those for freezing. Thawing procedures frequently call for a rapid thawing step. Heating the frozen product through container walls can only deliver limited heat flux due to the wall temperature limitation, and the thawing time may become unacceptably long.

Containers with vertical heat transfer surfaces may enhance the natural convection in the liquid phase with resulting intensification of thawing. The melting process is affected by solid-liquid density difference (the frozen mass is floating after a sufficient volume of liquid phase forms) and the currents caused by liquid convection (Yoo and Ro 1991). The convection in the liquid phase can be increased by recirculation of formed liquid phase via an external loop—turbulization of liquid, formation of liquid jet(s) interacting with the frozen mass (Bhansali and Black 1996), or by inducing a relative movement of solid and liquid phases. If possible, placing larger containers in a shaking bath may accomplish stirring of the liquid outside the containers (intensification of heat transfer) and agitating the internal liquid phase located between the container wall and frozen mass via their mutual movement within the container. The shaking should accomplish a multiple effect: turbulize liquid in the bath outside the container, turbulize liquid inside the container, cause relative motion between the solid and liquid phases inside the container. The two first phenomena increase heat transfer across the container wall, whereas the third one may lead to ablation of thawed product. Larger containers may have a heating jacket instead of being placed in a bath—then only the two last phenomena are involved. The movement of container causes internal turbulilization of liquefied product due to the mutual movement of solid and liquid phases. Such an approach may significantly accelerate the thawing rates. The heat transfer can be increased by the use of coils/ducts with extended finned heat transfer surfaces located inside the container. Details of the thawing system with recirculation of thawed liquid and the container design with finned heat transfer surfaces can be found in works by Wisniewski and Wu (1992, 1996).

Thawing of Products Frozen in Bags

The freezing of biological materials is frequently accomplished using flat plastic bags placed on trays in a blast freezer. The thawing step may be the reverse of freezing; recirculated air is heated to a certain temperature, and the latent heat of phase change during thawing is delivered by forced convection to flat bag surfaces from moving air. As in the case of freezing, heat transfer coefficients are low, and the process is slow. Flat bags with large top and bottom surface areas and small thicknesses are advantageous for such thawing.

Thermal conduction will play an additional role in the thawing; bag contents/supporting tray contact may also play roles in heat transfer. However, the tray may be separated from the frozen material not only by a bag wall but also by a layer of already melted liquid. In such a case, the heat transfer mechanism is different from that of freezing (i.e., conduction and natural convection in the liquid phase participate in an overall heat transfer). Conditions for Benard convection (Koschmider 1993; Lage et al. 1991) may form between the bottom heated flat wall and the cold bottom of the floating frozen mass.

A similar situation occurs when the trays are heated internally by a recirculating heat transfer fluid (i.e., a layer of liquid forms between the tray and the frozen material). If the frozen bags are stored in cabinet freezers and removed for the thawing step, a special thawing apparatus can be constructed to accelerate and control thawing by delivering heat by conduction to the top and bottom of the bag. Multiple collapsible trays/plates with internally recirculating heat transfer fluid at controlled temperature can be used to press the bags and deliver heat from two sides. Due to the solid/liquid density difference in aqueous solutions, one may anticipate more efficient thawing occurring at the upper bag surface due to solid proximity to the heat transfer surface, although the natural convection in the liquid below the frozen mass may also enhance melting of the lower solid surface. For large processed volumes, the size of apparatus and handling of many bags (loading/unloading) can pose operational problems. A continuous, contact-thawing apparatus may be devised using a single or double conveyor belt.

The addition of forced convection in the liquid phase may significantly accelerate the thawing process (Wisniewski and Wu 1992, 1996; Wisniewski, 1998 a, b). In the case of bags, the convection in melted liquid may be induced by pressing the flexible bag walls in a predetermined pattern or by a shaking/orbital motion of the whole bag-containing device. As in the case of vials or bottles, the bags may be immersed in a water bath with agitation, and a shaking or orbital motion may be imposed upon the bags to enhance the heat transfer inside and outside of the bag.

Thawing of Granules in Large-Scale Operations

Frozen granules can be fed in a controlled way into a warming/dissolving buffer bath to accomplish a rapid thawing processing step. A combination of the granules' feed rate and the heating rate of the buffer may permit rapid thawing in a controlled way. Such rapid thawing at controlled temperatures will prevent recrystallization of ice and resulting undesirable cell damage. During the thawing process, intensive agitation should be provided to maintain granules and cells in suspension and to provide system homogeneity and a sufficient intensity of heat transfer. Intensive agitation also removes the thawed solution from the vicinity of the granule surface, thus providing ablation-like conditions, which intensify the melting process. If cells enter the freezing machine as a concentrated suspension (after a centrifugation or a microfiltration step), any subsequent dilution during thawing may restore the original cell concentration or may adjust the concentration to a level suitable for further processing of cells. A continuously operating thawing system, with two metered streams (the liquid buffer and the solid granules) entering an agitated flow-through, heating thawing reactor could also be designed. Extended heat transfer surfaces in the form of coils, heated agitator, or both may be added to the jacket heating. The continuous thawing may, however, be limited to applications, where very large product quantities are processed.

The bed of granules may also be thawed using recirculation of already thawed liquid through the bed. A convective melting within the bed would occur (Plumb 1994). The bed thickness cannot be too high if the initial granule temperatures are very low; for example, at the beginning of the process, the warm liquid may cool down to the freezing temperature while passing through the initial zone of the bed and freeze over the granules in the other part of the bed with a potential for clogging the voids. The solid phase should be close to the melting temperature in such a system to provide continuous melting.

Thawing as a Part of the Freezing/Granulation Process

Referring to Case Study 3.1, granules can be thawed in an agitated contact heat transfer vessel similar in design to the freezing/ granulation apparatus described earlier. The heat is delivered by conduction through the jacket and, in larger devices, also through the agitator surfaces (heat transfer fluid at controlled temperature flows through the jacket and through the hollow shaft and agitating paddles/ribbons). Such a method of thawing does not introduce any additional buffer to the material; thus, there is no change in composition between the frozen and thawed products. During the thawing step, the heat transfer mechanism changes from the wall-granular agitated material through the wall-two-phase mixture (solid/liquid) to the wall-liquid in a final stage. The wall-agitated granular material heat transfer can be analyzed using available literature data (Muchowski 1988; Schlunder 1982). This period will be short since the appearance of liquid changes the heat transfer mechanism—that is, the main heat transfer will be from wall to liquid and then from liquid to granules. Agitation of such a slurry intensifies heat transfer due to mutual movement of granules and liquid phase. The heat transfer wall-slurry could be approximated by data on heat transfer of agitate non-Newtonian two-phase mixtures (Uhl 1966), but there is an additional factor of melting at constant temperature. The slurry temperature may rise when the volume occupied by frozen granules diminishes and most of the mass is in a liquid form. Heat transfer during the end period when liquid is heated will follow the heat transfer in agitated liquids (Newtonian or non-Newtonian depending on a product) and would depend on agitation parameters (agitator design, rotational speed) and on the gap between the agitator blade tip and the heating wall (Mueller 1988a; Uhl 1966). Wall temperature should be maintained at a level acceptable for the processed biological product (e.g., for processing bacterial suspensions, the wall temperature may not exceed 37° to 40° C. At slow agitation, an overall heat transfer coefficient may be within the range of 20–80 $\frac{W}{m^2 \cdot {}^\circ C}$. Intensive agitation of the liquid/solid mixture induces turbulent motion and increases heat transfer not only between the mixture and heated walls but also between the liquid and thawing granules. Additional ablation effects may take place on granule surfaces due to movement of liquid. At rapid agitation, the overall heat transfer coefficient may be within the range of 200–500 $\frac{W}{m^2 \cdot {}^\circ C}$. As a result, the thawing process can be accomplished rapidly. The mixing elements of the agitator (paddles, ribbons, bars, etc.) should cross the slurry surface during each rotation since, after a certain amount of liquid forms, the remaining granules will float at the surface due to the density difference. The rotor moving parts should force the granules into the bulk liquid.

Thawing runs were performed in the freezing apparatus using a jacket heated with water at 37° C, and results were positive. Rapid agitation provides good conditions for heat transfer due to intensive volumetric mixing and proximity to the wall of the moving agitator blades. Figure 3.22 shows an example of the thawing run using the agitated processor for thawing granules. The heat for melting was provided mostly from the heated jacket and, to some degree, by mechanical agitation. During the first period (I), granules are heated from the processor walls. The warming rate (a/b) depends on the wall temperature, intensity of agitation, and the heat transfer surface in contact with the agitate bed. The warming rate is critical since ice recrystallization occurs within the certain temperature range and this range has to be crossed rapidly to minimize the effect. After the temperature reaches the vicinity of 0° C, granules begin to melt and the melting process continues at the temperature plateau (near 0° C) with a continuous energy delivery into the agitated slurry (II). After the granules melt and the liquid phase forms, the temperature increases (period III of liquid heating).

As in freezing, thawing takes place in the mass of material at the temperature plateau. The difference is a heat transfer through the walls (i.e., there are temperature gradients within the agitated mass). These gradients depend on the intensity of agitation and the proximity of agitator blades to the heated walls. An increase in the area of heat transfer surfaces may approach conditions of volumetric heating and thawing. An increase in the heat transfer surface beyond the jacketed walls further intensifies and shortens the thawing step. Contact processor designs with heated rotors (using the heating surfaces in the form of paddles, tubes, or disks) and jackets can be used for a continuous thawing of granulated frozen materials. They ensure vigorous agitation of material and typically provide very large heating surfaces. As during freezing, most of the melting operation may be performed at the temperature plateau near 0° C

Figure 3.22. Thawing of granules in the agitated processor with heat supply from the jacket. *T* is the temperature, and *t* is the time, I is the granule heating period at the granule warming rate = $a/b = \left(\dfrac{\Delta T}{\Delta t}\right)$. II is the period of granule melting at the temperature plateau near 0° C. III is the period of warming the liquid phase.

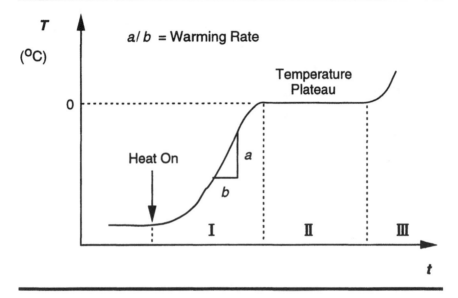

for the whole mass of material. At the beginning, the granule interior remains at its original storage temperature, and it takes some time to warm all of the granules up to approximately 0° C.

Melting of the solid, stationary spheres in the warmer fluid can be analyzed using an approach proposed by McLeod et al. (1996). However, in this case, the frozen granules are porous, the whole mass is agitated, and the fluid composition is the same as the granules. The melting process is much more complicated due to the relative movements among granules and between the granule and fluid and associated ablation and forced convection effects. The small-size granule (with a low Biot number) would ensure reaching such a temperature equilibrium in a short time and, therefore, crossing the range of ice recrystallization rapidly. This thawing method may provide the most uniform thawing environment for the whole mass among all thermal thawing techniques. The volumes of thawed granules may reach several hundred liters or more per batch. Continuous thawing could be accomplished using a trough-like device with heated walls and a single or multirotor agitator/conveyor. Such a device, although technically feasible, may not be recommended for extended periods of operation due to the possibility of local sticking of processed material to the walls. This material would then stay much longer in the machine than the rapidly passing bulk. Cleaning-in-place of agitated thawing devices is feasible due to an intensive, highly turbulent liquid agitation and a possibility of maintaining optimal cleaning temperature. Requirements for the shaft seal design are similar to those for the freezing machine.

Thawing in Containers with Finned Internal Heat Transfer Surfaces (Case Study 2)

The finned heat transfer surfaces enhance heat transfer in the case of thawing since they conduct heat more efficiently than the solidified product and the liquid phase that forms during thawing. Fins effectively conduct heat deeply into the liquid and solid phases due to a difference in thermal conductivity between the aqueous solution and solid product and the fin material (e.g., the stainless steel thermal conductivity is approximately 30 times larger than that of water). An approach to melting using quiescent liquid phase (heat conduction only) may significantly reduce melting rates. Natural convection effects, particularly for the vertical heat transfer surfaces, may increase 4–5 times the melting rates compared to quiescent liquid rates. The presence of fins may significantly increase the melting rates when compared to a smooth surface; for example, for the cylindrical heating surface, the addition of fins could increase the melting rates by a factor of 2–4 depending on the fin geometry and fin material. The application of fins for heat transfer is clearly justified here due to the thermal conductivity difference and relatively low heat transfer coefficients (natural convection) between the liquid and extended heat transfer surfaces. Due to this thermal conductivity difference, the proximity of fin ends to heat transfer surfaces can also extend the fin's effective length as was the case in solidification (see Figure 3.14). After the liquid phase forms, natural convection currents occur and enhance heat transfer at the heat exchanger walls and fins. The solid-liquid interface becomes curvilinear with more intensive thawing occurring near the top (liquid cavity is larger than at the bottom). The phenomenon of curvilinear solid-liquid interface was observed in all configurations and was more pronounced in wider interfin cavities (designs with fewer fins) than in designs with narrower cavities. Due to the proximity of fins to the vessel's heated bottom and rapid formation of the liquid layer there, these vertical convectional currents form very early and accelerate the thawing process. The natural convection effects depend on the height of heat transfer surfaces (jacketed walls, the central bayonet, and fins) and the temperature of these heated surfaces. Therefore, the taller vessels, which are beneficial for freezing in the finned central bayonet configuration, are also beneficial for thawing due to the more pronounced natural convection effects and resulting higher heat transfer coefficients. The temperature of heated surfaces is frequently equal to the temperature at which the product stability studies were handled earlier (in a small scale). The thawing is more intensive (shorter times) when the wall temperature is higher since the natural convection is more intensive. Figure 3.23 shows melting of a solid mass of product with one wall of the container heated to demonstrate formation of curvilinear solid-liquid interface. The convection currents in a liquid phase are shown with directional arrows. The warmest liquid is at the top of the heated wall, and the melting is the most intensive there. The descending liquid cools at the solid-liquid interface, which remains at the melting temperature (0° C for ice crystals). The heat transfer coefficient and resulting heat flux through the liquid phase toward the melting interface depend on the heat transfer surface height (increases with the height increase). Depending on the height of heat transfer surfaces, a transitional or turbulent natural convection flow pattern may be approached.

Frequently the wall temperature is maintained within the 15–25° C range. This ensures sufficient temperature difference between the heated surface and the solid product interface (near 0° C) to cause significant natural convection circulation in the liquid phase. The wall temperature maximum is frequently dictated by product stability and cannot be increased. This imposes a limit on the heat transfer coefficients and on the overall heat flux (limit on a temperature difference as a driving force). The fin configuration causes the convection patterns to be close to the patterns in vertical cavities (with one wall heated and the other cooled) rather than to the patterns near free vertical walls. There is a rising current along the heated surface and a descending current along the solid product-liquid interface. The central zone between these currents may remain relatively stagnant and a distinctive temperature distribution pattern may be observed

Figure 3.23. Thawing of the frozen mass with one wall heated. A curvilinear solid-liquid interface forms due to the presence of natural convection in the liquid phase. Arrows show direction of natural convection currents in the liquid phase. Q_h is the heat flux through the wall, T_h is the temperature of the heating fluid, T_m is the melting temperature at the solid product surface, and H is the height of heat transfer surface.

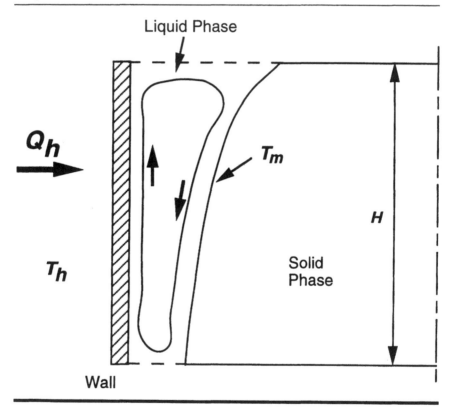

between the heated and product surfaces; that is, steeper changes can be seen at the interfaces, and the temperature profile in the central zone may remain relatively flat. The temperature distribution in the frozen mass changes on the principle of transitional heat transfer with the surface of the frozen mass kept near 0° C. Due to the division of the frozen mass into multiple compartments between the fins, the surface/volume ratio of the frozen product can reach high values soon after the beginning of thawing after the liquid phase forms. Since the dimensions of the frozen pieces are small, product exposure to the temperature range of ice recrystallization may be relatively short, and the product exposure time to molten eutectic may be limited.

Since the temperature changes along the fin length, complex patterns of natural convection currents develop in the liquid phase. The cavity shape is also dynamically changing. In the case of aqueous products, the frozen mass would float between the heat transfer surfaces remaining at the top of the liquid phase-filled cavity. This accelerates melting since the warmest liquid also rises to the top due to the natural convection along the vertical heated surfaces. This makes an analysis of the thawing process difficult, and in this case, a final experimental verification was required for various geometries. It has been observed in practice that the thawing pattern (convection currents and shape of the solid-liquid interface) may vary slightly depending on whether the vessel is placed on a horizontal or tilted floor. A significant heat transfer enhancement and melting acceleration can be achieved by agitation of the liquid phase. The liquid phase can be recirculated through an external heat exchanger using a recirculation pump. The returning liquid can turbulize the liquid phase in the vessel (increase in melting rate at the solid-liquid interface, disturbance of the temperature gradient patterns in the liquid phase, intensification of heat transfer across the liquid phase, and heat transfer increase at the heating surfaces). The returning liquid can also be directed against the frozen product surface with a significant increase in local melting rate due to the combination of forced convection and ablation effects. Since the product is floating in the liquid phase, such streams of returning liquid can cause movement of the whole frozen block, further increasing heat transfer and melting rates at the whole solid-liquid interface. In general, a relative movement between the solidified material and the liquid phase would significantly increase the thawing rate since it turbulizes liquid at the melting surface and at the heat transfer surface. Such a movement can be accomplished by a vibratory or shaking movement of the whole container. Agitation of the liquid phase may not be accomplished by traditional stirrers. Internally heated agitator, although feasible, makes design complicated. Special frame stirrers located close to heating surfaces may perform well after enough liquid forms between the solid phase and heat transfer surface. An agitator embedded in the solid phase and moving in an oscillatory mode may induce relative movement between the liquid and melting solid. A relatively simple solution is to induce a movement of the whole container, by a forced

vibration or a shaking or oscillatory movement. Significant liquid agitation and thawing intensification can be observed in such a design.

When compared to the thawing in a simple cylindrical container, the thawing in the presence of fins is significantly accelerated. Introduction of forced convection in the liquid phase shortens
the thawing time even further. The presence of fins and forced convection in the liquid phase are the important means to accelerate
thawing since the temperature difference (a heat transfer driving
force) is subject to limitation. The temperature of heat transfer surfaces should not reach a level causing product degradation and, in
practice, may not be higher than the temperature at which the
product stability tests were conducted during the initial formulation development. Development of the large-scale freezing and
thawing process and the final system design should consider this
product temperature limitation as one of the important parameters.

Thawing in Bags Using Plates with Extended Heat Transfer Surfaces (Fins) (Case Study 3.3)

The frozen product in bags can be thawed using the plates with
extended heat transfer surfaces (fins) in a configuration similar to
the one described for freezing (see Figure 3.20, page 266). The bag-
plates assembly taken from low-temperature storage was placed in
the environmental chamber with the temperature setpoint of 18° C,
and an extra fan was furnished for the chamber to increase air
recirculation. The thawing of product in the plate-bag assembly
was firstly tried in a horizontal position. Recirculation of air at
velocities close to 2 m/sec along the finned plates caused intensive
heat delivery from heated air to fins and the flat sides of the plates.
The melting of a frozen mass in a bag first occurred at the sides in
contact with the plates. The subsequent thawing was proceeding
in an asymmetrical fashion; the frozen block as at the upper plate
due to the density difference, and the natural convection currents
occurred between the bag wall in contact with the lower plate and
the bottom of the frozen mass. In such a configuration (a horizontal rectangular cavity with the upper side cooled and the lower side
heated), fluid stability is disturbed, and a cellular convection pattern develops with multiple cells with local liquid recirculation patterns. In the free liquid surface systems, the formation of convective (Benard) cells is surface-tension-dependent (Koschmieder
1993). Since the thermal buoyancy effect is secondary in such a

pattern, further investigation of the phenomenon is needed with a uniform layer of ice or frozen product floating at the top and the bottom wall being heated. The upper side of the frozen mass may not have good physical contact with the heated bag wall due to the presence of air bubbles, which reduce heat transfer.

Placement of the bag-plates assembly in a vertical position caused a different thawing pattern. The natural convection patterns developed in the liquid phase between the bag wall at the heated plate and the frozen mass, with the rising current at the heated wall and the descending current at the frozen mass. The upper part of the frozen mass was melting faster than the rest, causing the mass to rise and an apparent increase in the liquid volume at the bottom of the bag. The process was frequently unstable, with the frozen mass shifting sideways since it was not supported within the bag. When the distance between plates is small (thawing of thin bags), the natural convection in the liquid phase may not be fully developed and the thermal conduction through the liquid phase may dominate thawing heat transfer. The fins configuration on the plate can be vertical or horizontal depending on the airflow pattern (i.e., the airflow must be along the fins). Introduction of movement (rocking or shaking motion) to the assembly structure caused mutual liquid-solid movement within the bag and accelerated the melting rate.

The relative movement of liquid and solid in a thin container enclosed between the heat transfer plates is more difficult to accomplish and thawing intensification may be less pronounced, due to dimensional restrictions. The plates may be moved against each other (changing parallel configuration into a reciprocating slanted plate configuration) to cause longitudinal liquid flow from one end of the flexible container to another. Good contact between the container wall and the plate must be maintained.

The vertical position of the bag with fins positioned vertically was also utilized for thawing, using heating by air natural convection on the fin side. Frosting occurred on the fins' surfaces as the beginning of thawing. This phenomenon occurs when the plates are in contact with the frozen product, which is at a temperature below 0° C. The calculations for heat transfer on the fin side (based on air natural convection) may consider this event, but the frosting may not last long. Later, only some droplet condensation on the fins' surfaces could be observed since the plates were in contact with the liquid product at a temperature near 0° C and rising. Additional heat transfer by water vapor condensation on the fins may be considered, but the estimate should include information on air humidity in the room where thawing takes place. Figure 3.24 shows thawing between the finned plates in vertical configurations using heat transfer by natural convection of the surrounding room air.

The control of thawing rates can be performed by programmed changes in air temperature (controlled heating of recirculating air). Due to the small bag thickness and large heat transfer surfaces, the lag between the change in air temperature and the thawing product response is relatively short. The response lag also depends on the mass of finned plates, which can be significant. The air temperature may not exceed the temperature of earlier product stability studies; the available temperature difference as a heat transfer driving force will be limited. High velocities and uniform distribution of recirculating air may aid in obtaining high heat fluxes by creating airflow conditions feasible for high heat transfer coefficients at finned surfaces.

SUMMARY

Freezing of biological materials in the biopharmaceutical and biological manufacturing operations may provide a very convenient means to add operational flexibility; extend product life; and decouple upstream (fermentation and cell culture steps), downstream (product recovery and purification), and filling and finishing processes. Large volumes of raw fermentation broths can be frozen and stored for future processing and product recovery. Concentrated cell pastes (after filtration or centrifugation) could be frozen and granulated to be processed at the same or a remote facility. The large fermentation plant could operate on a campaign basis, and its product can be frozen, stored, and processed at one or many downstream processing plants with a capacity that cannot match directly the capacity of the fermentation plant working on a continuous basis. The frozen product can be easily shipped in the frozen state, adding flexibility to the concept of multi-site manufacturing. Contract manufacturing steps can be performed at various specialized manufacturers (an example: bulk production at the company plant and filling and finishing at the contract manufacturer site).

Figure 3.24. Thawing of a biological product in a bag compressed between two plates with extended heat transfer surfaces (fins) on the air side. Fins are arranged vertically to facilitate air natural convection. The direction of air currents between the fins are shown as dashed arrows.

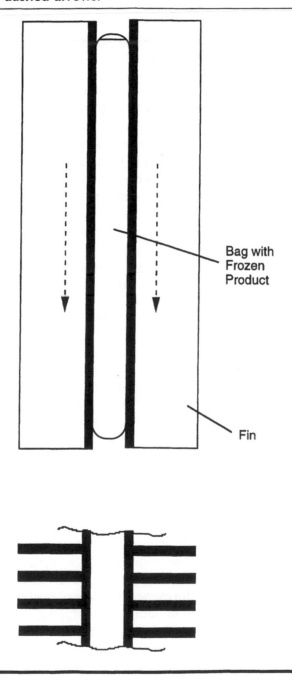

Bag with Frozen Product

Fin

Two groups of large-scale processes have been described: freezing and thawing of biological solutions (proteins and peptides), and freezing and thawing of cell suspensions (fermentation broths, cell pastes, cell fragments). In general, the large-scale cryopreservation of viable cells is the most difficult process, whereas handling killed cells and protein/peptide solutions may be less demanding.

The logistics of freezing operations involving large numbers of bags or bottles can be very complicated and labor intensive. Product losses (broken containers) and batch mix-ups may occur. The processing scheme may also require a large floor space and a significant capital investment. Rapid processing in a compact apparatus of large volumes of biological products can be an attractive alternative.

The considered range of processes for freezing cells and cell components has been narrowed to the following concepts: dripping product droplets into liquid nitrogen, droplet spray cooling in low-temperature gas, contact cooling in thin layers (such as in compressed bags), and cooling and granulation by agitation of the whole mass with cryogen delivery. The last method was selected due to its flexibility in pursuing various cooling protocols. For slowly settling cells, designs of containers with extended heat transfer surfaces were also considered.

Freezing of the product to a strictly determined temperature under controlled freezing conditions permits extended product storage without degradation; i.e., product expiration dates can be extended by a storage time in the frozen state. Thawing can be accomplished just prior to fill and finish operations with thawed quantities following actual market demands. If validated, the shelf life of the product may count then from the date of fill and finish and not from the date of freezing the bulk.

Freezing and Granulation of Large Quantities of Solutions and Suspensions

Due to the prototype character of the system design used for freezing large volumes of biological material, there were many initial concerns and uncertainties regarding procedure selection, design features, and anticipated performance. Review of available designs resulted in the new concept of building a prototype system with the application of liquefied gas as a freezing agent. The typical granule size ranges were between 1–4 and 1–2 mm, depending on the product. High agitation rates, when combined with uniform delivery of a cooling agent, may provide high homogeneity of agitated liquid and bed of particles and uniform cooling and freezing conditions of biological material. The method using the principle of initial water freezing-out at the temperature plateau may be suitable for the freezing of cells that can well survive the processing at low cooling rates (such as yeast or bacteria).

This rapid, controlled method of freezing biological materials may also be utilized to affect microorganisms in a predictable way. The freeze-granulation system can provide a wide and controlled range of temperature change rates not only to optimize conditions for live cell survival but also for selective damage to cells.

The freeze-granulation system features may be very beneficial in commercial applications in biotechnology, pharmaceutical, food, and other industries where precise control over the cooling and freezing process in large volumes is required. The prototype system has performed well with the bacteria and yeast suspensions. Other types of liquid or pasty materials may be frozen and granulated in the system under well-controlled freezing and cooling conditions. The volumetric seeding of ice crystals at a temperature plateau is novel, and a patent on its application to various processes is pending.

After testing the pilot freeze-granulation system, major design concerns were the seal design, exhaust filters design, control of particle size distribution, lump formation, control of cooling rate, agitation of material, geometry of a scaled-up system, and the concept of cleaning-in-place (CIP). Freezing rates were investigated using computer models for the freezing of spheres with changing temperature environment.

It was recommended that a double mechanical seal be used that can withstand thermal shocks, be cleanable, and also be steam sterilizable in-place. Due to high viscosity of some concentrated cell suspensions, such a seal should not have any deep, difficult-to-clean, narrow crevices where material could accumulate. Seal area around the shaft should remain wide open and be easily accessible to cleaning and rinsing solutions. Any possible pockets for condensate accumulation during steam sterilization must be avoided.

Low-temperature exhaust filters were built to the author's specifications after testing their component materials for mechanical properties at a temperature of $-80°$ C. Lump formation was resolved by rotor design optimization and optimization of the operating parameters, such as delivery rate of liquefied gas, rotor speed, and operation of the system jacket. The final freeze-granulation system has performed well in freezing of a variety of materials such as cells, microorganisms, cell components, biological liquid solutions and suspensions. The system is well automated and can deliver reproducible results. Effective cleaning-in-place procedures were developed and have provided positive results even after freezing of viscous and sticky materials.

The work on delivery of liquified gas involves analysis and testing of spraying liquid carbon dioxide and nitrogen. The use of a combination of liquefied carbon dioxide and nitrogen spraying may allow attainment of temperatures below those of glass transition of solutes and proteins or ice recrystallization at a controlled freezing rate. Application of liquid nitrogen as a secondary cooling agent may provide an opportunity to change the cooling rate rapidly after reaching a predetermined temperature level. Such a system is also suitable for performing two- or multiple-step cooling procedures, since any holding period at predetermined temperatures can be implemented by reduction of the gas delivery rate to maintain a constant temperature level for any required length of time. Use of multiple nozzles with individual valves may allow precise control of cooling and freezing rates.

The agitation rate was optimized to minimize mechanical stress imposed on mixed material by the rotors and, at the same time, maintain mixing levels adequate for particular processing steps. Speed and direction of the rotors' movement could change during the run to ensure optimal processing conditions.

The process dynamics and control have not posed any serious problems. Thorough agitation and product mass homogeneity are a key to system performance. The processes with temperature hold periods are dependent on the quality of mixing and performance (dynamics) of the cryogen delivery. As a result of on/off operation, the capacity of the nozzles may be reduced if gaseous phase forms in the liquid cryogen delivery systems. If necessary, the delivery systems should be vented from the formed gas to maintain them filled with the liquid phase only.

The jacket operation facilitated cooling of liquid product prior to the run. Its operation has also been used in an effort to minimize mechanical stress on the product and minimize product losses. The freezing system also can be used for thawing purposes by applying a heat transfer agent at an elevated temperature to the jacket.

The freeze-granulation system allows operation at various cooling rates even in a single run by changing the delivery of cryogenic fluid(s) and by applying a heat transfer agent to the jacket. Experiments with simultaneous delivery of cryogen and heating the walls of the apparatus using the jacket over extended periods of time have been conducted. Such a feature may be used in design of processes with nonlinear cooling protocols (Pitt 1992). For a well-agitated bed, the process dynamics and controllability over the desired profile of temperature change would mostly depend on performance of the cryogen delivery system(s).

Heat and mass transfer calculations and balances can be based on the delivery rate of cryogens and product characteristics. Free extra- and intracellular water are the most important parameters in calculating cooling and freezing balances. The cooling capacities can be estimated from latent heats of sublimation (carbon dioxide) or evaporation (nitrogen). Local heat transfer coefficients under the condition of spraying cryogens into agitated liquid or bed of granules remain to be investigated using a rigorous methodology. Agitation power input can be estimated from the motor power rating and an actual current. Heat gains from the room would depend on insulation parameters and usually are of secondary importance. Energy balance should consider the cold gas leaving the system—an estimate can be made using gas flow rate and its temperature.

The concept of agitated freezer/granulator can be applied in a wide range of scales, e.g., from bench scale laboratory apparatus to large industrial systems with a volume of hundreds of liters. Scale increase to thousands of liters is possible.

The recommended cooling rates for bacteria, yeast, and fungi (in the range of 1–10° C/min) have been easily attained in the freeze-granulation system during the process phases following granule formation. Higher cooling rates could be accomplished as well. This feature makes this system advantageous when compared to the dropping of liquids into liquefied gas where there is no practical control over freezing and cooling rates. The observed temperature plateau near 0° C was measured within the entire volume of freezing, agitated mass (thermocouples at different locations in the apparatus showed similar temperature trend patterns). This indicates a volumetric freezing-out of water prior to a critical point of transition with the mass breaking into granules. Thermal history of cells may be considered similar across the whole mass of material up to this point. Volumetric freezing at the beginning of the process involves formation and growth of ice crystals in extracellular space at almost constant temperature. This initial water removal by freezing-out and cryoconcentration of solutes and cells occur without a temperature drop (no cooling rate, as described in the literature, is involved). The process can be remotely compared to a volumetric, continuously seeded crystallization with suspended crystals. Water is being frozen out in extracellular space, and cells may be losing water osmotically at a controlled rate (depending on length of the temperature plateau) without approaching temperature levels of intracellular ice formation. Since the length of the temperature plateau time can be controlled (depending on delivery of cryogenic agent), an exposure of cells to concentrated solutes may be minimized.

This part of the process cannot be directly compared to literature data derived from freezing of product samples in vials or test tubes with external cooling across the wall. Only after the granules are formed do certain similarities exist, i.e., the freezing front moves through the semiliquid phase from the surface of a sample toward the center. But even in such a case, the cells and solutes have already been preconcentrated, the suspended ice crystals are present across the whole remaining liquid phase, the cells already have lost part of their water, and danger of intracellular ice formation is significantly reduced. Typical procedures for freezing in vials may not create conditions for such phenomena to occur. The vial freezing usually involves cooling through the wall without any agitation with a possibility of dendritic crystal growth from the vial walls toward the sample center. Freezing in vials may also involve a brief but significant undercooling of the whole volume.

The length of cell exposure to concentrated solutes at no or a very small cooling rate (during the temperature plateau) can be controlled by the delivery rate of the cryogenic agent. The delivery rate of cryogenic agent also determines the subsequent cooling rate during the temperature decrease period. The whole process could be completed within 20÷40 minutes. The operational cooling rates could reach 1–10° C/min and are uniform for the whole bed of small granules, which would be difficult to accomplish by other methodology. Higher cooling rates can be achieved by increasing liquid nitrogen or carbon dioxide spray delivery rates—the gas exhaust system can be rated to handle large volumes of gas if high cooling rates are anticipated for certain future products.

The removal of water from cells during the temperature plateau by freezing of extracellular water decreases the amount of free water inside the cells but does not affect the bound water. This intracellular bound water may freeze later at much lower temperatures.

The three principal granule freezing methods possess distinctive freezing patterns. In the case of immersion of liquid droplets of product into liquid nitrogen, the solidification is very rapid, there may be a large droplet under-cooling prior to solidification, the freezing front moves quickly, and very large numbers of fine ice crystals form. The cooling rate is very difficult to control. The cells are susceptible to intracellular ice formation under such a freezing regime. In the case of spray cooling by surrounding cold gas, the freezing rate is slower and depends upon the heat transfer of the droplets of gas. Some control of the cooling rate is possible by a change in cooling gas temperature, but an issue of drop handling over extended time at low cooling rates may become a problem. In the case of mixing solid cooling agent (carbon dioxide) particles with a freezing mass, there are multiple ice crystals formed within the whole volume of material. After reaching the freezing temperature level and certain minimum supercooling, the ice crystals may be nucleated by carbon dioxide particles or liquid nitrogen droplets. When the number of crystals increases, the viscosity of the frozen mass suddenly increases and the granules form. The delivery of solid carbon dioxide or liquid nitrogen may continue until reaching the desired material temperature, or it may stop and liquefied nitrogen only may be sprayed onto the agitated bed. Freezing of the remaining water within the granule interiors depends then on a heat flux through the granule surface. Higher heat transfer coefficients at the surface (and faster cooling) can be obtained by spraying liquid nitrogen rather than solid carbon dioxide due to high heat transfer for liquid nitrogen boiling versus contact and sublimation phenomena for carbon dioxide. There is also a larger temperature difference between granule interior and granule surface in the case of liquid nitrogen spray.

A variety of products can be frozen and granulated using this method. The important feature may be an ability to rapidly freeze large volumes of viable cells with a high level of cell viability preservation. The viable cell freezing and granulation may require a special design of internal agitation and an optimized operational mode (for example, changing the agitator speed during the process). Other products such as heat-killed cells, disrupted cells, cell fragments and components (for example, inclusion bodies), protein solutions, aqueous buffers, tissue suspensions, and plant materials (for example, a disrupted plant tissue) can also be frozen, granulated, and thawed using this method. The preservation of product in a granulated form not only permits great operational flexibility (the granules can be rapidly thawed for further processing in easily measured quantities) but also facilitates manufacturing scheduling in multi-product plants. For example, the granulated frozen product can be produced in a 2-month operation to be used during the following year while the plant is producing other product lines. Frozen product transfer between manufacturing plants can be easily performed. For example, the cells can be produced in the fermentation facility, concentrated in a filtration system or in a centrifuge, frozen and granulated, and shipped to the downstream processing plant.

The method can also be used for freezing protein solutions. Conditions for freezing/granulation of protein solutions may vary from conditions of freezing/granulation of cell suspensions—the presence of a large number of very small ice crystals may or may not be beneficial for proteins if there is a strong interaction between the ice surface and protein. The composition and concentration of solutes may determine how the proteins can be affected by the presence of ice crystals. The protein formulations containing protecting agents have been successfully frozen and granulated (for example, solutions with sucrose concentrations from 1–4 percent to more than 10 percent can be frozen/granulated). The technique can also be used for rapid cooling of granular materials.

Freezing of Large Volumes of Biological Solutions Using Extended Heat Transfer Surfaces

The freezing and thawing in jacketed containers with internal heat exchangers using extended heat transfer surfaces (fins) can provide very efficient means of aseptic processing of biological liquid and slurry products. The freezing and thawing processes can be accomplished in a rapid, uniform, and controlled way.

The fins act as effective heat transfer devices in both freezing and thawing steps. They are welded to the central bayonet heat exchanger and closely approach the vessel walls. In the freezing process, the small gap between the vessel-cooled wall and the fin tip fills with the frozen product, which forms a thermal bridge. After the bridge formation, the fin begins to conduct heat on both ends (the fin's thermally efficient length almost doubles). After fins are covered with the frozen product, they act as cold walls providing a cooled cavity with the freezing fronts moving to its center. The freezing time of the cavity is shortened due to the corner effect between fins at the bayonet surface. The freezing and thawing steps can be performed under conditions minimizing product exposure to adverse conditions. In the freezing process, the control of movement of the freezing front (the mushy zone) is a key to provide controlled product exposure to increased concentrations of solutes in the liquid phase and uniform product distribution across the frozen mass.

The freeze-thawing containers can be clean-in-place (CIP) using typical biopharmaceutical CIP cycles and sterilized-in-place (SIP) with steam. The product can be filled (sterile) into the container and, after freezing, kept stored under aseptic conditions (no seal breaking of closed containers). The thawing can also be accomplished under aseptic conditions, and the sterile product can be fed directly to the subsequent processing steps (such as filling and lyophilization).

The container design can be as large as 250–350 L and in a form close to a pharmaceutical tank design. Larger containers have a different configuration of the internal heat exchangers' and fins' interaction patterns. The containers may be sized to a batch size and, therefore, no batch division and later pooling is possible.

The cryopreservation in sealed containers with internal finned heat transfer surfaces is less costly than bulk lyophilization and may provide superior product quality after reconstitution. Product activity losses are much lower than after bulk lyophilization. In many cases, there may be no losses after the freeze-thaw cycle. Cryopreservation in large containers facilitate operation of multiproduct plants and adds flexibility to the manufacturing scheduling. The frozen product can be transferred between the bulk manufacturing and final filling and finishing facilities.

Thawing of Large Quantities of Solutions and Suspensions

An additional benefit of the freeze-granulation method is shaping the product into granules. Such a final form facilitates not only material handling and storage but also allows the thawing process to be run in a controlled way, for instance, by pouring frozen granules into a heated and agitated liquid. Such a method may allow running the thawing process at a rapid pace, which has been the recommended method of cell thawing by many protocols of cell culture collections (Malik 1987).

The freeze-granulation apparatus was also successfully used for thawing of frozen granules. The heating jacket was used to deliver energy for the product's latent heat of melting. Bed agitation maintains a high heat transfer coefficient between the wall and product. A typical limiting factor in maximizing the heat transfer is the wall temperature (e.g., when the product cannot be overheated). Further increase in the heat flux can be accomplished by use of the heated agitating rotors; the heat transfer area will increase, and high heat transfer coefficients will be possible due to the rapid relative movement between the product and heat transfer surfaces. For small granule sizes, such a thawing approach can provide close approximation to the volumetric thawing of the whole large product volume. For small granules (small Biot numbers), the entire granule can be rapidly heated to near 0° C and thawed completely. As during freezing, a temperature plateau for the whole bed can be observed near the 0° level.

The thawing process using extended heat transfer surfaces (fins) can be very efficient due to a difference in thermal conductivity between the liquid product and the fin material and due to a natural convection developing in the liquid phase early in the process. The natural convection with the associated temperature gradients causes more intensive melting of the top product layers, e.g., where the warmest liquid contacts the product.

The thawing step can be accelerated by introduction of forced convection into the initially formed liquid phase. The heat transfer increase can be caused by turbulence and by ablation at the melting solid surface. The turbulence in liquid phase can be introduced by liquid recirculation or by vessel movement. In the thawing process, a rapid thawing rate and rapid passing of the ice recrystallization temperature zone can be critical to product quality. Heat transfer intensification and volume compartmentalization allow rapid thawing and equalizing of temperature gradients within the frozen mass.

Enhanced Freezing in Small Containers

The traditional approaches to freezing of large quantities of biological solutions and suspensions typically apply to the division of large product volumes into small containers (for example, bags or bottles) and freezing them in blast freezers or liquid baths. The important factors for the freezing step are the container geometry, the cooling agent temperature (including its history during the process), and the external heat transfer coefficient. These parameters can be optimized to provide cooling and freezing conditions favorable for the processed biological material. The multiplate contact freezers for bags or flat containers can deliver high heat transfer coefficients, and the freezing process can be rapid and well controlled providing there are no air pockets in the heat flux path. However, such contact freezers can be expensive if a large number of plates is required. The heat transfer coefficients (gas-product container) in the blast freezers are usually low. The author has developed a proprietary design (patents are pending) of the plates with enhance heat transfer surfaces on the gas side to increase the heat transfer coefficient to the container by a factor of at least 2.5 to 3. Such a solution brings the freezing conditions closer to freezing in the plate contact freezers.

Freezing of biological products in cabinet and blast freezers can be enhanced by use of plates with extended heat transfer surfaces (fins). The product can be compressed between the plates or placed on horizontal plates/shelves. Flat containers (bags, trays) are preferred for rapid freezing. In the cabinet freezers with stagnant air or low air velocities, the finned plates should be placed vertically to facilitate natural convection. In the blast freezers with high air velocities, the plates should be placed with airflow moving along the fins.

Plates with extended heat transfer surfaces can provide high freezing rates impossible to obtain in the traditional cabinet or blast freezer applications (where the heat transfer occurs between gas and container surface). This enhancement in freezing rates is possible due to high effective heat transfer coefficients (a contact-type heat transfer) between the flat side of the plate and the wall of the product container. The increase in the effective heat transfer on the flat side of the plate is possible because of a large fin surface area on the gas side.

Efficient freezing and plate cooling capacity utilization depend upon the contact between the product container and the flat side of the plate. Therefore, the product container is compressed between the plates. The contact quality also depends upon the surface of the product container. This surface should be smooth without any corrugations. The container should be filled and positioned to avoid air pockets on the product side in the areas in contact with the plates. For example, the bags should be placed vertically with the air pocket (if unavoidable) moving to the bag top. Any corrugations on the container surface or air pockets in the vicinity of plates will hinder the heat transfer and reduce freezing rates. Use of double bags will reduce freezing rates compared to the smooth, single-wall bag.

The spacing between the plate assemblies is important to ensure sufficient air velocities and airflow distribution (convective heat transfer) and also to ensure low flow pressure drop if axial fans are used (they are sensitive to backpressure). Single, parallel rows of plates will be better than plates in series. The frame design to hold the plate-bag assemblies can also be important since it may affect the air distribution (airflow-guiding baffles may be attached to such a frame). The air velocity distribution should be checked for the system internals (supporting frame and plate assemblies). If one cannot measure the air velocity distribution inside the freezer, use an array of thermocouples (using very thin wires) in different points between the plate and the bag in selected locations. Temperature readings may indicate uniformity of freezing at different locations of the plates within the freezer. It is important to be sure that the flat sides of the plates are parallel (uniform thickness of the compressed bag).

The concept of a plate with extended heat transfer surface can also be applied to freezing similar to freezing on liquid-cooled shelves (e.g., for contact freezing in vials or in trays). The shelf can be placed in a cabinet or blast freezer with the flat top surface and the finned surface facing down. The vials or trays can be placed on the top, and the gas-shelf heat transfer will take place mostly through the fins. The conditions of heat transfer by natural convection may not be optimal for typical fin lengths, and a forced convection (air recirculation by fans) is recommended.

Enhanced Thawing in Small Containers

The thawing step involving small containers (bags, bottles) typically faces the problem of a limited temperature of the heating agent (product cannot be overheated). Under such restrictions, the thawing can be accelerated by an increase in heat transfer accomplished by an increase of heat transfer coefficient outside the container (increase air velocity and turbulence in the blast freezers; recirculate or agitate liquid in the liquid baths). Further acceleration of thawing can be achieved by introduction of a liquid-solid movement inside the container. The heat transfer can be enhanced by forced liquid convection in the liquid phase. Further thawing enhancement may occur if forced convection in the liquid phase is intensified and ablation effects at the solid-liquid interface occur.

The thawing rates can be enhanced by using fins, i.e., the plates with extended heat transfer surfaces can transfer much more energy into the product than in the case where the product container alone is exposed to still or moving air. The plates may be placed in a position facilitating natural or forced convection in air, depending on the thawing method. The natural convection configuration may be used under ambient conditions in a processing room. The forced air convection configuration requires air recirculation by fan(s) and may be done in an enclosed cabinet with controlled air temperature (an electrical or steam heater is placed in the recirculating air path). Convection in the liquid phase plays a significant role during thawing and may be enhanced by moving the whole assembly (creating a rocking or shaking movement).

The process enhancement using finned plates has been successful on a small and medium scale. The scale-up of this method is limited since large-scale operations would require filling a large number of containers (bags), placing them between plates, moving the plate assemblies into freezers, and, after freezing, removing the assemblies and putting them into storage. Thawing involves airflow by natural or forced convection at controlled temperature (air heating). Finally, the containers should be emptied and pooled into a single vessel. Filling and emptying the containers may be done manually or automated. Depending on the process stage, a clean room or isolator may be required for filling/emptying steps. The logistics of such an operation is complex, and processing of large product volume requires careful evaluation.

REFERENCES

AIRCO Industrial gases data book. 1982. Murray Hill, NJ.

Anchordoguy, T. J., J. F. Carpenter, J. H. Crowe, and L. M. Crowe. 1992. Temperature-dependent perturbation of phospholipid bilayers by dimethylsulfoxide. *Biochim. Biophys. Acta* 1404:117–122.

Angell, C. A. 1995. Formation of glasses from liquids and biopolymers. *Science* 267:1924–1935.

Apekis, L., P. Pissis, and G. Boudouris. 1983. Dielectric study of polycrystalline ice Ih by the depolarization thermocurrent method: The peak at ~220K. *J. Phys. Chem.* 87:4019–4021.

Awonorin, S. O. 1989. Film boiling characteristics of liquid nitrogen sprays on a heated plate. *Int. J. Heat Mass Transfer* 32:1853–1864.

Bar-Cohen, A. 1979. Fin thickness for an optimized natural convection array of rectangular fins. *J. Heat Transfer* 101:564–566.

Bartell, L., and J. Huang. 1994. Supercooling of water below the anomalous range near 226 K. *J. Phys. Chem.* 98:7455–7457.

Bergles, A. E. 1988. Augmentation of boiling and evaporation. In *Heat exchanger design handbook,* vol. 2, edited by E. Schlunder, 2.7.9.1–2.7.9.5. New York: Hemisphere Publ. Corp.

Bhansali, A., and W. Black. 1996. Local, instantaneous heat transfer coefficients for jet impingement on a phase change surface. *J. Heat Transfer ASME* 118:334–342.

Burmeister, L. 1993. *Convective heat transfer.* New York: J. Wiley.

Chandra, S., and S. D. Aziz. 1994. Leidenfrost evaporation of liquid nitrogen droplets. *J. Heat Transfer* 116:999–1006.

Charm, S. E. 1981. *Fundamentals of food engineering.* 3d ed. Westport, CT: AVI.

Cheremisinoff, N. 1986. Mixing of granular and loose solids. In *Encyclopedia of fluid mechanics,* vol. 4, edited by N. Cheremisinoff. Houston: Gulf Publishing Co.

Choi, K. J., and J. S., Hong. 1991. Experimental study of enhanced melting process under ultrasonic influence. *J. Thermophysics* 5:340–346.

Clemens, J., and C. Saltiel. 1996. Numerical modeling of materials processing in microwave furnaces. *Int. J. Heat Mass Transfer* 39:1665–1675.

Coakley, W. T. et al. 1977. Disruption of microorganisms. *Adv. Microbial. Physiol.* 16:279–341.

Coleman, C. J. 1990. The microwave heating of frozen substances. *Appl. Math. Modeling* 14:439–443.

Constant, T., C. Moyne, and P. Perre. 1996. Drying with internal heat generation: Theoretical aspects and application to microwave heating. *AIChE J.* 42:359–368.

CRC handbook of chemistry and physics, 1995. 77th ed. Edited by D.R. Lide. Boca Raton: CRC Press.

de Alvis, A., and P. Fryer. 1990. The use of direct resistance heating in the food industry. *J. Food Engineering* 11:3–27.

deLoecker, R., and F. Pennickx. 1987. Biochemical and functional aspects of recovery of mammalian systems from deep sub-zero temperatures. In *Temperature and animal cells,* edited by K. Bowler and B. Fuller, Symp. Soc. Experiment. Biology, No 41, 407–427.

de Noel, J. G., F. M. Klis, J. Priem, T. Munnik, and H. van den Ende. 1990b. The glucanase-soluble mannoproteins limit cell wall porosity in *Saccharomyces cerevisiae. Yeast* 6:491–499.

Deb, S., and S.-C. Yao. 1989. Analysis on film boiling heat transfer of impacting sprays. *Int. J. Heat Mass Transfer* 32:2099–2112.

Deshpande, S., H. Bolin, and D. Salunkje. 1982. Freeze concentration of fruit juices. *Food Technology,* May:68–82.

Dharma-Wardana, M. 1983. Thermal conductivity of the ice polymorphs and the ice clathrates. *J. Phys. Chem.* 87:4185–4190.

Diller, K. R., M. E. Crawford, and L. J. Hayes. 1985. Variation in thermal history during freezing due to the pattern of latent heat evolution. *AIChE. Symp. Ser. 81,* 245:234–239.

Drach, V., N. Sack, and J. Fricke. 1996. Transient heat transfer from surfaces of defined roughness into liquid nitrogen. *Int. J. Heat Mass Transfer* 39:1758–1762.

Duluc, M., M. Francois, and J. Brunet. 1996. Liquid nitrogen boiling around a temperature controlled heated wire. *Int. J. Heat Mass Transfer* 39:1758–1762.

Dumas, J. P., M. Strub, and F. Broto. 1990. Heat transfer during the freezing of undercooled liquids dispersed within an emulsion. In *Heat transfer* 1990, vol. 3, edited by G. Hetsroni. Proc. 9th Intl' Heat Transfer Conf., 45–50.

El Moctar, A., H. Peerhossaini, and J. Bardon. 1996. Numerical and experimental investigation of direct electric conduction in a channel flow. *Int. J. Heat Mass Transfer* 39:975–993.

El Moctar, A., H. Peerhossaini, P. LePeurian, and J. Bardon. 1993. Ohmic heating of complex fluids. *Int. J. Heat Mass Transfer* 36:3143–3152.

Fahy, G. M., D. I. Levy, and S. E. Ali. 1987. Some emerging principles underlying the physical properties, biological actions, and utility of vitrification solutions. *Cryobiology* 24:196–213.

Farrant, J., C. Walter, H. Lee, and L. McGann. 1977. Use of two-step cooling procedures to examine factors influencing cell survival following freezing and thawing. *Cryobiology* 14:273–286.

Fletcher, L. 1988. Recent developments in contact conductance heat transfer. *J. Heat Transfer* 110:1059–1070.

Frost, W. 1975. *Heat transfer at low temperatures.* New York: Plenum Press.

Fukusako, S. 1990. Thermophysical properties of ice, snow, and sea ice. *Int. J. Thermophys.* 11:353–372.

Geankoplis, C. 1983. *Transport processes: Momentum, heat and mass.* Boston: Allyn & Bacon.

Gerhardt, P., G. Murray, W. Wood, and N. Kreig, eds. 1990. *Methods for general and molecular bacteriology.* Washington, DC: Am. Soc. Microbiol.

Graham, L. L., T. J. Beveridge, and N. Nanninga. 1991. Periplasmic space and the concept of the periplasm. *TIBS,* September:328–329.

Han, R. H., T. C. Hua, and H. S. Ren. 1995. Experimental investigation of cooling rates of small samples during quenching into subcooled LN2. *Cryoletters* 16:157–162.

Hartmann, U., and M. W. Scheiwe. 1984. Film boiling heat transfer to liquid nitrogen: A comparison of transient and steady state measuring techniques. *Adv. Cryogenic Eng.* 29:307–314.

Hill, P., and K. Ng. 1996. Statistics of multiple particle breakage. *AIChE J.* 42:1600–1611.

Hua, T. C., E. G. Cravalho, and L. Jiang. 1982. The temperature difference across the cell membrane during freezing and its effect on water transport. *Cryoletters* 3:255–264.

Hughes, D. E. et al. 1971. The disintegration of microorganisms. *Meth. Microbiol.* 5B:1–54.

Hwang, C.-C., S. Lin, and L.-F. Shen. 1994. Effects of wall conduction and interface thermal resistance on the phase-change problem. *Int. J. Heat Mass Transfer* 37:1849–1855.

Irvine, T., and J. Taborek. 1988. Thermal contact resistance. In *Heat exchanger design handbook,* vol. 2, edited by E. Schlunder, 2.4.6.1–2.4.6.6. New York: Hemisphere Publ. Corp.

Jackson, S., and R. Whitworth. 1996. Evidence for ferroelectric ordering of ice Ih. *J. Chem. Phys.* 103:7647–7648.

Jacobson, L., and J. McKittrick. 1994. Rapid solidification processing. *Materials Sci. Eng.* R11:355–408.

Johari, G. P., A. Hallbrucker, and E. Mayer. 1987. The glass transition of hyperquenched water. *Nature* 330:552–553.

Jones, J. 1984. Microstructure of rapidly solidified materials. *Materials Sci. Eng.* 65:145–156.

Kaji, N., Y. Mori, Y. Tochitani, and K. Komotori. 1982. Electrodynamic augmentation of direct-contact heat transfer to drops passing through an immiscible dielectric liquid: Effect of field-induced shuttle migration between parallel plane electrodes of drops. In *Heat transfer* 1982, vol. 5, edited by U. Grigull et al., 231–236. Proc. 7th Intl Heat Transfer Conf., Munich. Washington, DC: Hemisphere Publ. Corp.

Kalhori, B., and S. Ramadhyani. 1985. Studies on heat transfer from a vertical cylinder, with or without fins, embedded in a solid phase change medium. *J. Heat Transfer* 107:44–51.

Kern, D., and A. Kraus. 1972. *Extended surface heat transfer.* New York: McGraw-Hill.

Kirsop, B. E., and A. Doyle, eds. 1991. *Maintenance of microorganisms and cultured cells: A manual of laboratory methods.* 2d ed. London: Academic Press.

Ko, Y., and S. Chung. 1996. An experiment on the breakup of impinging droplets on a hot surface. *Experiments in fluids* 21:118–123.

Koebe, H., A. Werner, V. Lange, and F. Schildberg. 1993. Temperature gradients in freezing chambers of rate-controlled cooling machines. *Cryobiology* 30:349–352.

Koschmieder, E. 1993. *Benard cells and Taylor vortices.* Cambridge: Cambridge University Press.

Kosky, P. G., and D. N. Lyon. 1968. Pool boiling heat transfer to cryogenic liquids (3 parts). *AIChE J.* 14:372–379, 380–383, 383–387.

Kruuv, J., and D. J. Glofcheski. 1992. Protective effects of amino acids against freeze-thaw damage in mammalian cells. *Cryobiology* 29:291–295.

Kurz, W., and D. Fisher. 1989. *Fundamentals of solidification.* Aedermannsdorf: Trans Tech.

Kurz, W., and P. Gilgien. 1994. Selection of microstructures in rapid solidification processing. *Materials Sci. Eng.* A178:171–178.

Lage, J., A. Bejan, and J. Georgiadia. 1991. On the effect of the Prandtl number on the onset of Benard convection. *Int. J. Heat and Fluid Flow* 12:184–188.

Lande, M., J. Donovan, and M. Zeidel. 1995. The relationship between membrane fluidity and permeabilities to water, solutes, ammonia, and protons. *J. Gen. Physiol.* 106:67–84.

Ledezma, G., A. Morega, and A. Bejan. 1996. Optimal spacing between pin fins with impinging flow. *J. Heat Transfer* 118:570–577.

Leslie, S., E. Israeli, B. Lighthart, J. Crowe, and L. Crowe. 1995. Trehalose and sucrose protect both membranes and proteins in intact bacteria during drying. *Appl. Environment. Microbiol.* 61:3592–3597.

Levine, H., and L. Slade. 1988. Principles of "cryostabilization" technology from structure/property relationship of carbohydrate/water systems, A review. *Cryoletters* 9:21–63.

Lim, J., and A. Bejan. 1992. The Prandtl number effect on melting dominated by natural convection. *J. Heat Transfer* 114:784–787.

Lyon, D. N. 1964. Peak nucleate-boiling heat fluxes and nucleate boiling heat transfer coefficients for liquid N_2, liquid O_2, and their mixtures in pool boiling at atmospheric pressure. *Int. J. Heat Mass Transfer* 7:1097–1116.

MacFarlane, D. R. 1986. Devitrification in glass-forming aqueous solutions. *Cryobiology* 23:230–244.

MacFarlane, D. R. 1987. Physical aspects of vitrification in aqueous solutions. *Cryobiology* 24:181–195.

MacFarlane, D., M. Forsyth, and C. Barton. 1992. Vitrification and devitrification in cryopreservation. In *Adv. low-temp. biology*, vol. 1, edited by P. Steponkus, 221–278. London: JAI Press Ltd.

MacGregor, S. 1991. Air entrainment in spray jets. *Int. J. Heat and Fluid Flow* 12:279–283.

Macklis, J. D., and F. D. Ketterer. 1978. Microwave properties of cryoprotectants. *Cryobiology* 15:627–635.

Macklis, J. D., F. D. Ketterer, and E. G. Cravalho. 1979. Temperature dependence of the microwave properties of aqueous solutions of ethylene glycol between +15° and –70° C. *Cryobiology* 16:272–286.

Malhotra, K., A. Mujumdar, and M. Okazaki. 1990. Particle flow patterns in a mechanically stirred two-dimensional cylindrical vessel. In *Drying* 89, edited by A. Mujumdar and M. Roques. New York: Hemisphere Publ. Corp.

Malik, L. 1987. Preservation of biotechnologically important microorganisms in culture collections. *Progress Biotechnol.* 4:145–186.

Marvillet, C. 1995. Heat transfer intensification in evaporators and condensers. In *Two-phase flows with phase transition*, edited by C. H. Sieverding, 1–73. Lecture series 1995–06, Rhode Sâint Genese, Belgium: von Karman Institute for Fluid Dynamics.

McLeod, P., D. Riley, and S. Sparks. 1996. Melting of a sphere in hot fluid. *J. Fluid Mech.* 327:393–409.

Merte, H., and J. A. Clark. 1964. Boiling heat transfer with cryogenic fluids at standard, fractional, and near-zero gravity. *J. Heat Transfer* 86:351–359.

Morega, A., A. Began, and S. Lee. 1995. Free stream cooling of a stack of parallel plates. *Int. J. Heat Mass Transfer* 38:519–531.

Muchowski, E. 1988. Packed and agitated beds. In *Heat exchanger design handbook*, vol. 2, edited by E. Schlunder, 2.8.1.1–2.8.3.9. New York: Hemisphere Publ. Corp.

Mudgett, R. 1986. Microwave properties and heating characteristics of foods. *Food Technology,* June:84–93.

Mueller, D. 1988a. Mechanically aided heat exchangers. In *Heat transfer equipment design,* edited by R. Shah, E. Subbarao, and R. Mashelkar, 351–361. New York: Hemisphere Publ. Corp.

Mueller, D. 1988b. Cryopreparation of microorganisms for electron microscopy. *Methods in Microbiol.* 20:1–28.

Omori, Y. et al. 1989. A novel method—a "freeze-blast" method—to disrupt microbial cells. *J. Ferment. Bioeng.* 67:52–56.

Omori, Y. et al. 1990. Application of the freeze-blast method to disruption of cultured plant cells. *J. Ferment. Bioeng.* 69:132–134.

Park, Y. S., and L. Huang. 1992. Cryoprotective activity of synthetic glycophospholipids and their interactions with trehalose. *Biochim. Biophys. Acta* 1124:241–248.

Peyayopanakul, W., and J. W. Westwater. 1978. Evaluation of the unsteady-state quenching method for determining boiling curves. *Int. J. Heat Mass Transfer* 21:1437–1445.

Pitt, R. E. 1992. Thermodynamics and intracellular ice formation. In *Adv. in low temp. biology,* vol. 1, edited by P. Steponkus, 63–99. London: JAI Press Ltd.

Plumb, O. 1994. Convective melting of packed beds. *Int. J. Heat Mass Transfer* 37:829–836.

Pringle, M. J., and D. Chapman. 1981. Biomembrane structure and effects of temperature. In *Effects of low temperatures on biological membranes,* edited by G. J. Morris and A. Clarke, 21–40. London: Academic Press.

Razelos, P., and B. Satyaprakash. 1993. Analysis and optimization of convective trapezoidal profile longitudinal fins. *J. Heat Transfer* 115:461–463.

Recommendations for the processing and handling of frozen foods. 1986. Paris: International Institute of Refrigeration.

Rupley, J. A., and G. Careri. 1991. Protein hydration and function. *Adv. Protein Chem.* 41:37–172.

Ryan, W., P. Collier, V. Satyagal, R. Sachdev, and F. Sumodjo. 1995. Characterization of a cryogenic pelletizer for preserving delicate biologicals. *BioPharm* 8, October:32–38.

Saito, A., H. Hong, and O. Hirokane. 1992. Heat transfer enhancement in the direct contact melting process. *Int. J. Heat Mass Transfer* 35:295–305.

Saito, A., S. Okawa, A. Tojiki, H. Une, and K. Tanogashira. 1992. Fundamental research on external factors affecting the freezing of supercooled water. *Int. J. Heat Mass Transfer* 35:2527–2536.

Sartor, G., A. Hallbrucker, K. Hofer, and E. Mayer. 1992. Calorimetric glass-liquid transition and crystallization behavior of a vitreous, but freezable, water fraction in hydrated methemoglobin. *J. Phys. Chem.* 96:5133–5138.

Sartor, G., E. Mayer, and G. P. Johari. 1994. Calorimetric studies of the kinetic unfreezing of molecular motions in hydrated lysozyme, hemoglobin, and myoglobin. *Biophys. J.* 66:249–258.

Schiffmann, R. 1986. Food product development for microwave processing. *Food Technology,* June:94–98.

Schlunder, E. 1982. Particle heat transfer. In *Heat Transfer* 1982, vol. 1, edited by U. Grigull, E. Hahne, K. Stephan, and J. Straub, 195–211. 7th Intern. Heat Transfer Conf., Munich. Washington, DC: Hemisphere Publ. Corp.

Schmidt, D., and M. Akers. 1997. Cryogranulation: A potential new final process for bulk drug substances. *Biopharm,* April: 28–32.

Schubert, D. 1987. Biophysical approaches to the study of biological membranes. In *Biological membranes,* edited by J. B. Findlay and W. H. Evans, 241–280. Oxford: IRL Press.

Sherrington, J., and P. Oliver. 1981. *Granulation.* Amsterdam: Elsevier.

Somboonsuk, K., J. T. Mason, and R. Trivedi. 1984. Interdendritic spacing: Part I. Experimental studies. *Metall. Trans.* 15a:967–975.

Souzu, H. 1992. Freeze-drying of microorganisms. In *Encyclopedia of Microbiology,* vol. 2, edited by J. Lederberg, 231–243. San Diego: Academic Press.

Spieles, G. et al. 1995. The effect of storage temperature on the stability of frozen erythrocytes. *Cryobiology* 32:366–378.

Sterbacek, Z., and P. Tausk. 1965. *Mixing in the chemical industry.* Oxford: Pergamon Press.

Sutton, R. 1992. Critical cooling rates for aqueous cryoprotectants in the presence of sugars and polysaccharides. *Crybiology* 29:585–598.

Takahashi, T. 1983. Electric charge separation during ice deformation and fracture under a temperature gradient. *J. Phys. Chem.* 87:4122–4124.

Timmerhaus, K. D., and T. M. Flynn. 1989. *Cryogenic process engineering.* New York: Plenum Press.

Uhl, V. 1966. Mechanically aided heat transfer. In *Mixing: Theory and practice,* edited by V. Uhl and J. Gray, 279–328. New York: Academic Press.

Vance, D. E., and J. Vance, eds. 1985. *Biochemistry of lipids and membranes.* Menlo Park: Benjamin/Cummings.

Vance, D. E., and J. Vance, eds. 1991. *Biochemistry of lipids, lipoproteins, and membranes.* Amsterdam: Elsevier.

Veziroglu, T. N., M. A. Heurta, and S. Kakac. 1976. Exact solutions for thermal conductances of planar and circular contacts. In *Thermal conductivity* 14, edited by P. G. Clemens and T. K. Chu, 435–448. New York: Plenum Press.

Vorotilin, A., A. Zinchenko, and V. Moiseyev. 1991. Cell injury at the stage of thawing. *Cryoletters* 12:77–86.

Weibel, E. K. 1987. The Biofreezer, a new contained freezing equipment in biotechnology. *Appl. Microb. Biotech.* 27:46–49.

Wisniewski, R. 1988. Design objectives for aseptic seals. In *Bioprocess Eng. Symp.* 1988, edited by D. E. DeLucia, T. E. Diller, and M. Prager, 11–21. New York: ASME.

Wisniewski, R. 1989. Anticipated effects of seal interface operating conditions on biological materials. In *Bioprocess Eng. Symp.* 1989, edited by T. E. Diller, R. M. Hochmuth, and Y. I. Cho, 87–96. New York: ASME.

Wisniewski, R. 1992. Unpublished results.

Wisniewski, R. 1996. Unpublished results.

Wisniewski, R. 1998a. Developing large-scale cryopreservation systems for biopharmceutical products. *Biopharm* 11(6):50–60.

Wisniewski, R. 1998b. Large-scale cryopreservation of cells, cell components, and biological solutions. *Biopharm* 11(9):42–61.

Wisniewski, R., and V. Wu. 1992. Large-scale freezing and thawing of biopharamaceutical drug product. Proceedings of the International Congress: *Advanced Technologies for Manufacturing of Aseptic and Terminally Sterilized Pharmaceuticals and Biopharmaceuticals,* 17–19 February, Basel, Switzerland.

Wisniewski, R., and V. Wu. 1996. Large-scale freezing and thawing of biopharmaceutical products. In *Biotechnology and biopharmaceutical manufacturing, processing, and preservation,* edited by K. Avis and V. Wu, 7–59. Buffalo Grove, IL: Interpharm Press, Inc.

Wolfe, J., and P. L. Steponkus. 1983. Tension in the plasma membrane during osmotic contraction. *Cryoletters* 4:315–322.

Wolfe, J., Z. Yan, and J. M. Pope. 1994. Hydration forces and membrane stresses: Cryobiological implications and a new technique for measurement. *Biophys. Chem.* 49:51–58.

Yoo, J., and S. Ro. 1991. Melting process with solid-liquid density change and natural convection in a rectangular cavity. *Int. J. Heat and Fluid Flow* 12:365–374.

Zabrodsky, S., and H. Martin. 1988. Fluid-to-particle heat transfer in fluidized beds. In *Heat exchanger design handbook,* vol. 2, edited by E. Schlunder. 2.5.5.1–2.5.5.7. New York: Hemisphere Publ. Corp.

Zeng, X., and A. Faghri. 1994. Experimental and numerical study of microwave thawing heat transfer of food materials. *J. Heat Transfer* 116:446–455.

Zhang, C.-T., and K.-C. Chou. 1996. Beat motion in DNA double helix and a mechanism of energy exchange between its two strands with microwave frequency. *Chemical Physics* 206:271–277.

4

METHOD SELECTION CONSIDERATIONS FOR LONG-TERM PRESERVATION OF MAMMALIAN CELLS AND MICROORGANISMS

Carmen M. Wagner

Wyeth-Lederle Vaccines and Pediatrics

INTRODUCTION

Microorganisms and mammalian cells are fundamental source materials for biopharmaceutical manufacturing. Long-term preservation, storage, and recovery of these cells are critical to ensure long-term availability of commercial products. The goal is to preserve and store cells without compromising purity, function, and genetic cell makeup. Furthermore, the recovered cells must consistently, reproducibly, and reliably provide an initial inoculum that will show optimum growth performance and production yields.

Chapter Focus

This chapter focuses on method selection for long-term cell preservation and presents some points to consider for addressing the technical, administrative, regulatory, and business issues associated with control of preserved cells. The key technical consideration is the selection of a suitable method that, at a minimum, considers application needs and addresses sample requirements. Administratively, it helps to institute policy and procedures to outline roles and responsibilities, specifying what needs to be done during seed bank development, in what sequence, and by whom. Instructions should also specify the content and format of the final seed development report, who should approve it, and where it should be filed.

As to regulatory considerations, commercial applications require that preservation be done in compliance with government regulatory requirements. In today's global business environment, this often means complying with worldwide global regulatory expectations. Regulatory requirements can be addressed through the development of a complete preservation program documentation, including the creation of inventory systems called seed or cell banks, documentation on historical seed information, seed characterization, and records for routine use. These expectations are defined in regulatory documents that cover the organization, management, and control of preserved specimens (ICH 1995; FDA 1985, 1992, 1993, 1995, 1996, 1997; WHO 1987). FDA documents can be obtained by fax. See Table 4.1.

In regard to business considerations, a well-thought-out preservation program, with proper controls and documentation during seed development and routine manufacturing, ensures the continuous availability and quality of final product. For business, technical, administrative, and regulatory reasons, the program should be well planned, carefully executed, and thoroughly documented. This chapter provides information on how to do that.

The Methods

Freezing and freeze-drying are the methods addressed in this chapter. Even though both methods can successfully preserve mammalian and microbial cells, application and/or sample requirements may point to one method as being more suitable than the other. Freezing and freeze-drying have been extensively used in biopharmaceutical manufacturing to preserve live cells intended for use in production and testing, and to preserve cells and their by-products that are used as reagents or product intermediates. In the remainder of this chapter, the focus will be on preservation of cells. The words "cells," "culture(s)," and "specimen(s)" will be used interchangeably to refer to biological source materials of either microbial or mammalian origin.

Table 4.1. FDA Documents That Can Be Obtained by Fax: Guidelines, Guidance, Points to Consider and Other Documents.

Hard Copy	FAX ID	Document Date	Title	Pages
D0372	0372	02/28/97	PTC* in the Manufacture and Testing of Monoclonal Antibody Products for Human Use	47
D0336	0336	12/22/96	PTC on Plasmid DNA Vaccines for Preventive Infectious Disease Indications	36
D0262	0262	01/02/96	Draft Addendum to the PTC in Human Somatic Cell and Gene Therapy	18
D0236	0236	08/22/95	PTC in the Manufacture and Testing of Therapeutic Products for Human Use Derived from Transgenic Animals	20
D0139	0139	07/12/93	Draft PTC in the Characterization of Cell Lines Used to Produce Biologicals	42
D0126	0126	04/06/92	Supplement to the PTC in the Production and Testing of New Drugs and Biologicals Produced by Recombinant DNA Technology: Nucleic Acid Characterization and Genetic Stability	9

Continued on next page.

Continued from previous page.

Hard Copy	FAX ID	Document Date	Title	Pages
D0124		03/01/92	PTC in the Manufacturer of In Vitro Monoclonal Antibody Products for Further Manufacturing into Blood Grouping Reagent and Anti-Human Globulin	16
D0115	0115	08/27/91	Draft PTC in Human Somatic Cell Therapy and Gene Therapy	21
D0090		08/22/89	PTC in the Collection, Processing, and Testing of Ex Vivo Activated Mononuclear Leukocytes for Administration to Humans	21
D0048	0048	04/10/85	Draft PTC in the Production and Testing of New Drugs and Biologicals Produced by Recombinant DNA Technology	14
D0038		06/20/83	PTC in the Manufacture of In Vitro Monoclonal Antibody Products Subject to Licensure	5

*Points to consider.

Available from the Center for Biologics Evaluation and Research Office of Communication, Training and Manufacturers Assistance, IIFM-40, 1401 Rockville Pike, Rockville, MD 20852–1448. 301–827–1800, 1–800–835–4709, FAX Information System 1–888–CBER–FAX or 301–827–3844 10–Mar–98

318

The technical aspects of freezing and freeze-drying have been extensively covered in the literature (Adams 1991; Ashwood-Smith 1980; Mazur 1970; Snowman 1991) and will not be covered in detail here. It is appropriate, however, to review briefly the two methodologies.

Freezing, or cryopreservation, is one of the simplest methods for preserving biotechnology specimens at temperatures that usually vary from –5° to –196° C. The upper temperatures in this range, –20° to –30° C in ordinary freezers, are suitable for storage of some microorganisms for 1–2 years. However, freezing at ultracold temperatures of –140° to –196° C (or deep-freezing) is preferred for long-term storage of many microorganisms and most, if not all, mammalian cells. Deep-freezing requires the placement of pure cultures in a suspending medium, freezing, and storing them for use up to several years later. This is an excellent way to adequately preserve metabolic and morphological properties, and cell viability and recovery are also usually good.

Freeze-drying, or lyophilization, is a multistep process that results in a cakelike residue that should be completely soluble and easy to rehydrate in a suitable liquid medium (Mellor 1978). Freeze-drying usually prevents shrinkage and minimizes chemical changes. The freeze-drying process is influenced by a number of factors, including the nature and age of the culture, the cell concentration, the nature of the suspending fluid, and the drying temperature used. Freeze-drying is the most economical and reliable method for long-term storage for those organisms that can withstand dehydration.

A major advantage of freeze-drying is the lack of requirement for specialized storage conditions or equipment. Additionally, freeze-dried cultures can be shipped and distributed without needing special storage conditions. One disadvantage is that dehydration can be mutagenic. Other concerns include the potential selection of contaminants or variants resistant to freeze-drying, loss of plasmid or plasmid function, seed contamination during dispensing, product cross-contamination, and potential genotypic changes, particularly during the drying stages of the freeze-drying process. (Ashwood-Smith 1976). The terms "freeze-drying" and "lyophilization" will be used interchangeably in this book.

Chapter Organization and Contents

Following this introduction, the reader will find a section on definitions, an overview of seed stock management, and key considerations in developing a master plan for method selection. Emphasis is given to regulatory compliance and the identification of activities necessary to satisfy GMP requirements, including documentation. Safety considerations are also addressed. The chapter ends with examples of procedures used for preserving microorganisms and mammalian cells.

DEFINITIONS

Freezing. The process of preserving specimens by exposing them to low or ultralow temperatures, typically from −5° to −196° C. Ordinary laboratory freezers allow storage at −20° to −30° C. Some microorganisms and mammalian cells are also stored for short periods at −70° C. Alternatives to ordinary freezing include storage in ultralow temperatures in mechanical freezers or colder temperatures in liquid nitrogen or nitrogen vapor phase environments (−140° to −196° C).

Freeze-Drying or Lyophilization. A multistage process used to achieve stable preservation of a wide variety of biological and pharmaceutical specimens. During freeze-drying, water vapor is removed directly from the frozen material by sublimation. The process consists of three major stages: freezing, sublimation, and desorption.

Short-Term Storage. In the context of this chapter, short-term storage refers to periods no longer than 6 months.

Long-Term Storage. In the context of this chapter, long-term storage refers to periods longer than 6 months. Some specimens have been stored for several years without negative impact on performance.

Master Seed Bank (MSB). Also known as Master Cell Bank (MCB) or Master Seed Bank System (MSBS), the MSB can be defined as a collection of cells derived from a single tissue or cell type. Cells giving rise to the MSB should be of uniform composition. The MSB is usually divided into 2 levels, the Master Seed Stock (MSS) and the Working Seed Stock (WSS). Some

MSBs also contain a third level, the so-called Submaster Seed Stock (SSS), from which the WSS is derived. In this chapter, the name MSB or MSBS will be used. The most common arrangement is the 2 level MSB, where the MSS is the origin of the WSS.

Master Seed Stock (MSS).   This level of the MSB contains the seed vials that are the origin for the SSS or the WSS. The MSS is the fundamental source material for product manufacturing and thus must be properly secured for the life of the product. The WSS should always be derived from the associated MSB.

Submaster Seed Stock (SSS).   The second level of the MSB may be the SSS, which is derived from one or more vials of the MSS. Not all MSBs have a submaster level. The submaster can add another level of security to the preservation of the specimen, but it does require an increase in the number of cell passages. Most often, this level is not part of MSBs developed for commercial applications and thus will not be emphasized in this chapter.

Working Seed Stock (WSS).   The WSS is usually derived from one or more vials of the MSS. If the SSS exists, then the WSS is originated from the SSS. The MSS is expanded by serial subculture up to a passage number selected by the manufacturer and approved by the regulatory agencies. The expanded culture is pooled and distributed into small aliquots and frozen for future use. The WSS is also known as the Manufacturer's Working Seed Stock (MWSS) or Manufacturer's Working Cell Stock (MWCS), since it is the seed stock used to initiate biopharmaceutical production.

Passage Number.   Once mammalian cells and microorganisms are established in culture, they require periodic media change or subculturing. These periodic media changes and serial passages are usually determined by the rate of cell growth, observed pH changes in the media, depletion of nutrients, and/or observed morphological changes in the culture. The number of times the cultures are subcultured or passed represents the passage number for that culture.

Seeding Density.   Most cells have a defined density for both seeding (starting the culture) and maximum culture density. That simply means that a minimum number of cells is required to allow the culture to thrive, stimulating growth and cell function (concentration necessary for seeding), and that beyond a certain cell concentration (maximum density that allows effective growth), the culture will not do well.

Cell Doubling.   Bacterial and mammalian cell growth is usually measured by an increase in cell numbers, not necessarily an increase in the size of the individual cells. Normally, one cell divides into 2 cells, 2 cells divide into 4 cells, and so on. When the arithmetic number of cells in each generation is expressed as a power of 2, the exponent expresses the number of cell doublings that have occurred.

Adventitious Agent.   Microorganisms that can contaminate cell cultures. These contaminating organisms include bacteria, fungi, mycoplasma, and endogenous and adventitious viruses.

MANAGEMENT OF SEED STOCK INVENTORY: THE MASTER SEED BANK (MSB)

Long-term preservation of cells is crucial to biopharmaceutical manufacturing because of the reliance on live source materials for production and testing. This long-term preservation program must assure stable and reproducible specimen recovery with minimum changes to cell morphology, physiology, and genetic makeup. Consequently, specimen preparation must be carefully planned, the preservation and storage method wisely chosen, and the preservation process accurately and thoroughly documented to ensure the quality of the specimen, process reproducibility, and overall product quality. Once all this is accomplished, specimen storage must be organized in an inventory system that ensures the security and control of the preserved specimen. The inventory system of choice is seed banks.

In commercial applications, it is common to store preserved specimens in small aliquots in the form of seed banks. This facilitates specimen traceability and good inventory management. In many ways, the term "master seed bank" invokes the image of small aliquots of mammalian cell cultures. However, the concept has also been applied to organizing the inventory of microorganisms and plant, yeast, mammalian, or insect cell cultures in small or large aliquots. Such an inventory system simply provides an effective way to organize and control the long-term storage of any preserved specimen maintained as pure cultures.

Seed Bank Development

The overall concept of a MSB System can be explained by the expansion of the initial seed into the MSS, then to the WSS, accompanied by appropriate quality control and documentation (Hay 1988). Briefly, the initial seed is expanded by serial subculture up to a certain passage number, at which point the cells are pooled and small aliquots are frozen as the MSS culture. The MSS can then be used as an internal reference standard to assure the maintenance of the phenotypic and genotypic characteristics of the production stock.

The next level, the WSS, is derived from at least one vial of the MSS. The MSS is used to create the WSS, as needed. The WSS can be accessed routinely to start production. If the WSS is accidentally destroyed, it can be easily replaced by expansion of the associated MSS. Replacing the master seed stock is not automatic and usually requires further FDA or foreign boards of health approval. It is also more time consuming and costly.

Different approaches can be used to develop Master Seed Banks. The approach described in Figure 4.1 uses one vial of the MSS to give rise to the submaster seed stock (SSS), which then gives rise to the WSS. This process flow can apply to microorganisms or mammalian cells. As stated previously, most commercial seed banks only have two levels: the MSS and the WSS.

The MSS and the derived manufacturer's WSS must have enough aliquots to ensure batch reproducibility and preservation of identity, purity, and genetic stability throughout the life of the product. This can be accomplished by controlling specimen handling, limiting the number of passages, and whenever possible, starting with cells at a low doubling level or passage.

Characterization of the Seed Bank

All seed levels should be well characterized to ensure consistency in passage number and genetic stability for all aliquots. Validated methods should be used to support testing during development and maintenance of the seed bank. Seed characterization should comply with regulatory expectations. FDA and WHO guidelines contain information on recommended characterization testing (FDA 1985, 1992, 1993, 1997; WHO 1987).

Seed Bank Security

The safety of the master seed is critical in commercial applications, and availability of the seed material must be guaranteed throughout the life of the product. To ensure security of the seed bank, it is recommended that preserved material be kept in locked,

Figure 4.1. Process flow for the development of a Master Seed Bank. Three levels (MSS, SSS, and WSS) are shown. In practice, most commercial seed banks only have two levels, the MSS and the WSS.

Development of a Master Seed Bank System

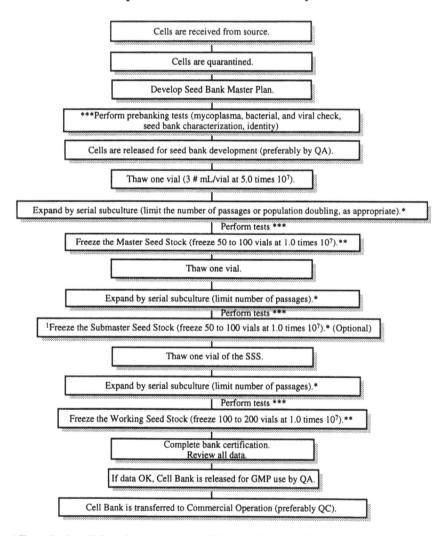

* The application will dictate how many passages will be needed. In general, allow as few passages as possible.

** The ultimate number of containers frozen also is dependent on how much culture can be expanded without causing changes in performance.

*** Test requirements will depend on cell source and application.

access-controlled freezers or controlled storage areas (for lyophilized specimens). It is also important to store preserved specimens in more than one location to avoid accidental loss of product. This is especially true in the case of mechanical freezers, due to the possibility of equipment failure or power loss. Inventory record logs should be maintained for all seed stocks. This can help determine when another WSS will have to be initiated to ensure continuity of production.

Validation

Another important issue is the validation of equipment and methods used in association with the seed bank development. Validation is further discussed later in this chapter.

Benefits

The Master Seed System, with the master, submaster (optional), and the working seed cultures, can be used to establish a safe source of starting material for product manufacturing and quality control testing. Use of this type of inventory system provides the best assurance that the same type of source material, with identical properties, will be consistently used. This, in turn, helps ensure the uniformity and reproducibility of the production stock and the final product vial.

KEY CONSIDERATIONS FOR METHOD SELECTION

The method is the central piece of the preservation program, but the method alone is not sufficient to ensure the program's success. The appropriate method selection must be accompanied by suitable documentation, method validation, equipment validation, plan for backup storage, and other considerations that will help ensure the long-term quality of the preserved material.

To start method selection, one should first define the outcome (i.e., what the application calls for). In other words, one needs to perform a needs assessment. If a familiar method is already available, the selection process can be simplified. Otherwise, one will need to identify one or more methods that may be suitable to the application and then consider the needs of the specimen, the needs of the application, and the method capabilities in order to select the best option. Questions associated with needs assessment are further defined in Table 4.2.

Table 4.2. Method Selection: Needs Assessment

What is the application?

What is the criticality of material being preserved?

How long do you expect to store the preserved specimen?

What are the special specimen requirements?

What is the impact of preservation on the specimen?

Is the main purpose of the preservation to preserve function or morphology?

Are cryoprotectants recommended? What is the appropriate choice?

Are there special culturing requirements? What are the optimum cultural conditions?

Is the optimum stage of growth for processing the organism or cell known?

What is the process scale? How will specimens be used after preservation?

What are the safety and GMP considerations?

What are the cost considerations?

What are the method/procedure limitations?

The development of seed banks and choice of preservation method may be approached differently by different companies. The approach presented here follows a formal methodology to select the best option and ensure the compatibility of the method, sample, and application in compliance with regulatory requirements.

The Specimen and the Application

The method must be compatible with the specimen characteristics and the intended use. In addition, the method design must fit the desired outcome. For example, the selected method may call for special sample preparation through concentration by evaporation, incorporation of cryoprotective agents, or filtration in order to increase recovery following thawing. These treatments, however, may not be suitable for the application because they may be harmful to the specimen. If this is the case, alternatives need to be investigated to identify the proper sample preparation conditions that fit the specimen requirements.

Another challenge is illustrated by an application that requires ambient storage because low-temperature storage is not easily available. In this case, storage at 25° C may be the only realistic option for the application in question but may not be appropriate for the specimen. In the same application, the key outcome may be high output of by-product, not high viable recovery of frozen specimen. Considering both requirements, a method calling for low-temperature storage would not be the best choice for the application and should not be considered. And, since high viable cell recovery is not required, a method such as freeze-drying can adequately address the application needs. Freeze-drying would also satisfy storage requirements for ambient temperature. This example illustrates the need to consider the sample's special requirements, the application, and the desired outcome. A compromise is often needed in order to accommodate all requirements.

Equipment and Reagent Needs

The sample and application requirements will, in many ways, dictate the choice of method. This, in turn, will dictate the kind of equipment needed for preservation and storage. Early considerations of equipment and reagents are important from a scientific perspective and also for budget planning and evaluation of needed resources. For example, if it is determined that specimens should be freeze-dried, the equipment needed will be costly and will require special expertise for its validation, operation, and maintenance. Early considerations for testing requirements are also important since some testing may require new equipment not available within the organization.

Current Good Manufacturing Practices (cGMPs) should be followed at every step, from equipment and reagent selection to validation and use. Additional comments on equipment and reagents follow.

Freezers

The choices include ordinary freezers and ultralow-temperature freezers. These can be programmable or mechanically controlled-rate freezers. Unlike storage at –80° C in ordinary mechanical freezers, liquid nitrogen storage often requires controlled-rate freezing equipment for controlled, slow-rate freezing of specimens. It is believed that controlled-rate freezing of microorganisms and mammalian cells is more effective than rapid freezing.

All freezers used to store preserved specimens should be alarmed and, preferably, connected to emergency generators in case of malfunction or general power failure. Liquid nitrogen freezers should have alarmed sensors that warn of low levels of liquid nitrogen. It is critical to monitor the liquid nitrogen levels to ensure that temperatures do not rise above –130° C, since that can negatively impact the stability of microorganisms in long-term storage.

Liquid nitrogen freezers require more attention and maintenance than ordinary and mechanical ultralow freezers. Unless automated connections are available for liquid nitrogen replacement, the user must schedule routine checks to ensure that a consistent supply of liquid nitrogen is available. These types of freezers also require special safety handling practices as described later in this chapter.

Freeze-Dryers

Small freeze-dryers can be utilized in the laboratory, whereas large-capacity equipment is commonly used for commercial applications. The equipment and supplies available from different vendors may offer different conveniences, but they should operate on similar principles. There are basically two types of freeze-dryers: the shelf and the centrifugal. Centrifugal freeze-dryers can be more technically demanding for inexperienced workers. Some freeze-dryers cannot support large-scale processes and are not built for commercial, heavy-use applications. During initial evaluation, it is important to consider the volume of the specimen and the type of organism in order to select the proper equipment.

Freeze-drying presents its own special challenges. For example, special precautions should be taken to avoid contamination of equipment and product during the freeze-drying operation (Barbaraee and Sanchez 1982). Validation of sterilization cycles and cleaning of the lyophilizer should be completed before the equipment is used for lyophilizing product. Another challenge involves scaling up the lyophilization process. The product to be lyophilized should be studied on a small, pilot scale. It is also desirable that the same type of lyophilizer used in pilot scales be used for production scale operations, if possible, to minimize problems during technology transfer. This is rarely the case, however, and the lack of equipment comparability is often the cause of many difficult challenges during technology transfer of lyophilization processes from development to commercial scale. In any case the production equipment should be capable of replicating the requirements established during the pilot scale runs.

Freezing Containers

Glass or plastic vials, ampoules, flat trays, bottles, flasks, and bags can be used to freeze or freeze-dry small or large volumes of a specimen. The choice of container will depend on the application, the specimen characteristics, the method selected, and the volume to be frozen. If bulk material is stable in plastic, it may be more convenient and practical to set up storage in large plastic bottles or bags. This may also facilitate handling and shipping of preserved materials. If glass vessels are used, they must be of adequate mechanical strength and resistant to thermal shock in order to survive the freezing procedure and storage.

Stoppered or capped vials seem to be the most appropriate containers for freeze-drying small volumes, but ampoules can also be used. Plastic, screw-capped vials are often used to freeze small volumes. Glass ampoules, both trimmed and funnel tipped, can be used for cell lines and animal viruses. Both types can be flame sealed. Bags may be the container of choice for bulk applications because they are easier to sterilize and transport than large bottles and may reduce the expense of package validation. For freeze-drying, they may work better than open trays since they facilitate handling and are less likely to introduce contamination. Bags should definitely be part of the initial evaluation to determine if they fit the application in question. Bags are usually available in different sizes and are often made up of multilayers of polyethylene and gas barriers to prevent oxygen or CO_2 contamination by permeation. They are usually sterilizable and disposable, and they can help prevent contamination due to improper cleaning of reusable containers (as in the case of reusable polypropylene or Teflon bottles). Contamination can also be prevented by designing bags with multiple ports for inlet, for outlet, and for sampling. Bags and ports should be capable of withstanding sterilization.

They should also be capable of holding the desired temperature during shipping, even if that requires additional insulation packaging. For all containers, it is important to ensure that the sealing is done in a clean, protected environment to prevent contamination.

Storage Boxes

Storage boxes should be made of construction materials that can ensure the integrity of the preserved specimen for long periods of time. Cardboard boxes may cause problems during storage in ordinary freezers because contaminating mold may grow vigorously due to high humidity levels. Stainless steel storage boxes, which are more durable and less likely to promote microbial growth, can be used in ordinary or liquid nitrogen freezers.

Reagents and Materials

Raw materials, freezing/thawing media, buffering solutions, and cryoprotectants are needed to support the preservation program. Primary and backup suppliers of buffers, cryoprotectants, culture media, and other reagents and materials should be qualified, and a program should be in place for qualification of crossover from old to new lots. In addition, assays used to qualify reagents and materials should be validated and documented in a written protocol, approved by designated functions. Expiration dating should be assigned to all reagents and materials based on in-house data or dating specified by the manufacturer.

Additives and Cryoprotectants

The preservation of animal cells and microorganisms has become common practice partly because of the availability of agents that protect cells and help minimize damage caused by freezing and thawing. The phenomenon of cryoprotection is not yet completely understood, and neither are the mechanisms of injury they are supposed to prevent. Nevertheless, cryoprotectants have been used successfully to help prevent significant injury to many cryopreserved biological materials.

Cryoprotectants can be classified as penetrating and nonpenetrating agents. Penetrating cryoprotectants are usually associated with protection of living cells against slow-freezing injury, while nonpenetrating agents are thought to protect frozen material during rapid rates of freezing and thawing.

Glucose, sucrose, bovine serum albumin, sodium glutamate, dextran, mannitol, glycerol, and dimethyl sulfhoxide (DMSO) have all been used as cryoprotectants. The cryoprotective properties of glycerol were first demonstrated in 1949 (Polge 1949). Since that time, the list of available cryoprotectants has grown, and research has helped clarify the special applications of individual cryoprotectants (Farrant 1980; Meryman 1977).

Glycerol and DMSO are the two cryprotectants most commonly used in liquid nitrogen freezing of microorganisms and mammalian cells. DMSO is usually used at final concentrations of 5 percent v/v. It can be sterilized with a Seitz filter and added to a cell mixture to achieve the final concentration of 5 percent v/v. DMSO does not work well in freeze-drying applications since it tends to concentrate to toxic levels. Glycerol is often used at 10 percent v/v.

Cryoprotectants should be of reagent grade and should undergo quality control testing. Preferably, they should be stored in small aliquots to prevent multiuse of the same storage container. This helps prevent contamination and oxidative breakdown. Sterility of cryoprotectants can be ensured by filtration or autoclaving, depending on the sensitivity of the cryoprotective agent to heat. For example, glycerol can be autoclaved, whereas DMSO should be filtered.

Cryoprotectants have been covered extensively in the literature (Fahy 1986; MacFarlane and Forsyth 1990; Meryman 1971) and are addressed in more detail in other chapters in this book.

Regulatory Compliance

Preserved specimens used in the manufacture of products for human use should comply with the GMP regulations and guidelines specified by the U.S. FDA, boards of health from countries outside the United States, and the World Health Organization, as appropriate. Activities that support regulatory compliance include those associated with quality assurance, quality control, and validation. Figure 4.2 illustrates the author's model for a Total Quality Assurance (QA) Program associated with specimen preservation. The Total QA Program emphasizes regulatory compliance and control of facilities, equipment, utilities, processes, and all documentation. It starts with activities associated with product research and continues through product development, scale-up, and technology transfer to operations for product commercialization.

Figure 4.2. Illustration of an approach to ensure the total quality assurance of seed bank development.

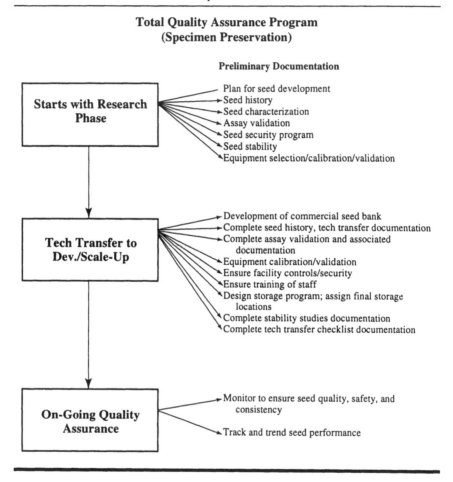

The quality *assurance* section in this chapter addresses all activities associated with general GMPs and documentation, whereas quality *control* activities are associated with laboratory testing. The validation section addresses facility, equipment, utilities, assays, and process validation. Method validation includes those assays used to characterize the MSB and is associated with control of raw materials and reagents.

Quality Assurance

The development of the MSB should comply with government regulations and company policies regarding GMP, auditing, and documentation. Thorough traceability through accurate documentation is good business practice and is a GMP requirement. Some key quality assurance considerations needed to control the seed bank follow.

Prevention of Viable and Nonviable Contamination During Freeze-Drying. Industrial freeze-dryers used for biological products should be designed so that the chamber opens into a slightly pressurized, Class 100, aseptic environment to prevent microbial contamination and intrusion of particulates. An alternative includes a sterile air curtain to protect the chamber door and the product. The freeze-dryer chamber is sterilized by gas treatment (such as vaporized hydrogen peroxide gas) or by steam under pressure. Product protection should be provided during loading and unloading.

Product Stability. Stable reagents, stable seed material, and stable product intermediates will most likely lead to stable final products. Starting with a stable seed is of the utmost importance to assure the availability of a stable final product.

Specimen Consistency and Stability. Procedures used to preserve specimens for subsequent storage and handling will influence the genetic stability, recovery, and function of the cell population. Typically, the freezing and thawing process has a negative impact on stability. Therefore, large volumes should be divided into manageable aliquots to prevent repeated cycles of freezing and thawing that can potentially reduce viability and can affect functional performance. Real time stability studies should be part of the seed bank characterization for commercial applications and should start in the development phase of the product life cycle. A number of vials should be put aside and thawed at different time points (for example, 1, 3, 6, 9, 12, 24, and 36 months) to evaluate seed performance upon thawing following prolonged storage. Vial contents should be checked for viability, allowed to grow and tested that seed characteristics are still acceptable.

Transport of Preserved Biological Specimens. Bulk materials are often shipped to other company sites or other locations. They are often maintained frozen or at 2–8° C in bottles, vials, or bulk containers/bags and, by quality assurance and federal regulation, are expected to be shipped under equivalent conditions. It is important to plan up front the conditions needed for storage and shipping and to develop procedures for ensuring the appropriate conditions during transport. These procedures must be approved by appropriate personnel and should conform to government regulations. Procedures also should be on file in the responsible department. Users are advised to work with suppliers of bags and other containers to ensure that the proper transport equipment (insulated packages) will ensure cost-effective and validatable shipments.

The Documentation System. Documentation should consist of production batch records, SOPs, change control forms, technology transfer documents, and other supporting documentation. At a minimum, the company should assure that:

- The MSB documentation clearly supports the characterization of the preserved specimen, including identity, purity, seed history, traceability, and genetic stability.

- Storage conditions are controlled and documented at all locations.

- Documentation is available to demonstrate the absence of contamination, including process validation studies for inactivation and removal of adventitious agents and other contaminants during the seed development and simulated seed expansion for production, if appropriate.

- All failures to meet specifications are fully documented in an investigation report or form, as appropriate.

- Written and approved procedures to monitor and control contamination of the master seed bank and production working seed stock are available.

- Proper documentation for all bovine-derived material regarding issues with Transmissible Spongiform Encephalopathies (TSE) are in place.

- Rejection criteria for thawed specimens are included in procedures.

- All documents are easily accessible.

During routine manufacturing of commercial products, the company must ensure that:

- Batch records fully reflect written procedures and commitments made during the filing.
- Growth promotion and functional integrity of stock are supported by data.
- Use of the working seed bank is consistent with validated processes regarding methods for recovery and for expansion of preserved cells.
- No critical changes to the process have occurred to compromise the validated removal/inactivation of contaminants.
- New working seed development does not deviate from license commitments.
- Transport of frozen materials abides by conditions equivalent to those of storage.
- Test reagents, raw materials, and other components are within their expiration dating when used in the seed propagation for product manufacture.
- All raw materials are controlled and released by QA.
- All containers of the seed bank are consistently treated according to specific procedures during storage and use. For example, once removed from storage, containers should not be returned to stock to avoid problems from repetitive freezing and thawing.

Quality Control

Quality control (QC) of the preserved specimen is another important piece of the total product quality program. The quality control laboratory plays an important role in the confirmation of the quality, safety, and efficacy of product. Communication with the QC laboratory must start early in the planning of the preservation program to ensure the availability of validated equipment, methods, and qualified reagents when seeds are transferred to QC. The quality of the seed bank is closely associated with the accuracy of the data that is generated by development and transferred to QC. This, in turn, is closely associated with the quality of the reagents, standards, and the validity of the methods used to generate the data. Specific activities associated with quality control testing include:

- Control of raw materials, the culture media, and test reagents
- Method and lab equipment validation
- Development and approval of rationale for all test specifications
- Auditing of outside contract testing laboratories and review of protocols and test results, if appropriate
- Access to all generated documentation in support of seed development
- Documentation of assay transfer from Development to QC

Examples of quality control documentation are shown in Figure 4.3 and Figure 4.4. For companies doing business worldwide, method validation and validation documentation must comply with worldwide regulatory expectations.

Testing

Testing activities include seed characterization and testing of raw materials, components, and other laboratory reagents. Raw materials and components must be tested by Quality Control and released by Quality Assurance before they can be used in the preparation of GMP master seed bank and product manufacture.

All laboratory reagents and reference standards must be well characterized and properly stored to ensure their purity and stability. Selection of characterization methods should be based on technical assessment of specimen properties, application requirements, and overall method performance, including limits of detection or sensitivity.

Validation

As stated previously, validation considerations are an integral part of the total product quality. A master validation plan should be in place and should include validation of equipment, utilities, facilities, processes, assay methods, container-closure systems, assay standards, reagents, and test controls as appropriate. A brief explanation of each validation category follows:

Equipment/Utilities. All equipment and utilities used to develop the MSB and to store preserved specimens must be validated and

---

Figure 4.3. Example of inventory log sheet to track/record information on use of seed vials that are part of the seed bank.

Freezer #: _____   Cell Type: _____

Initial # of Vials Frozen: _____

Vials Removed	Vials Remaining	Operator Initial/Date	Comments

---

monitored. In addition, the laboratory equipment, instruments, and utilities used for testing or to support validation efforts must be calibrated and validated.

Guidance on the validation of liquid nitrogen freezers is available in an excellent article by Simione and Karpinsky (1996). Validation and other GMP considerations for lyophilization equipment are widely covered in the literature (Snowman 1991; Trappler

---

Figure 4.4. Example of document that can be used to summarize information on the master seed bank development. This information, together with the batch sheet for seed bank development, can become part of the technology transfer summary.

Master Seed Bank Development

Source of Material _____

Volume _____   Circle one: MSS, SSS, WSS

Passages   P#:   _____

               Date: _____

Freezing   Concentration: mg / mg/ vial   No. of vials frozen: _____

               Date: _____

               Storage Box #: _____

Testing Yes   ☐   Type of Testing: _____

               No   ☐   _____

                                                _____

                                                _____

Processed by: _____   Date: _____

Reviewed by: _____   Date: _____

    QA Released by: _____ Date: _____

1989; Williams and Polli 1984). Table 4.3 summarizes the important components of the validation of a lyophilized product.

Facility. The facility used to develop the seed bank should be built, validated, and routinely monitored for environmental bioburden to prevent seed contamination. Access to the facility should be restricted to authorized personnel only. The air quality and room surfaces should meet acceptable Class 100 standards (ASTM 1994). Alternatively, isolation technology can be used. If isolators are used, they must be validated, and monitoring procedures must be put in place to ensure the integrity of the work environment within the isolator (Coles 1998, Wagner and Akers 1995).

Process. Process validation makes good business sense and is a GMP requirement. For mammalian cell lines, the process should be validated for viral removal and inactivation. Additional information on validation of cell lines used to produce monoclonal antibodies and recombinant DNA products may be found in the *Points to Consider in the Manufacture and Testing of Monoclonal Antibody Products for Human Use* (FDA 1997) and the *Points to Consider in the Production and Testing of New Drugs and Biologicals Produced by Recombinant DNA Technology* (FDA 1985). In addition, useful information can be found in the "Supplement to the Points to Consider . . . Recombinant DNA Technology: Nucleic Acid Characterization and Genetic Stability" (FDA 1992). A guideline is also available for *Characterization of Cell Lines Used to Produce Biologicals* (FDA 1993). All of these documents are published by the Food and Drug Administration (FDA), Center for Biological Evaluation and Research (CBER).

Methods. Methods utilized in cryopreservation or freeze-drying should be validated. Others requiring validation include, but are not limited to, methods used to demonstrate the integrity of container-closure systems, to ensure the purity and overall integrity of the frozen specimen, and to control raw materials, reagents, and standards. The European ICH Harmonized Tripartite Guideline on analytical testing provides guidance on validation of analytical procedures (1994, 1997).

Container-Closure Systems. The integrity of container-closure systems is essential to ensure the integrity of the frozen specimen during freezing and storage. Some important considerations for the validation of the container-closure system include freedom from

Table 4.3. List of Considerations for the Validation of a Lyophilized Product.*

Preparation of Container-Closure System

Vial washing with WFI: particle, chemical, and microorganism removal
Dry heat sterilization: endotoxin removal (depyrogenation)
Closure washing: chemical and particle removal
Sterilization: steam, VHP, or other

Compounding

Control of temperature, pH, etc.
Mixing: speed, time, potency, and consistency

Equipment

Filtration System

Sterilization
Integrity testing: bubble point and forward flow
Bacterial retention studies

Aseptic Filling

Check weight
Media fills
Environmental monitoring

Lyophilization

Cleaning and sterilization (machine only)
Mechanical systems: leak test, vacuum pump, ice capacity, compressors, and
    shelf temperature uniformity
Product freezing: load configuration and temperature controls
Primary drying (frozen)
Secondary drying
Cake: uniformity, height, color, and reconstitution

After Lyophilization

Stoppering vials

Class 100 or better environment
All loads and vial sizes validated
Environmental humidity controlled

Capping of vials.

Leakage tests: normal and elevated temperature
Bacterial challenge
Package integrity

*If isolators are used, additional considerations apply (Wagner and Akers 1995).

viable and particulate contamination and freedom from endotoxins. Absence of leaks, impermeability to gas transfer, and maintenance of sterility throughout the shelf life of the product are also required.

The Documentation System

Documentation is the essence of GMPs, and it is also good business practice. It is important to consider documentation needs early in the life of the project to satisfy business needs and meet regulatory requirements. Early assessment of documentation requirements can help the process of resource and budget planning. Documentation is the key for recording activities associated with quality assurance, quality control, and validation. Accurate and easily accessible records should be available to document the history, the traceability of the specimen, and any other activity associated with the preservation program. This is particularly important if the material is being used to support clinical trials or commercial product manufacturing and needs to comply with cGMPs.

Documentation is also one of the main components of an effective technology transfer program. The technology transfer documents should be identified early to establish the content and desired format. Figure 4.5 presents an example of a document that can be used to assist with the technology transfer. The format for all documents may differ according to company policies and style, but the topics covered should be similar. If the technology transfer is associated with a Pre-Approval Inspection (PAI), Table 4.4 provides guidance for PAI preparation.

Safety Considerations

Four major aspects of safety are addressed: the safety of the seed stock, operator safety regarding procedures and equipment, safety associated with shipping of frozen specimens, and biohazard waste handling.

Safety of the Master Seed Bank (MSB)

There are two major aspects of safety associated with the seed system. One relates to the physical protection to prevent loss of specimens. The other is associated with maintenance of the integrity of the seed, such as protection against contamination with adventitious agents or seed cross-contamination.

Figure 4.5. Example of a document that can be utilized to record all key information needed for an effective technology transfer.

Product File/Code	Cell Line Transferred	

The following information is needed to transfer seed stock into QA/QC Laboratory. Please supply all listed information in the transfer package, initial, and date the appropriate column. N/A any items that are not applicable.

Source Laboratory: _____ Contact Person: _____
Receiving Laboratory: _____ Contact Person: _____

Checklist for Bacterial/Cell Line Seed History

Information Requested	Information or Notes
Cell Line	
Name: Code Number:	
Characterization: gram reaction, microscopic description, antibiotic sensitivity, plasmid designation, recombinant, etc.:	
Cryopreservative:	
Storage conditions:	
Required equipment, lab environment:	
Plating (agar) growth medium:	
Liquid growth medium:	
Special nutrition requirements:	
Growth conditions (temperature, atmosphere, rpm, etc.):	
Identity method:	
Seed History Qualification	
Seed source (ATCC, clinical isolate, etc.):	
Seed identification / Name:	
Flow diagram of seed history:	
Seed expansion strategy:	
Master seed backup available? Storage locations(s):	
Documentation	
References (e.g., SOP #'s):	
Review of cell growth, storage, ID protocols in both labs by (names):	
Training documentation:	
Data (e.g., seed stock) storage location(s):	

Documentation of associated information and location:
  Equipment calibration/validation:
  Vendor qualification(s):
  Raw material/reagents qualification:
  Assay validation documentation:	

Figure 4.5. Continued.

Literature references:	
Safety	
Safety and biosafety information:	
Vaccination requirement (if any):	
Acceptance Criteria	

Acceptance criteria for transfer:
    Requirements summary:
    Data summary:	

Conclusion:	
Laboratory qualification statement:	
Signatures	
Source laboratory approval:	Name_____Date_____ Title_____
Receiving laboratory approval:	Name_____Date_____ Title_____
Quality assurance approval:	Name_____Date_____ Title_____

Release of seed for manufacturing (select one, initial, and date. Write comments, if appropriate.):

Approved_____   Denied_____

Comments:

_____

_____

The MSB, primarily the MSS, has to be physically secured to ensure the long-term availability of the source material used to manufacture the commercial product. The security program should satisfy business and cGMP requirements. The usual approach includes local storage at the plant site and one or more remote storage locations. Many companies take advantage of expert services, such as the one provided by American Type Culture Collection (ATCC), and use their facility as one of the remote locations for storage of seed bank vials.

The seed system should also be protected from intraspecies or microbial contamination throughout its development, storage, and maintenance. Seed stocks should be well characterized and should be tested periodically to ensure the proper identity of the specimens and the absence of contamination. Viral safety evaluation of biotechnology products derived from animal cell lines is a regulatory requirement (ICH 1998).

Recently, Transmissible Spongiform Encephalitis (TSE) safety has gained increased attention in Europe and the United States and some

Table 4.4. Preparation for a Pre-Approval Inspection - A Checklist.

	Check/Comments
• Is the history and general characterization of the cell bank easily accessible? Is it complete?	
• Is an Environmental Monitoring program in place for the cell banking laboratory?	
• Is a facility cleaning program in place for the cell banking laboratory?	
• Has the effectiveness of cleaning reagents been validated?	
• Is access to the laboratory restricted with a locked door? Is access to storage unit(s) restricted, and is unit locked?	
• Is air HEPA filtered?	
• Is 100% of the air exhausted?	
• What is the air exchange rate?	
• What is lab classification?	
• Are surfaces built for easy cleaning?	
• If a mammalian cell bank: Is routine testing for adventitious agents part of the quality control program?	
• Does the MSB qualification include validation of viral elimination?	
• Are characterization methods validated and documented?	
• Is all the equipment validated? Is validation reveiwed by QA?	
• Are SOPs available and approved?	
• If isolators are used, have they been validated? Is validation documented?	
• What Environmental monitoring program is available for the isolators?	
• Are sterilization logs available for the isolators?	
• Is the rationale for choosing the method in a written format? Is it readily available?	
• Is an isolator system used? If so, is it validated?	

very restrictive guidelines regarding the use of bovine-derived reagents could come into effect in the future. The World Health Organization has recently issued a consultation report (WHO 1997). In the United States, as required in the 21 CFR 207.31, manufacturers of biological products are required to provide information regarding the sources and control of any bovine- or ovine-derived materials. The MSB should be free of bovine- or ovine-derived materials to avoid concerns with potential TSE contamination of the final product.

Operator Safety

Handling of Specimens. Microorganisms and human cell cultures should be handled with care to ensure the protection of the culture as well as the protection of the operator. This is especially true for tumor cell lines because of the known presence of oncogenes. The operator should be informed of the potential biohazards associated with each specimen and should know the safety precautions associated with it. Material Safety Data Sheets (MSDS) should be available to describe the safety precautions for all chemical reagents. An equivalent of the MSDS should be available for all cell cultures. It is prudent to handle all specimens in a vertical, unidirectional flow biosafety cabinet or isolator system to protect the operator and to protect the specimen against microbial contamination. It is also advisable to handle one type of specimen at a time to avoid the risk of cross-contamination between 2 or more products.

Equipment. Special attention should be paid to handling liquid nitrogen. Cryogenic storage can present many opportunities for accidents if appropriate safety precautions are not used. Glass or plastic ampoules and vials can be used to store frozen samples in liquid nitrogen or in the vapor phase. Pinhole leaks in ampoules have been known to allow the penetration of liquid nitrogen into the ampoule, causing it to explode upon warming. Vials are usually safer but can also become a hazard if liquid nitrogen accumulates at the cap-vial interface due to an inadequate seal. The accumulated liquid nitrogen can spray and cause burns when the vial comes into contact with warmer temperatures. Leakage may also cause the contents of the vial to contaminate the work environment and the operator. Storage of vials in the vapor phase eliminates the risk of penetration of liquid nitrogen into the vial and thus prevents the danger of explosion and personnel contamination from broken vials and ampoules. Further, using the pull-seal technique for sealing glass ampoules may help reduce the risk of pinhole leaks in the sealed tip. If using ampoules, use special safety precautions when opening them to avoid the possibility of skin cuts.

Another matter that requires attention is the extreme cold temperature of liquid nitrogen that can cause rapid freezing damage to exposed human tissues. Personal protection—such as face and neck shields, insulated gloves, and special aprons or lab coats—should be worn when handling liquid nitrogen. Skin and eye exposure to liquid nitrogen's extreme cold temperatures must be avoided.

Liquid nitrogen freezers should be placed in well-ventilated areas to prevent asphyxiation of the operator. During fill operations and other use, liquid nitrogen can quickly displace the room air and cause oxygen deprivation to the operator. If several freezers are concentrated in one area, it is advisable to install an oxygen monitor that can immediately warn operators if nitrogen gas infiltrates the area and air quality deteriorates.

Freeze-dryer safety must also be considered. Special precautions should be taken to avoid contamination of equipment and product cross-contamination during freeze-drying. Validation of sterilization cycles, cleaning validation, and personnel training should be complete before equipment is released for routine use.

Shipping

As stated in Appendix D of the CDC/NIH booklet on *Biosafety in Microbiological and Biomedical Laboratories* (1993), "The interstate shipment of indigenous etiological agents, diagnostic specimens, and biological products is subject to applicable packaging, labeling, and shipping requirements of the Interstate Shipment of Etiologic Agents (42 CFR Part 72)." These guidelines can also be applied to internal transportation of specimens. Proper labeling and packaging can help avoid accidents and personnel/environment contamination. Packaging and labeling requirements for interstate shipment of etiological agents are shown on Figure 4.6.

Waste Handling

Safe waste handling is an important component of the overall safety program. Waste generated from handling cultures of infectious agents and other human-derived cells or tissues is usually considered to be regulated biological or medical waste and

Figure 4.6. Packaging and labeling requirements for interstate shipment of etiologic agents based on 42CFR, Part 72. From: CDH/NIH, 1993. *Biosafety in Microbiological and Biomedical Laboratories.*

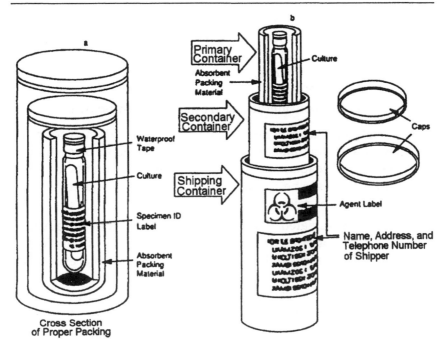

Figure 4.6 (a) and 4.6 (b) diagram the packaging and labeling of etiologic agents in volumes of less than 50 mL in accordance with the provisions of subparagraph 72.3 (a) of the regulation on Interstate Shipment of Etiologic Agents (42 CFR, Part 72). A revision has been proposed that may result in additional package labeling requirements, but this has not been issued as of the publication of this third edition of BMBL.

For further information on any provision of this regulation, contact:
Centers for Disease Control and Prevention
Attn: Biosafety Branch Chief
Mail Stop F–05
1600 Clifton Road N.E.
Atlanta, GA 30333
Telephone: (404)639–3883
FAX: (404)639–2294

Note that the shipper's name, address and telephone number must be on the outer and inner containers. The reader is also advised to refer to additional provisions of the Department of Transportation (49 CFR, Parts 171–180) Hazardous Materials Regulations.

must be managed accordingly. Procedures should be available to guide employees on how to decontaminate and dispose of any broken glass vials or general labware used during seed development and storage. This also applies to handling solid waste such as absorbent paper, gloves, pipette tips, tubes, flasks, culture dishes, and sharps used during manipulations of the seed stock material. Further guidance is available in the CDC/NIH biosafety manual (CDC 1993) and in the WHO *Laboratory Safety Manual* (1993). For large-scale work, guidance is available in the work of Maigetter et al. (1990).

General safety considerations are summarized below:

- Safety information should be available for all microorganisms and mammalian cells. MSDS must be in place for all chemical reagents. MSDS-equivalent documents can be used to describe all safety precautions associated with a given cell or microorganism.

- Issues associated with the presence of ovine or bovine-derived materials in a manufacturer's process should be addressed early in the project. It is advisable to prevent use of such materials in the MSB preparation and/or storage to expedite European regulatory approvals. Further details on safety recommendations are available (WHO 1997).

- The company's Institutional Biosafety Committee (IBC) should review the safety issues associated with all recombinant-derived organisms. Nonrecombinant organisms also should be reviewed by a biosafety committee. Biosafety levels should be assigned to all organisms. Recommendations for biosafety levels are given in the CDC/NIH biosafety manual (CDC 1993).

- Personal protection is a must when handling liquid nitrogen. Gloves, a face shield, and a lab coat or apron help protect skin against freezing burns from liquid nitrogen.

- Ampoules or vials must be removed carefully from the liquid nitrogen freezer and opened in a laminar flow biosafety cabinet. Storage should be such that containers are easily accessible and easy to retrieve.

- Adequate ventilation must exist when inserting or removing ampoules from the liquid nitrogen freezer.

PRESERVATION PROCEDURES

Several methods have been used for short-term preservation of biological specimens, including subculturing, immersion in mineral oil, drying, and ordinary freezing of microorganisms. However, long-term preservation and storage of microorganisms and mammalian cells require more specialized methods such as deep-freezing or freeze-drying. These two methods have become popular choices due to improvements in refrigeration systems and the greater availability of liquid nitrogen and commercial lyophilizers. This section provides examples of freezing and freeze-drying procedures that can be used to preserve mammalian cells and microorganisms. Following some general comments, Part I will address mammalian cells; Part II, bacterial preservation; Part III, procedures used for preserving virus-infected cells; and Part IV, other support procedures.

General Comments

Preservation in liquid or vapor phase nitrogen or lyophilization are the preferred methods for specimens that must be stored for long periods. The choice between lyophilization and ultralow freezing depends on the specimen characteristics and the application.

Many freezing procedures place the specimen immediately in the temperature that will be used for storage, but this approach is not suitable for all types of specimens. Sensitive specimens—such as some mammalian cells, certain viruses, bacteria, algae, and protozoa—may need to be frozen in two stages. In the first stage, the specimen is cooled slowly at a rate of 1° to 2° C/min. The specimen is held at this initial temperature temporarily to allow for some dehydration and cell shrinkage but briefly enough to prevent intracellular ice formation. In the second stage, the specimen is quickly brought to the final storage temperature of −70°, −140°, or −196° C. Programmable and mechanically controlled-rate freezers are available and can be used for applications that require well-controlled and reproducible processes (such as for GMP applications).

Regarding thawing, the opposite is usually true. Most specimens are helped by quick thawing followed by cell wash for quick removal of cryoprotectants from the suspension medium. It seems that rapid thawing combined with optimal cooling can enhance the chances for successful recovery after ultralow temperature preservation. For small volume containers, immersion in a shaker water bath at 37° C for a few minutes is sufficient to promote quick thawing.

During method selection and during seed bank characterization and qualification, viability assessment should be performed to establish the baseline before freezing. Viability should also be tested after thawing to assess method performance. One or more containers should be checked to ensure the quality of the material against preestablished specifications. The number of containers tested will depend on the size of a lot. It is common to use one vial for each seed lot of less than 50–60 preserved containers. For larger lots, 3 or more vials—from the beginning, middle, and end of the fill—can be pulled for testing to ensure better representation of the quality of the whole lot. Each lot of preserved specimens should be tested at different stages of development to demonstrate the maintenance of identity, purity, and genetic properties. Once the seed bank is established, batch-to-batch consistency should be verified with respect to the starter culture's uniformity and stability, according to predetermined specifications. Table 4.5 summarizes the testing associated with a mammalian cell bank development. Further information on freezing and freeze-drying procedures is available in the literature (Hay 1989; Simione and Brown 1991). Microorganisms and mammalian cells are addressed separately to emphasize some of the special requirements for each of these specimens.

Special Hints for Successful Outcome

- Start with healthy cultures, usually near the end of the log growth phase.
- Identify optimum growth conditions for cells to be preserved.
- Identify optimum stage of growth for harvesting organisms. This will vary for different organisms, but in general, late log phase is most suitable for preservation.
- Be aware that, in general, cells such as spores with very little water content are injured by low temperatures.
- It is believed that freezing bacteria in the presence of growth medium generally permits good recoveries.
- Be aware that, in general, poor viability yields may still be adequate for perpetuation of microbial cultures. Standards

Table 4.5. Development of the Master Seed Stock for Bacterial or Mammalian Cell Line.

Phenotypic Characterization

Microscopic evaluation
Isoenzymes studies

Genotypic Characterization

DNA fingerprinting
Southern Blot Analysis of total DNA
Polymerase Chain Reaction (PCR)

Molecular Characterization

Sequence analysis
Restriction endonuclease mapping

Functional Performance

"Use test" characterization: checking for reliable and reproducible production of by-product

Assessment of Contamination

Checking for
Mycoplasma
Viruses
Bacteria

for recovery may not be as rigid for microorganisms as they are for mammalian cell cultures.

- Where yields are particularly poor, the addition of cryoprotectants—such as DMSO or glycerol—before freezing is beneficial.

- In the storage of freeze-dried material, the residual moisture content (proportion of residual water to dry substance) appears to be a vital stability parameter.

- Different organisms subjected to apparently identical conditions of freeze-drying may show different survival rates.

- In most cases, microorganisms are most sensitive to drying during the logarithmic phase of their growth cycle.

- Microorganisms highly sensitive to freeze-drying may do better if they are left to rehydrate for a few minutes before subculturing. This might allow for the initiation of repair mechanisms and may improve recovery.

Evaluation of Method Performance

To consider the method successful, an adequate number of cells must initiate growth under ordinary conditions. In some cases, cells will only grow on complete media; in others, cells will only grow under special conditions. The effectiveness of the procedure can be assessed by evaluating specimen performance following recovery. This can be done by looking at viability/reproducibility, functional/metabolic stability, and genetic stability. A brief description of each follows.

Viability, Growth, and Reproduction

The preservation method can be assessed by the percentage of viable cells recovered (viability) and by the ability of the inoculum to grow and reproduce in the recovery medium. The degree of injury varies with the specimen and method used. According to the literature, cell injury and death are frequent occurrences during preservation; the question is, how can both be minimized or prevented altogether (Choate 1967).

Routine qualitative and quantitative microbiological methods can provide information regarding the success of recovery of microorganisms. Qualitative methods show the absence or presence of specified organisms, whereas quantitative methods enumerate the number of microorganisms present. Both approaches are recognized by the various pharmacopoeias. In the simplest approach, a qualitative assessment of viability can be made by transferring the contents of a thawed vial to a selected growth medium, with subsequent examination of growth or no growth. More precise and quantitative assessment can be achieved by a direct measurement of cell numbers by plate counts following serial dilution. Plate counts can be done by the pour or spread plate method, as described in Part IV in this chapter. For mammalian cells, viability and growth can be assessed by direct measurement of cell counts in a hemocytometer or by performing growth evaluation experiments in specified growth medium. These are illustrated later in this chapter. For both types of specimens, selection of the proper medium is essential to avoid results that are misleading. In other words, ensure that the medium is not the cause of the lack of growth.

The size and complexity of the organism can significantly impact survival after lyophilization or deep-freezing; however, size and complexity are not the only determinants of success. For example, viruses are smaller and less complex than bacteria but have been shown to be generally more sensitive to lyophilization. This could be due to the fact that viruses need to associate themselves with a host cell and be infective to survive. It should be noted, however, that among viruses there seems to be a correlation between size and survival. Small viruses can be less sensitive to lyophilization conditions than large viruses.

It must be emphasized that other aspects of the preservation process can add to cell injury and thus impact the performance of the preserved specimen. For example, methods for washing and harvesting cells during preservation and thawing can cause metabolic or structural injury. Cold shock from subjection to rapid chilling is also a possible cause of injury. Many protocols routinely utilized in microbiology or cell culture laboratories call for spinning cells at g forces in the range of 5 Kg to 15 Kg, which in centrifuge tubes usually translates to high hydrostatic pressures. If proper care is not exercised upon thawing, these washing procedures can cause stress and impact the performance of the preserved specimen. This lack of performance may not be directly related to injury caused by freezing but may be due to improper recovery practices. Such practices include using solutions removed from the refrigerator and not allowed to warm up or washing cells that have been warmed up in a refrigerated centrifuge or cooled centrifuge rotor. Users are cautioned to exercise care with such practices to avoid adding unnecessary stress to preserved specimens, thereby compromising performance.

As already related, viability can be a good indicator of successful preservation, but it is not the only attribute of a successful preservation program. Loss of viability can be less critical if functional performance is the key desired outcome.

Functional Performance

Preservation effectiveness is often assessed in terms of functional performance. Functional performance should be considered when selecting a method, since it has been shown that in some cases methods preserve the viability but not the functional performance of the specimen. This was demonstrated by Sharp (1984), who showed that organisms that exhibited high survival after preservation could lose the ability to produce the expected end product. The same has been observed in mammalian cell culturing, where recovery of viable cells is high, but the ability to produce a given by-product is lost or significantly reduced (Wagner, unpublished observation).

The foregoing examples illustrate that biochemical or biological functional performance can also be a good indicator of successful preservation for both microorganisms and mammalian cells. Functional performance, together with viability, can help provide sensitive and quantitative measures for effective quality control of preserved specimens. A small decrease in viability in the first year may not be a reason for serious concern. However, loss of function soon after freezing may indicate that the specimen's functional recovery may be unacceptable in a few years. In a production setting, where the life of a product is on average 5 years, such rapid loss of function can be a reason for concern, even if viability remains high.

Genetic Stability

One of the primary goals of a preservation program is to ensure that the genetic composition of the progeny is the same as that of the original culture. This is primarily an issue for mammalian cell cultures where extensive evaluation of the genetic profile is undertaken. In the case of microorganisms, complete genetic assessment is rarely done, and some genetic variation may go undetected. In practice, this may not be a concern because it has been shown that most auxotrophic mutants are not as hardy as the parent prototrophs, and a small number of mutants would probably not survive preservation conditions. In spite of this theory, it is important to strive for methods that prevent development of mutants. The good news is that growth and reproduction tend to be arrested under extreme conditions of pH, desiccation, or temperature. The bad news is that even one of the popular methods of choice—freeze-drying—has been associated with inducing mutations in microorganisms (Ashwood-Smith and Grant 1976). In general, however, the conditions normally present during freezing and freeze-drying are not likely to lead to genetic instability and/or survival of any mutants. As a whole, in spite of these potential problems, freezing and freeze-drying methods are still the best choices for preserving biological specimens.

Part I: Procedures for Preservation of Mammalian Cells

General Comments

The source seed material used in product manufacture may very well be the most important component of the production system since without it there is no product. Seed preparation is the first stage in a series of aseptic manipulations that culminate with a final sterile product, and the seed must remain stable and reproducible for the life of the product. Perhaps these are some of the reasons that have lead regulatory agencies worldwide to see the starting production inoculum as an important first step in ensuring the quality and safety of the final product.

The quality of biopharmaceuticals starts with a well-developed and well-controlled MSB. One of the major challenges of mammalian cell seed banks is to assure the absence of contamination. The complexity of the system, the many variables involved, and the multitude of steps needed to develop the seed bank provide many opportunities for contamination. The initial culture, the media components, and other raw materials are all potential sources of contamination. In addition, the equipment, the facilities, and the personnel may also put the seed at risk. The methodology used must address these challenges, and the bank must be controlled for endogenous and adventitious contamination. Very thorough safety testing is normally carried out to ensure the absence of adventitious agents and other contaminants in mammalian cell cultures.

A general procedure for preserving mammalian cells in liquid nitrogen follows:

Freezing of Animal Cell Culture in Liquid Nitrogen

1. Select exponentially growing cultures.
2. Prepare freezing medium by adding 5 percent glycerol or up to 10 percent DMSO (by volume) to freshly prepared culture medium.
3. Harvest cells and wash by centrifuging cells at ambient temperature. Resuspend in the freezing medium so that

cell concentration is approximately 2.0×10^6 to 1.0×10^7 viable cells/mL.

4. Dispense 1 mL of cell suspension into ampoules and seal.

5. Test ampoules for leaks.

6. Freeze.

    a. For manual freezing:

    Place at −20° C for 4 to 6 hours.

    Move to −70° C overnight.

    Transfer to liquid nitrogen freezer within 18 hours of freezing.

    b. In controlled-rate freezers:

    Program freezer to drop from ambient temperature to −20° C at a rate of 1° to 2° C/min.

    Hold for about 4 to 6 hours.

    Transfer to −70° C or liquid nitrogen freezer for permanent storage.

Note: The process should be validated for each application.

Recovery of Frozen Cells

1. Quick-thaw in a 37° C water bath.

2. Gently pipette cells into fresh sterile growth medium.

3. Rinse vial once with medium to ensure that all cells are transferred to fresh medium.

4. Centrifuge cells for 5 min at low g force, sufficient to pellet cells, usually 800–1200 rpms.

5. When centrifuge stops (do not use brake), remove supernatant and resuspend cells with fresh medium.

6. Plate at about 0.5 to 1.0×10^6/mL in a T-25 flask to establish cells.

7. Proceed with cell expansion as dictated by procedures.

Part II: Procedures for Preservation of Bacteria

General Comments

Bacteria can be divided into 3 major groups: gram-negative vegetative cells, gram-positive vegetative cells, and spores. Freezing and freeze-drying methods are widely used for long-term preservation of bacterial cultures used for research, development, manufacture, and quality control, including preservative effectiveness testing of final products. Both methods have been used with different degrees of success, but neither has resulted in 100 percent recovery of viable specimens. Many bacteria do well with freezing at –40° to –70° C for long periods of time, but those that are highly resistant to dehydration, such as spores, can withstand lyophilization treatment and can be stored freeze-dried. In addition, vegetative organisms such as *Streptococcus pyogenes* and *Staphylococcus aureus* have also been shown to be particularly resistant to freeze-drying injury (Heckly 1978). Freeze-dried inocula have been shown to be stable at ambient temperatures and to retain viable cells for as long as 10–20 years if handled correctly. Even though many laboratories successfully preserve bacterial specimens at –70° C, many experts believe that long-term preservation of bacteria is more effective at low temperatures. In a paper published in 1992, Simione recommended that master seed cultures be stored at low temperatures (e.g., in liquid nitrogen), whereas working seed cultures can be maintained frozen or freeze-dried.

Many bacterial components vary regarding their antigenic composition and level of expression. The understanding of these characteristics is essential in order to select the appropriate methodology for preserving the inocula. If the seed stocks do not maintain the required viability, stability, and functional properties, undesired changes may result and the starter cultures may acquire unwanted characteristics. These unwanted characteristics may go unnoticed, and their by-products may be passed on to process intermediates and final product. The development of an inventory system that includes suitable initial characterization, routine monitoring, and quality control systems facilitates the early detection of any changes during development and maintenance of the seed stock.

A full range of testing should be done to characterize the bacterial strain as well as the master and working seed stocks. Specific testing should be instituted as part of the routine quality control and monitoring of the production starter cultures. In the case of recombinant seeds, testing is usually more extensive and should include complete characterization of the vector itself with full details on the preparation of the nucleic acid and the genes involved. The types of tests required to characterize and maintain the seed bank will vary depending on the source material itself, the intended use of the final product, and the host/expression system. (A list of possible characterization tests are previously shown in Table 4.5.)

Ordinary Freezing of Bacteria

1. Prepare slant cultures.
2. Pick one or more colonies and prepare broth cultures.
3. Dispense culture into tubes or vials.
4. Store frozen in ordinary freezers.

Note that cryoprotectants are usually not used in this method. This method is not suitable for long-term preservation for most microorganisms.

Bacteria Used for Vaccine Production

The consistent production of bacterial vaccines is highly dependent on the existence of a reliable and reproducible source material and process. In the author's opinion, this can be achieved in part by developing well-characterized and well-documented seed systems to ensure good traceability and reproducibility of the starting production inoculum. Vaccine manufacturers can make use of this approach to ensure control of their source materials and better satisfy regulatory requirements. The basic process, shown in Figure 4.7, can be used for all bacterial vaccine systems. The master culture should originate from a well-documented and well-characterized bacterial cell source. Further details associated with regulatory requirements for manufacturing biologicals from bacterial seed stocks can be found in several regulatory documents published by the FDA/Center for Biological Evaluation and Research, European Regulatory Agencies, and WHO. (FDA 1995, 1992, 1993, WHO 1987, ICH 1995)

Freezing Bacterial Vaccine at −70° C

1. Grow bacteria on an appropriate solid medium.
2. Pick a single colony of the bacteria (single colonies are preferred).
3. Inoculate 2 mL of medium and grow at 37° C in a shaker to suitable OD (optical density).
4. Dilute 1 mL of culture in 100 mL of medium.

Figure 4.7. General process flow for the preservation of bacteria.

Preservation of Bacterial Vaccines

Culture Preparation and Freezing

↓

Inoculate media

↓

Incubate culture

↓

Harvest culture

↓

Prepare culture suspension in medium with cryoprotectant

Aliquot into rubber-stoppered vials or cryotubes

| Transfer vials to freeze-dryer | or | Transfer to liquid nitrogen | or | Transfer to freezer ($-70^0 \pm 5^0$ C) |

---

5. Plate 10 mL of dilution onto agar. Remove 1 mL and add to 1-L bottle.

6. Incubate with shaking at 37° C for 7 hrs or sufficient time to grow culture to a suitable OD.

7. Aliquot bacterial suspension into 1-mL cryotubes.

The appropriate method of selection and the development of effective quality control procedures are essential to maintain and manage the preserved specimen to ensure the continuous manufacturing of quality as well as safe and efficacious vaccines.

Liquid/Gas Phase Nitrogen Freezing and Storage

Most bacteria do fine at –70° C storage for many years, but bacterial cultures can also be stored in a liquid nitrogen freezer. Depending on the criticality of the specimen, and if cost is not a major obstacle, one may want to consider the benefits of liquid or gas phase nitrogen storage at least for the MSS aliquots.

Part III: Procedures for Preservation of Virus-Infected Cells

General Comments

Viruses may be regarded as simple, small, living organisms or, a highly complex network of aggregated chemical substances. Viruses contain nucleic acid and a protein coat, sometimes enclosed by an envelope made of lipids, proteins, and carbohydrates. Viruses range from 20 nm to 14,000 nm in length. The nucleic acid is either RNA or DNA. For a virus to multiply, it must invade a host cell and use the cell machinery to produce the viral chemicals needed to reproduce. Viruses that infect bacteria are called phages.

Viruses can be preserved as isolated viral particles or as part of the host cell. By-products of viral cultures, such as viral antigens, may also be preserved for use in diagnostic products as laboratory reagents for research and as pharmaceuticals. Preservation of virus-infected cells is somewhat challenging since the cell and the virus requirements must be considered. A procedure to freeze virus-infected mammalian cells follows.

General Procedure to Freeze Virus-Infected Mammalian Cells

1. Record the cell density and viability of the culture to be frozen.

2. Prepare the freezing medium using the following percentages:

   RPMI 1640   70 percent

   FCS   20 percent

   DMSO   < 10 percent

   The medium should be freshly prepared for each use.

3. Freeze the specimen at a concentration of approximately 1.0×10^7/mL. If long-term freezing is desired (greater than 3 months in liquid nitrogen), freeze 1.0 mL of this concentration in a plastic cryotube. If short-term freezing is the objective, the specimen may be frozen in 1.5-mL glass vials to be stored for less than 3 months at −70° C.

4. To determine the cell culture volume needed to reach a concentration of 1.0×10^7, use the following formula:

$$\frac{\left(Total\ cells \right)}{1 \cdot 10^7} = \text{mLs needed to make a } (1 \times 10^7) \text{mL concentration}$$

5. Prepare 10 mL of freezing medium by pipetting 7 mL of RPMI 1640 into a 50-mL centrifuge tube. Add 2 mL of FCS and approximately 1 mL of DMSO to the medium. Mix thoroughly.

6. Pipette the volume needed to reach the right concentration. Prepare a total of 7 centrifuge tubes.

7. Centrifuge all tubes at 1200 rpm for 7 min.

8. When centrifuge stops, decant supernatant and discard.

9. Forcefully eject 1 mL into a centrifuge tube. If needed, lightly vortex the tube to help break the pellet. Do not use a pipette to break the pellet.

10. After the pellets are resuspended, transfer the specimen to cryovials (NUNC or equivalent) for long-term freezing, or use glass vials for short-term freezing. Volumes are not to exceed 1.0 mL/vial.

11. Label all vials with the lot number and date. Record the following information:

   - Density and viability of the frozen culture
   - Whether long- or short-term freezing was done
   - Total number of vials frozen
   - Lot number assigned to the specimen
   - Freezer number where stored
   - Freezer location

Procedure to Recover Virus-Infected Cells from Frozen State

1. Turn on the water bath to warm up to 37° C. Monitor and record the temperature.

2. When the water bath temperature has reached 37° C, place the vial(s) in the floating rack just long enough to thaw the vial contents. Do not leave the vial in the water bath any longer than necessary, and do not immerse the vials to avoid potential contamination.

3. Pipette the contents of the vial(s) into a 15-mL centrifuge tube. Add culture medium to a total volume of 14 mL. Rinse vial(s) with remaining medium.

4. Centrifuge the resuspended culture at 1000 rpm for about 7 min.

5. When the centrifuge has stopped, remove the supernatant and resuspend the pellet in 10 mL of culture medium. Lightly vortex to help resuspend the pellet, if necessary.

6. Transfer the contents of the centrifuge tube to a T-75 flask containing 20 mL of culture medium. Wash the centrifuge tube to transfer all cells to the flask.

7. Check the final cell concentration. Record the cell concentration.

8. Incubate for 24 hrs.

9. After 24 hrs, split the culture 1:2 and perform a cell count.

10. Recovered cells can be expanded according to specific manufacturing or laboratory procedures.

Part IV: Other Support Procedures

Other basic microbiology and cell biology procedures are used to support the seed development of both microorganisms and mammalian cell lines. Microbiological methods are either quantitative techniques used to count microorganisms or qualitative methods that assess the purity of the specimen. Some of these procedures include assessment of cell viability, absence of contamination, and cell function. This section provides examples of support procedures that help ensure the quality of the seed bank.

Basic Microbiology Procedures

The growth of microorganisms can be measured by direct or indirect methods. Commonly used direct methods include plate counts, membrane filtration, and direct microscopic counts. Indirect methods include measurements of turbidity, metabolic activity, and dry weight. More extensive information on measurements of bacteria can be found in general microbiology books or laboratory manuals. Only 3 methods are briefly described here: plate counts, direct microscopic counts, and turbidity.

Plate Counts. A plate count is the most frequently used method for direct measurement of viable bacteria. The process takes at least 24 hours and may be inadequate for quality control applications when time is an issue. To ensure that the original inoculum is in the appropriate range (e.g., 10–100 cells per mL), a bacterial suspension can be diluted by serial dilution. Briefly, the bacteria are suspended in 100 mL of sterile water, and the bacterial suspension is passed through a retaining filter membrane. The filter is placed in a petri dish with a pad that has been soaked in liquid nutrient medium. The incubated colonies are allowed to grow into visible colonies that can then be counted.

Whether using neat or diluted inocula, two methods can be applied: spread plate and pour plate. In the spread plate method, the microbial colonies grow only on the surface of the medium in the plate. In the pour method, the microorganisms are mixed with warm, *not hot,* agar medium. The agar is poured, and the microbial colonies grow in and on the solid medium.

If the bacterial cell suspension has been diluted, the final calculations for the number of viable bacteria should take into account the dilution factor. The following formula can be used to calculate the total number of bacterial per mL:

number of colonies counted × reciprocal of sample dilution = number of bacteria/mL

Direct Microscopic Counts. Bacteria can be counted with the help of a set of specially designed slides called a Petroff-Hausser cell counter. These slides have special wells of known depth and squares of known surface area. The volume of fluid over each large square is 1/1,250,000th of a mL. The bacterial suspension is added to the squares, and the average number of bacteria in the squares is multiplied by the factor (1,250,000) to find the final bacterial count per mL.

The main advantage of this method is the speed. If a quick estimate of cell number is needed, this direct microscopic count method can be used in place of a plate count. The disadvantages are that motile bacteria are difficult to accurately assess by this method, and dead cells may be inadvertently counted, causing an inaccurate assessment of viable cells.

Turbidity. A quick assessment of bacterial numbers also can be achieved by the turbidity method, wherein the amount of light striking the spectrophotometer detector is inversely proportional to the number of bacteria. In this technique, the more bacteria in the suspension, the less light is transmitted; light tends to hit the bacteria, be scattered, and not reach the detector in the spectrophotometer.

Cell Biology Procedure

Determination of Viability by Trypan Blue Exclusion. As presented in the first part of this chapter, viability should be performed before and after preservation. For mammalian cell lines, the number or percentage of viable mammalian cells can be determined by staining with trypan blue. Viable cells exclude the dye, whereas dead cells take up the dye. After staining with trypan blue, cells should be counted almost immediately since viable cells also can take up the dye after a few minutes of exposure to dye. Also, since trypan blue has a great affinity for proteins, elimination of serum from the cell diluent helps avoid interference, ensuring a more accurate determination of viability. A brief description of the procedure follows.

1. On the day of use, mix 4 parts of 0.2 percent trypan blue with 1 part of a 5× saline solution.

2. To one part of the trypan blue saline solution, add 1 part of the cell suspension (1:2 dilution).

3. Load cells into a hemocytometer and immediately count the number of viable (unstained) cells and the number of dead (stained) cells. Record the results for each count separately. For appropriate accuracy, it is recommended to count a total of about 200 cells.

4. Calculate the number of viable cells:

$$\text{viable cells/mL} = (\text{av. \# of viable cells in large square}) \times 10^4 \times 1/\text{dilution}$$

or

$$\% \text{ viable cells} = \text{\# of viable cells} / \text{\# of viable cells} + \text{\# of dead cells} \times 100$$

Note that erythrosin B can also be used in dye exclusion tests (100 mg erythrosinB/100 mL of an isotonic phosphate buffered solution).

SUMMARY/CONCLUSION

In summary, the quality of the preserved specimen is crucial for the success of the biopharmaceutical process. Quality can be achieved by using a well thought-out, organized approach to seed development and storage, including careful method selection. Selection of the best method should be based on not only the application and specimen requirements but also the method capabilities. Whatever the method selected, the preservation and recovery processes should ensure the return of the specimen to a reliable and predictable level of performance, usually with minimal or no morphological change and minimal cell death.

In addressing method selection, it is important to point out that successful recovery is not always assessed solely on the basis of cellular viability. In actuality, since some cell death is often present, functional metabolic attributes and genetic stability are frequently the critical indicators of a successful preservation program.

The preservation and storage of specimens used in the manufacture of products for human use should comply with good manufacturing practices and guidelines. Quality assurance, quality control, and validation principles should be considered early in the master plan for the preservation program to ensure an effective, well-documented, and compliant Master Seed Bank.

REFERENCES

Adams, G. D. J. 1991. Freeze-drying of biological materials. *Drying Technology* 9:891–925

Ashwood-Smith, M. J. 1980. Preservation of microorganisms by freezing, freeze-drying and desiccation. In *Low temperature preservation in medicine and biology,* edited by M. J. Ashwood-Smith and J. Farrant. Tonbridge Wells, UK: Pitman Medical.

Ashwood-Smith, M. J., and E. Grant. 1976. Mutation induction in bacteria by freeze-drying. *Cryobiology* 13:206–213.

ASTM. 1994. *Standard guide for design and maintenance of low-temperature storage facilities for maintaining cryopreserved biological materials. E 1564–93.* Annual Book of ASTM Standards, 1480–1481.

ASTM. 1994. *Standard guide for inventory control and handling of biological material maintained at low temperatures,* E 1565–93. Annual Book of ASTM Standards, 1482–1483.

Barbaraee, J. M., and A. Sanchez. 1982. Cross-contamination during lyophilization. *Cryobiology* 19:443–447.

Centers for Disease Control and Prevention/National Institutes of Health. 1993. *Biosafety in microbiological and biomedical laboratories,* NIH 88–8395. Washington, DC: U.S. Government Printing Office, U.S. Department of Health and Human Services.

Choate, R. V., and M. T. Alexander. 1967. The effect of the rehydration temperature and rehydration medium on the viability of freeze-dried *Spirillum atlanticum. Cryobiology* 3:419–422.

Coles, Tim. 1998. *Isolation technology: A practical guide.* Buffalo Grove, IL: Interpharm Press.

Fahy, G. M. 1986. The relevance of cryoprotectant "toxicity" to "cryobiology." *Cryobiology* 23:1–13.

Farrant, J. 1980. General observations on cell preservation. In Low temperature preservation in medicine and biology, edited by M. J. Ashwood-Smith and J. Farrant. Tonbridge Wells, UK: Pitman Medical.

Food and Drug Administration. Office of Biologics Research and Review, Center for Drugs and Biologicals. *Points to consider in the production and testing of new drugs and biologicals produced by r-DNA technology.* Rockville, MD: Center for Biological Evaluation and Research.

Food and Drug Administration. 1992, 1995. *Supplement to the points to consider in the production and testing of new drugs and biologicals produced by recombinant DNA technology: Nucleic acid characterization and genetic stability.* Rockville, MD: Center for Biological Evaluation and Research.

Food and Drug Administration. 1993. *Points to consider in the characterization of cell lines used to produce biologicals.* Rockville, MD: Center for Biological Evaluation and Research.

Food and Drug Administration. 1996. *Points to consider on plasmid DNA vaccines for preventive infectious disease indications.* Rockville, MD: Center for Biological Evaluation and Research.

Food and Drug Administration. 1997. *Points to consider in the manufacture and testing of monoclonal antibody products for human use.* Rockville, MD: Center for Biological Evaluation and Research.

Grout, B., M. McLellan, and J. Morris. 1990. Cryopreservation and the maintenance of cell lines. *TIBTECH* 8:293–297.

Hay, R. J. 1978. *Preservation of cell-culture stocks in liquid nitrogen.* Vol. 4, no. 2. Rockville, MD: ATCC.

Hay, R. J. 1988. The seed stock concept and quality control for cell lines. *Analytical Biochem.* 171:225–237.

Hay, R. J. 1989. Preservation and characterization. In *Animal cell culture: A practical approach,* edited by I. Freshney, 71–112. Oxford: IRL Press.

Heckly, R. J. 1978. Preservation of microorganisms. *Adv. Appl. Microbiol.* 24:1–53.

International Conference on Harmonization. 1994. *Guideline on validation of analytical procedures: Definitions and terminology.* Geneva, Switzerland.

International Conference on Harmonization. 1997. *Note for guidance on validation of analytical procedures: Methodology.* Geneva, Switzerland.

International Conference on Harmonization. Q5A. 1998. *Viral safety evaluation of biotechnology products derived from cell lines of human or animal origin.* Geneva, Switzerland: ICH Secretariat.

International Conference on Harmonization. Step 4. 1995. *Analysis of the expression construct in cells used for production of r-DNA derived protein products.* Geneva, Switzerland: ICH Secretariat.

Kruse, P. F., and M. K. Patterson, eds. 1973. *Tissue culture: Methods and applications.* New York: Academic Press.

MacFarlane, D. R., and M. Forsyth. 1990. Recent insights on the role of cryoprotective agents in vitrification. *Cryobiology* 27:345–358.

Maigetter, R. Z., F. J. Bailey, and B. Miller. 1990. Safe handling of microorganisms in small and large-scale BL-3 fermentation facilities. *BioPharm*, February:22–29.

Mazur, P. 1970. Cryobiology: The freezing of biological systems. *Science* 168:939–948.

Mellor, J. D. 1978. *Fundamentals of freeze-drying.* New York: Academic Press.

Meryman, H. T. 1971. Cryoprotective agents. *Cryobiology* 8:173–183.

Meryman, H. T., R. J. Williams, and M. St. J. Douglas. 1977. Freezing injury from solution effects and its prevention by natural or artificial cryoprotection. *Cryobiology* 14:287–302.

Polge, C., A. U. Smith, and A. S. Parkes. 1949. Revival of spermatozoa after vitrification and dehydration at low temperatures. *Nature* (London) 164:666.

Sharp, R. J. 1984. The preservation of genetically unstable microorganisms and cryopreservation of fermentation seed cultures. *Adv. Biotech. Process.* 3:81–109.

Simione, F. P. 1992. Key issues relating to the genetic stability and preservation of cells and cell banks. *J. Parenteral Sci. and Technol.* 46:226–232.

Simione, F. P., and F. M. Brown. 1991. *ATCC preservation methods: Freezing and freeze-drying.* Rockville, MD: ATCC.

Simione, F. P., and J. Z. Karpinsky. 1996. Points to consider before validating a liquid nitrogen freezer. In *Validation practices for biotechnology products, ASTM STP 1260,* edited by J. K. Shillenn. Rockville, MD: American Society for Testing and Materials.

Snowman, J. W. 1991. Freeze-drying of sterile products. In *Sterile pharmaceutical manufacturing,* vol. 1, edited by M. J. Groves, W. P. Olson, and M. H. Anisfeld, 79–108. Buffalo Grove, IL: Interpharm Press, Inc.

Trappler, E. H. 1989. Validation of lyophilization. *Pharm. Tech.,* January:56–60.

Wagner, C. M. Personal observations (unpublished).

Wagner, C. M., and J. E. Akers, eds. 1995. *Isolator technology: Applications in the pharmaceutical and biotechnology industries.* Buffalo Grove, IL: Interpharm Press, Inc.

Williams, N. A., and G. P. Polli. 1984. The lyophilization of pharmaceuticals: A literature review. *J. Parenteral Sci. and Technol.* 38 (2):48–59.

World Health Organization. 1987. *Acceptability of cell substrates for production of biologicals.* Technical report series 747:1–29. Geneva, Switzerland: WHO.

World Health Organization. 1993. *Laboratory Safety Manual.* Geneva, Switzerland: WHO.

World Health Organization. 1997. *Report of a WHO consultation on medicinal and other products in relation to human and animal transmissible spongiform encephalopathies.* Geneva, Switzerland: WHO.

5

CRYOPRESERVATION OF MAMMALIAN CELL CULTURES

William H. Siegel

BioWhittaker, Inc.

INTRODUCTION

This chapter will provide industry workers with practical guidance for the successful cryopreservation of mammalian cells and will present ideas and options to improve cryopreservation results.

The primary reason for freezing cells is risk management. Cell lines maintained only in continuous culture are at risk for contamination and accidental death from use of defective growth medium and from equipment failure. Cryopreservation provides insurance that cell stock can be recovered when adverse or unexpected events occur. Second, cryopreservation will provide time for evaluation and further discovery of new applications for cells or cellular products and can maintain a consistent passage number. It also can be used to enhance cell properties and select subpopulations of cells.

Cryopreservation Provides Insurance Against

Contamination

Bacterial and fungal contaminations are a common and everpresent nuisance in cell cultures. Contamination can occur even if antibiotics are present in the medium. Moreover, continuous antibiotic use can mask low-level contaminants and cannot substitute for careful aseptic technique.

Mycoplasma contamination has been shown to alter nearly every cellular parameter (Hay et al. 1992). Estimates of the number of cell lines contaminated with mycoplasma range from about 10 percent upward (Hay 1996a). Most antibiotics are ineffective against mycoplasmas and merely suppress detection.

Virus contamination in cell cultures can occur from intentional cocultivation of two cell lines, one infected and one not. Viral infection can be naturally occurring or perhaps accidentally transmitted during virus propagation if cells supporting viral growth are not carefully segregated from uninfected cell stock. Virus-contaminated liquid nitrogen has been reported and has the potential for infection of cells stored in the liquid phase of liquid nitrogen freezers (Shafer et al. 1976).

The best-known case of cross-contamination of one cell line with another is widespread contamination with the HeLa cell line (Nelson-Rees et al. 1981). Results from a United States testing laboratory indicated that an average of 35 percent of cell lines tested for identity were not the species or precise identity that was expected by the submitting investigator (Hay 1996a).

Medium and Equipment Failure

Defective medium caused by omission of required growth supplements, impure water, or errors in medium preparation can result in rapid cell death. Incubators, roller bottle assemblies, and bioreactors can fail mechanically or due to power interruption and adversely affect cell growth.

Cryopreservation Provides Time

Storage of Cells That Cannot Be Used Immediately

Since it may not be practical to simultaneously screen all the clones that have been generated in large batches, storage of cells may be required. For example, the production of transfected cells and hybridomas will yield too many clones to screen at once. Furthermore, continued culturing of numerous clones until they can be tested may represent significant time, medium expense, and contamination risk. Another example of the need for long-term storage of cryopreserved mammalian cells, normally in a Master Cell Bank (MCB), is for the production and quality control testing of sequential lots of biopharmaceuticals.

Maintenance of Constant Passage Number

Cells with a high passage number may not behave in the same manner as cells with a low passage number. Some cells will senesce and die at a high passage number. Maintaining a constant passage number will minimize the risk of spontaneous genetic mutations (genetic drift) and also reduce the risk of changes in cellular expression (phenotypic changes). Genetic drift in continuous cell lines after many passages may be due to spontaneous chromosome rearrangement. Another mechanism is thought to be the failure of mutant cells to be eliminated from the main population of cells *in vitro*. Phenotypic changes in the expression of cell products can occur over time. An example would be the failure of a hybridoma to continue producing antibody.

Biodiversity Can Be Conserved

Cells from endangered species can be stored for future study or to await further discoveries that may improve the odds of survival of the species. Cell studies can also be done retrospectively.

Cryopreservation Provides Enhancement of Cell Properties and a Selection Method

Enhancement of Cellular Characteristics

Certain cellular characteristics can be enhanced by cryopreservation. For example, secretion of cytokines and interferon-γ was improved after freezing human mononuclear cells (Venkataraman 1995).

Subpopulation Selection

Freezing can be used to select subpopulations of cells. Transformed and nontransformed human lymphocyte populations were separated by different freeze-thaw regimens (Knight et al. 1972). Mouse T lymphocytes are more sensitive than B lymphocytes to freeze-thaw and hypotonic shock (Strong et al. 1974). The ability to reduce or eliminate human T cells from frozen preparations is of clinical importance for bone marrow transplantation (Strong et al. 1974).

Suggested Practice to Maintain Consistent Cell Stocks

Whenever a new cell line is acquired, frozen *seed stock ampoules* should be prepared immediately for future use. Seed stock cells are then used to prepare frozen *working stock ampoules,* and cells from this *working stock* are then used to prepare proliferating cells for production, quality control, or experimentation. Cells are maintained in continuous culture for only 8–12 weeks and then discarded. To prepare the next continuous culture for production, quality control, or experimentation, return to frozen *working stock ampoules.* This scheme minimizes selective pressure that may have been exerted unknowingly, eliminates any low-level contaminant that may have been acquired during the 8–12-week culture period, and maintains a constant beginning passage number. An approximate 12–16-week continuous culture period before returning to frozen stock has been suggested by others (Doyle and Morris 1996). Although there is some flexibility in the prudent interval of continuous culture before thawing another stock ampoule, it should be done with regularity.

Industrial validation of culture conditions to ensure cell stock consistency and stability should incorporate a version of the foregoing procedures in cell handling SOPs for the establishment of the MCB and the manufacturer's working cell bank (WCB). The MCB (seed stock) is defined as "a collection of cells of uniform composition derived from a single tissue or cell." The manufacturer's WCB (working stock) is derived from one or more vials of the MCB.

BRIEF THEORY OF CRYOPRESERVATION

Successful cryopreservation of mammalian cell lines occurs when the cooling rate is neither too rapid nor too slow. The cooling rate is *critical* (Leibo and Mazur 1971). Although the rate of thawing is also important, the cooling rate is far more important (Harris and Griffiths 1977).

Very rapid cooling produces cell injury believed to be caused by intracellular ice formation and the lack of cell size shrinkage as shown in Figure 5.1 (Farrant 1980; Leibo et al. 1970; Leibo and Mazur 1971). This injury is nearly always fatal to the cell upon thawing. Very slow cooling causes cell damage due to increased extracellular solute concentration resulting from ice crystal formation,

Figure 5.1. Effects of freezing rate on ice crystal formation.

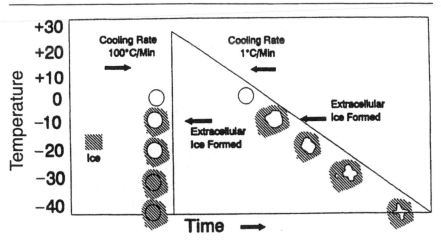

causing extreme osmotic shrinkage of the cell (Leibo et al. 1970; Leibo and Mazur 1971). Ideally, the cells should be cooled at a rate slow enough to avoid intracellular ice formation and fast enough to minimize damage from shrinkage and solute effects.

Cryoprotectant additives enlarge the range of cooling rates in which this can occur. There is no single ideal cooling rate. The optimal cooling rate for a particular cell line will vary with the cryoprotective additive used and its concentration (Leibo et al. 1970), although a cooling rate range of –1° C to –3° C/min is effective in freezing most mammalian cells (Macy and Shannon 1976; Morris 1995).

CRYOPROTECTANT ADDITIVES

Cryoprotectant additives attenuate slow freezing damage, but they are far less protective with rapid cooling rates (Farrant 1980; Ashwood-Smith et al. 1972). The use of cryoprotectant additives began after the observation in 1949 that survival of frozen fowl semen was enhanced by the addition of glycerol (Polge et al. 1949). The next milestone was the discovery of the cryoprotective properties of dimethyl sulfoxide (DMSO) (Lovelock and Bishop 1959).

Penetrating vs. Nonpenetrating

Penetrating cryoprotectant additives are capable of entering the cell. Common penetrating cryoprotectants include dimethyl sulfoxide (DMSO), glycerol, ethylene glycol, methanol, and dimethyl acetamide. DMSO may, in many cases, possess superior cryoprotective capacity (Porterfield and Ashwood-Smith 1962; Nagington and Greaves 1962; Paul 1975). Perhaps this is due to the ability of DMSO to exit the cell more easily than glycerol, thus causing less osmotic damage.

Nonpenetrating additives include polyvinylpyrollidone (PVP), hydroxyethyl starch (HES), dextrans (70,000 MW, Pharmacia), albumin, and polyethylene glycols. None of these agents has the protective capacity of DMSO (Ashwood-Smith et al. 1972).

Requirements for Cryoprotectants

Good solubility in water and low toxicity are two basic requirements for cryoprotectant additives. Good water solubility is required to minimize damage by solute effects. It is beneficial for the cryoprotectant to remain in solution longer as crystallization occurs (Farrant 1980).

Toxicity of two commonly used cryoprotectants, DMSO and glycerol, has been evaluated. Their toxicity varies with temperature, concentration, time of exposure, as well as with the particular cell being frozen (Greene et al. 1970). Lower temperatures (0°–4° C) and shorter exposures minimize toxic effects. Toxicity of 10 percent DMSO on human lymphocytes at room temperature for 30 min is minimal (Knight 1980). Similarly exposed mouse lymphocytes are more sensitive to DMSO, as measured by mitogen stimulation and a 50 percent reduction (vs. control) in ^{3}H-thymidine uptake (Thorpe et al. 1976). Mouse T lymphocytes are more sensitive than B lymphocytes to the osmotic stress of DMSO addition and removal (Knight 1980; Strong et al. 1974). It was reported that concentrations of 2.5 percent or less of DMSO or glycerol are not toxic to chick embryo fibroblasts after 18 hr in culture (Porterfield and Ashwood-Smith 1962).

Some investigators recommend that PVP be dialyzed against pH 7 buffer for 24 hours to remove toxic components or it can be dialyzed against water and then freeze-dried (Farrant and Ashwood-Smith 1980). However, evidence indicates that there is no difference between dialyzed and undialyzed PVP when freezing

Chinese hamster cells (Puck strain A) (Ashwood-Smith et al. 1972). Toxicity of PVP varies with the cell line being frozen and the purity of the particular batch of PVP used.

Concentration

DMSO and glycerol are used in the empirically determined concentration range of 5 percent to 10 percent. The optimal concentration will depend on the cell, the composition of the freezing mixture, and the cooling rate. Higher concentrations of DMSO and glycerol were more effective with slower cooling rates (Farrant et al. 1972; Leibo et al. 1970).

Quality

Reagent (Freshney 1994; Ferguson 1960) or spectrophotometric-grade glycerol should be employed (Farrant and Ashwood-Smith 1980). The use of glycerol that is more than one year old should be avoided (Freshney 1994), because oxidative by-products in the glycerol may be toxic (Shannon and Macy 1973). DMSO also should be at least reagent grade.

Sterilization and Storage

DMSO can be sterilized by filtration through a 0.22-μm Teflon FGLP filter (Millipore) that has been prewashed with methanol and then DMSO. Most other membrane filters are unsuitable because DMSO will dissolve them. Undiluted glycerol can be sterilized by autoclaving for 15 minutes at 121° C (Simione and Brown 1991). Both additives should be stored in single-use glass containers at 2° to 8° C and protected from light (Hay et al. 1992). DMSO freezes at +18° C. Sterile DMSO or freezing medium containing DMSO is available from ATCC (No. X-4), BioWhittaker (No. 12-132), Sigma Chemical Co. (No. D-2650), or Life Technologies (No. 11101-011).

SAFETY ISSUES
Liquid Nitrogen Hazard
Frostbite Risk (Skin)

Liquid nitrogen presents several risks (Simione and Brown 1991). The first, and most obvious, is frostbite at –196° C. Exposure to skin should be minimized. Cryoprotective gloves (VWR) should be worn.

Splash Risk (Eyes)

Safety glasses or a face shield should be worn to protect the eyes. One situation with a high splash risk is immersing a warm ampoule bucket into the liquid nitrogen freezer as one stands directly over it. This causes the nitrogen to boil vigorously and splash directly upward. Immersion of any warm object will cause boiling and splashing.

Asphyxiation Risk

Nitrogen is heavier than air, odorless, and tasteless. It can displace atmospheric oxygen and reduce oxygen concentrations to unsafe levels. The normal atmospheric oxygen level is 20.9 percent. The minimum "safe" level is 19.5 percent, as determined by government regulatory agencies. Reduction of oxygen levels to 16 percent causes disorientation; 12 percent, unconsciousness; and 5 percent, death by asphyxiation. The victim can be overcome quite suddenly and without warning, because there may be no symptoms prior to collapse and unconsciousness. Low-oxygen alarm sensors (Mine Safety Appliances) should be used in large-scale nitrogen freezer storage facilities and in rooms with large liquid nitrogen freezers.

Ampoule Explosion Risk

Liquid nitrogen can seep into poorly sealed ampoules during immersion storage. Thawing the ampoule, or even removing it from the storage freezer, can cause explosive expansion of the nitrogen. Even if the ampoule does not explode, it may spray the contents into the room as an aerosol. Explosion can be delayed and may occur even several minutes after thawing. Many researchers are surprised to learn that plastic ampoules will also explode. Exploding ampoules sound like rifle shots and can be nearly as deadly. A full face and throat shield (VWR) should be worn with protective gloves when working with ampoules that have been—*or may have been*—immersed in liquid nitrogen.

Biohazardous Material

Biohazardous material includes, but is not limited to, primary human material, cells exposed to or transformed by a primate oncogenic virus, and virus-containing and mycoplasma-containing cells. Biohazardous material should always be stored in plastic ampoules in the vapor phase of liquid nitrogen (Freshney 1994). Additional information on biosafety is available (U.S. Department of Health and Human Services 1993).

DMSO: A Powerful Solvent

Any contact with DMSO should be avoided. DMSO is capable of penetrating skin and rubber gloves and can carry toxic substances with it.

PREPARING CELLS TO FREEZE

The following provides guidance on preparing cells for a freezing process. For manufacturing applications, it is essential that the entire process of cell preparation, freezing, and thawing be fully documented and validated to ensure good recovery results and maximum reproducibility. These procedures would then be described in any application for a license in which the cells are used. Once a fully validated cryopreservation procedure has been developed, it must be followed precisely each time cells are frozen. The quality and integrity of the cell stock is vital to commercial product success.

Words of Caution

Freeze only one cell line at a time. This practice will *eliminate* the possibility of mixing one cell line with another and *reduce the risk* of introducing cells into ampoules that are incorrectly labeled.

 Never freeze *all* available cells. Maintain some cells in continuous culture until the frozen cells have been successfully thawed, grown without contamination, properly characterized, their functionality confirmed, and all has been documented.

Cell Condition before Freezing

Cells in the logarithmic or late logarithmic growth phase should be used. Avoid freezing cells that have reached the stationary phase. Adherent dependent cells should have had a fluid renewal 24–48 hours earlier. Most adherent independent suspension cultures should be harvested at a concentration of $4-8 \times 10^5$ cells/mL.

 Viability should exceed 90 percent in most cases. Starting with viability below 80 percent is inadvisable because it may indicate that the cells have reached the stationary growth phase. The viability percentage is evaluated by dye exclusion testing and microscopic examination. A volume of cells (0.1 mL) is mixed with an equal volume of dye, either 0.4 percent trypan blue (Sigma No. T-6146 BioWhittaker No. 17-942E) or 0.1 percent erythrosin bluish dye (Sigma No. E-9259). Viable cells with intact membranes exclude the dye and appear very bright under the microscope. Dead cells have leaky membranes, will take up the dye, and will appear dark in the microscope. The dyes are prepared in calcium- and magnesium-free saline. Erythrosin bluish dye is preserved with 0.05 percent methyl paraben (Sigma No. H-3647).

Ampoules

ATCC has switched from glass ampoules to plastic ampoules (Nunc, Inc., and Wheaton Scientific) for most cells as a convenience to the customer. Opening a plastic, screw-cap ampoule is easier than scoring the neck of a glass ampoule and breaking off the top. As a safety precaution, known biohazardous material should only be stored in plastic ampoules.

Contamination risks may increase when plastic ampoules are used instead of glass ampoules. If the seal is not tight, contaminants may enter the ampoule during storage in the liquid phase of liquid nitrogen or from the water bath used to thaw the cells. In addition, even with a tight seal, water from the water bath used to thaw the cells may cling to the neck of the ampoule and contaminate the contents upon opening. Ampoules with external threads may be more susceptible to this hazard. Plastic ampoules are not as thermally conductive as glass, which affects the freeze-thaw rate of the contents and could produce results different from those obtained with glass ampoules.

Glass ampoules are available with either unscored or prescored necks (Wheaton Scientific). Prescored ampoules are identified by a gold band around the neck. The prescored necks offer convenience but are more likely to break open if dropped. To open the prescored ampoule, wrap it in sterile cloth and break it into two pieces. Unscored ampoules should be broken in the same manner but need to be scored at the neck with a triangular glass file before breaking.

Glass ampoules are preferable for long-term storage of cells, because a hermetic seal can be achieved. Ampoules expected to be stored for the long term should either be immersed in liquid nitrogen or stored very near the surface of it. Poorly sealed ampoules in these locations have an elevated risk of explosion and contamination, as already noted.

Poor labeling can undermine the cryopreservation effort by causing confusion over the ampoule's contents and by causing noncompliance with GMP regulatory standards. Glass ampoules can be labeled with ceramic ink, using a straight pen, and placing the ampoules (empty) in a hot air oven at 120° C for one hour to anneal the ink. MicroMate™ labels offer adhesive designed to tolerate both liquid nitrogen storage and 37° C water bath thawing (Life Technologies, No.10696). These labels can be used on either glass or plastic ampoules. ATCC uses a computerized labeling device (Intermec Corp.). The labels (also Intermec Corp.) produced by this machine can be used on glass or plastic, adhere very well, have printing that is extremely indelible, and are impervious to most solvents.

Plastic ampoules sterilized by gamma irradiation are available (Nunc, Inc., and Wheaton Scientific). Empty glass ampoules (Figure 5.2), each covered by a glass cap (Wheaton Scientific), are best sterilized with exposure to dry heat by heating in an oven at 180° C for a minimum of 4 hours. Although autoclaving can be used instead of dry heat, dry heat is preferred because it drives off pyrogens that may be present. Elimination of pyrogens is necessary for cell lines but not for other cultures (Simione and Brown 1991).

Freezing Mixture Preparation

Most cell lines can be successfully harvested, resuspended in freezing mixture, and aliquoted into ampoules as follows:

- Harvest cells by centrifugation at 100 to 200 × g for 10 min.
- Resuspend the cell pellet in complete culture medium (culture medium with all supplements and serum at 10–20 percent) that has been further supplemented with 5 percent DMSO.
- Adjust cell density to $2-6 \times 10^6$ viable cells/mL.
- Dispense 1-mL aliquots into ampoules.

Although results with some cell lines may be improved by the optimization of certain variables (discussion follows), the foregoing steps have been found to be very effective.

Figure 5.2. Three labeled glass ampoules: left, filled and sealed; center, empty without cap; right, empty with cap, ready for sterilization.

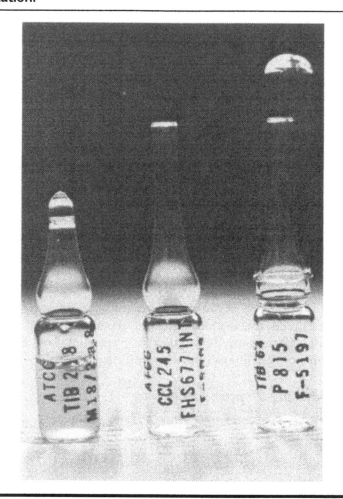

The number of cells frozen should be sufficient to permit adequate dilution of the cryoprotectant at recovery yet still maintain the cell density above critical levels (Freshney 1994). In general, a minimum inoculum into the culture vessel at recovery is required for the cells to recover quickly from the frozen state. Inoculation of fewer viable cells than the required minimum *for the particular cell* will result in very slow recovery or failure to recover. Using 2 million cells satisfies this requirement for most cells. One million cells *may* be sufficient, but using only 100,000 cells is likely to be problematic.

Freezing more cells may improve recovery results up to a point. Ferguson (1960) reported that 2 million cells gave significantly better survival rates than 10 million cells after 3 months of storage. Ferguson speculates that this may be due to the higher number of cells causing a difference in the freezing rate. Test ampoules had been placed directly into a –65° C mechanical freezer for freezing.

After centrifugation, the cells will occasionally remain in clumps that are difficult to count with a hemocytometer. If the clumps are small, reasonable estimates can be made of the number of cells in each visualized cluster. Clumping and cellular debris also can produce erroneous cell counts from a Coulter counter. For this reason, the hemocytometer may be preferable to the Coulter counter.

If clusters are too large to count, the following treatment may provide a remedy. The cells are washed 3 times in calcium and magnesium-free saline, Hank's salts solution (BioWhittaker, Sigma), or phosphate-buffered saline. (Elimination of the calcium and magnesium cations promotes single-cell suspensions.) The wash solution should be at 4° C. Aspirate vigorously with a pipette to generate a single-cell suspension. If clusters persist, aspirate the cells through a 22-gauge needle or cannula (blunt-tipped needle). A cannula is the safer alternative since needles always present the risk of accidental injection.

Serum at a concentration of 10 percent to 15 percent provides additional cryoprotection above that offered by DMSO alone (Ashwood-Smith et al. 1972; Ashwood-Smith 1980; Strong et al. 1974; Thorpe et al. 1976). It has been reported that higher concentrations of serum (50 percent to 90 percent) yield improved results (Martin 1994). Many investigators prefer a freezing mixture of 90 percent fetal bovine serum with 10 percent DMSO (Martin 1994). As already noted, most cells at ATCC are frozen with 10 percent to 20 percent serum. Although higher concentrations of serum in the freeze mixture can produce good results for small-batch freezes, it would be impractical on a large scale because of the serum expense.

Cells have been successfully recovered after being frozen in the absence of any serum or protein hydrolysate. After freezing in

4–10 percent DMSO with 0.1 percent (w/v) methyl cellulose (15 centipoise), the average viability upon thawing the 3 cells tested was 86 percent (Brown and Nagle 1965). The established cat kidney, human HeLa, and mouse L cells had each been grown in suspension in chemically defined medium without serum. A variety of murine tumor cells were successfully cryopreserved using the same freeze mixture (Waymouth and Varnum 1976). Inclusion of 0.1 percent (w/v) methyl cellulose (4000 centipoise) in the freeze mixture also has been found to attenuate the adverse effects of no serum (Ohno et al. 1988). Others have demonstrated that successful cryopreservation could be accomplished without serum (Leibo et al. 1970) or by using glycerol instead of DMSO (Evans et al. 1962). Hamster cells were frozen without serum but with 4 mM PVP and 1.0 M DMSO in Hank's balanced salt solution with 97 percent survival (Leibo and Mazur 1971). The protective effects of DMSO and PVP appear to be additive.

The cryoprotectant, typically DMSO, can be introduced to the cells in one of 2 ways. The cell pellet can be resuspended in medium containing the cryoprotectant. Alternatively, the suspension at twice the desired cell density can be mixed with an equal volume of medium containing twice the desired concentration of cryoprotectant. The addition of DMSO directly to the cell suspension is inadvisable because it causes release of heat sufficient to kill cells.

When freezing the cells for the first time, the most conservative approach should be taken. All additives and growth factors should be included. Morris (1995) suggests that *inclusion* of growth factors, such as Interleukin-2 (IL-2), may stabilize surface proteins and improve recovery results. However, inclusion of growth supplements in the freeze mixture may be unnecessary. Although they may be required for cell growth, the supplements (such as IL-2) are only available to the cells for a short time at reduced temperatures before the cells are frozen. I have routinely omitted IL-2 from the medium when freezing the cytotoxic mouse T cell—CTLL-2, ATCC TIB-214—without adverse effect. Elimination of expensive additives from the freeze mix can represent significant savings on large-scale freezes.

Elimination of zwitterionic buffers, such as HEPES and Tricine, is advocated to avoid hypertonic stress in the freeze mixture during cooling (Morris 1995). While this may improve results

with troublesome cells, I have observed that HEPES has no apparent adverse effect in cryopreserving the cells grown and frozen
with it.

Dispensing

Small-batch freezes (10 to 20 one-mL ampoules) can be dispensed
with a syringe or pipette. The cell suspension should be frequently
mixed to achieve a uniform cell density in each ampoule. Larger-
batch freezes are dispensed with a Cornwall syringe connected to
a reservoir of cells that is continuously stirred, such as a Celstir
unit (Wheaton Scientific), as shown in Figure 5.3. It is beneficial to
add an appropriate percent of sterile CO_2 in air to the reservoir to
maintain proper pH. Precise automatic dispensing can also be
achieved with the Wheaton Omnispense.

Ampoule Sealing

A plastic ampoule with a screw cap is easily sealed by screwing the
cap on tightly. However, storage at $-180°$ C in liquid nitrogen
vapor causes materials to contract so that the seal may no longer
be tight. The cap should not be excessively tightened, since that
can cause deformation of the ampoule gasket and lead to a poor
seal at both storage and thawing temperatures.

Sealing of glass ampoules can be done manually with a dual-arm
gas-oxygen torch. The flame is adjusted to a small size to avoid heating the ampoule contents. As the ampoule is rotated through the
hottest portion of the flame, the top of the ampoule is pulled away.
This pull-seal technique reduces pinhole leaks (compared to just
melting the top of the ampoule without pulling away the scrap).
Semiautomatic sealing devices rotate the ampoule and pull the scrap
away as the ampoule is sealed (Figure 5.4). The sealed glass ampoules
are placed on a cane (Shur-Bend Mfg.) and submerged in a 4° C solution of 1 percent methylene blue in 70 percent ethanol for 10 to 30
min (Freshney 1994). Pinhole leaks in ampoules will be revealed by
contents that have turned blue. These ampoules should be discarded.
This safety step is messy but necessary to reduce the risk of ampoule
explosion (refer to the earlier discussion of safety issues).

The cells should be frozen as soon as possible after the ampoules
are sealed. Ampoules containing cells are stored at 4° C until they
can be frozen. Reportedly, WI-38 cells suffered a significant loss of

Figure 5.3. Celstir unit with Cornwall syringe.

viability when stored at 4° C for more than 6 hours in a mixture containing either 5 percent glycerol or 5 percent DMSO prior to freezing (Greene et al. 1970).

Configuration Variations

Typically, cryopreservation is carried out with ampoules containing a one mL volume. Other configurations are possible and may be

Figure 5.4. Semiautomatic ampoule sealer.

more practical in certain circumstances. Ure et al. (1992) reported a simple and rapid method to preserve transfected embryonic stem cell clones in 24-well plates. Briefly, he added 0.25 mL of complete culture medium with 10 percent DMSO to each well of growing cells derived from EFC-1 cells. The plate was transferred to an insulated box and placed at –80° C to freeze. Thawing was accomplished by addition of 2.0 mL of complete culture medium that had been equilibrated at 37° C. The medium was aspirated and replaced after one minute or less, and the cells were then transferred to a CO_2 incubator. This procedure was 100 percent effective for 227 clones frozen on 13 separate plates—every clone recovered and grew after storage for up to 6 weeks at –80° C. This technique may also be applicable to quality control procedures.

Confluent adherent dependent cells (NIH-3T3, L929, Vero, IMR-90, MRC-5, HeLa, and primary rat liver parenchymal cells) have been frozen in 35-mm plates and in 24-well plates. Although cells bound to polystyrene microbeads and suspended in culture medium were frozen successfully, dextran microbeads were unsuitable, as were 96-well plates (Ohno 1996).

It is possible to freeze whole flasks of cells and aliquots larger than one mL. The cells are grown in flasks to late log phase, 10 percent DMSO is added to the medium, and the flasks are frozen at –70° C to –90° C in an expanded polystyrene box with a 15-mm wall (Ohno et al. 1991). Cultured human lymphocytes have been frozen in 5-mL, 10-mL, and 30-mL aliquots in 8-mL, 16-mL, and 35-mL specimen vials (Wheaton Scientific), respectively (Glick 1980).

Concerns about Differentiation Induction by DMSO

Some cell lines, such as HL-60, ATCC CCL-240, can be induced to differentiate by exposure to DMSO. One way to cryopreserve such cells is to avoid DMSO and use glycerol as the cryoprotectant. Another approach is to use DMSO for its superior cryoprotective properties. Differentiation induction of HL-60 requires 5–7 days of incubation with DMSO at 37° C (Collins et. al. 1978). The brief exposure to DMSO at much lower temperatures during the freezing process is unlikely to cause induction of differentiation.

COOLING

Rates

It is extremely important that the cooling rate be neither too slow nor too fast. The freeze program used with the microprocessor controlled-rate freezers at ATCC is as follows. After equilibration at 4° C for 10 to 15 min, the ampoule contents are cooled at 10° C/min to –2° C, followed by a cooling rate of –1° C/min down to –40° C. At –40° C, the temperature is reduced at 10° C/min until –90° C is reached. The cells are then quickly transferred to liquid nitrogen for storage.

The –1° C/min cooling rate has been advocated by others (Harris and Griffiths 1977; Glick 1980; Freshney 1994). A cooling rate range of –1° C to –3° C/min is effective in freezing most mammalian cells (Macy and Shannon 1976; Morris 1995). Although both the cooling rate and the thawing rate are important, the cooling rate has a more significant effect on recovery results (Harris and

Griffiths 1977). Optimal cooling rates can vary as much as 100-fold (Coriell 1979) depending on the cell type being frozen (Farrant et al. 1972) and on the composition of the freezing mixture (Leibo and Mazur 1971). Generally, the more permeable the cell membrane is to water, the faster the cooling rate that produces optimal survival (Farrant et al. 1972). Also, the lower the concentration of DMSO used, the higher the cooling rate that yields optimal DNA synthesis (Farrant et al. 1972). Dawson (1992) lists suggested protocols for freezing lymphocytes, granulocytes, rat myoblasts, and murine BALB/c 3T3. Glick (1980) outlines a method for freezing human lymphocyte cultures.

Methods

Liquid Nitrogen Controlled-Rate Freezer

Use of a controlled-rate freezer will provide the best results with the most consistency. The programmed-rate freezer is the only freezing equipment that allows precise counteraction to the latent heat of fusion released when crystallization occurs, usually between –2° C and –8° C. If rapid cooling does not occur at this point to compensate for the latent heat of fusion, sudden warming occurs with some ensuing cell damage. Various freezing routines can be tested with the assurance of good repeatability. A disadvantage is that this equipment is relatively expensive (Forma Scientific; Taylor-Wharton) compared to equipment requirements for other methods. However, the controlled-rate freezer is the method of choice for most industrial applications and would prove to be a good investment.

Mechanical Freezer (–80° C) with Insulated Box

A –80° C mechanical freezer can be used to freeze cells if insulation is present to reduce the rate of cooling. The ampoules are placed in a styrofoam box with walls and a cover that are about one inch thick. The box is placed directly into the –80° C freezer and left overnight. The ampoules are transferred to liquid nitrogen storage the following day. This regimen closely approximates a –1° C/min temperature drop, but this will not be as reproducible as programmed, controlled-rate freezing. Results also vary more widely with different cell types. Placing the cells into a –20° C freezer initially, either with or without insulation, is likely to yield fewer viable cells upon recovery.

Figure 5.5. Freezing device designed to be fitted into the neck of a liquid nitrogen freezer. The O-ring on the core controls insertion depth into the freezer.

Mechanical Freezer (–80° C) with Alcohol Bath

A very effective, inexpensive device is available for small freeze batches of 18 ampoules or less (Nalgene). This freezing container, "Mr. Frosty," consists of a reservoir containing 100 percent isopropyl alcohol and a rack that holds the ampoules. The alcohol does not come into contact with the ampoules. This device gives repeatable results at a cooling rate of approximately –1° C/min when placed into a –70° or –80° C freezer. It has been used successfully for cryopreservation of human peripheral blood mononuclear cells (Venkataraman 1992; Vingerhoets et al. 1995).

A variation of this method is the direct immersion of the ampoules into an alcohol–dry ice bath at –78° C. This method cools the cells at a much faster rate than the Mr. Frosty device. Reportedly, there is a dramatic reduction in the recovery rate with this method (Ferguson 1960). In general, this is a method of last resort that gives poor and unpredictable results.

Liquid Nitrogen Freezer Neck

An inexpensive device for 8 ampoules or less consists of a plug of expanded polystyrene that is inserted into the neck of the liquid nitrogen freezer. As the cool nitrogen gas escapes the freezer, it passes outside the plug and cools the contents (Nagington and Greaves 1962; McGann et al. 1972). A similar product for 8 ampoules or less is made by Union Carbide Industrial Gases, Inc. (Figure 5.5). A significant difference from the previous plug device is that the ampoules can be suspended at various levels within the plug by a retaining ring. This allows slightly more control of the freezing rate—that is, deeper insertion yields a more rapid freezing rate.

RECOVERY

Thawing

Thawing the frozen cells is an important step in the entire cryopreservation process. Poor thawing technique can severely reduce viable cell recovery. Typically, cells are thawed rapidly in a 37° C water bath and added to a flask containing complete culture medium. Variations on this procedure may improve recovery results depending on the particular cell.

Rapid warming is beneficial (Coriell 1979; McGann and Farrant 1976; Thorpe et al. 1976). The cells pass quickly through the –50° to 0° C zone where most damage is thought to occur (Coriell 1979). Others also advocate rapid warming (Leibo et al. 1970; Dawson 1992; Ferguson 1960; Morris 1995). The faster the warming rate, the better the rate of survival (Harris and Griffiths 1977). Slow thawing can reduce the survival rate by as much as 50 percent (Ferguson 1960). Small sample volumes facilitate rapid thawing.

Typically, the ampoules are removed from liquid nitrogen storage and placed directly into a 37° C water bath. The water bath should be covered as a safety precaution against exploding ampoules. Personnel handling the ampoules should wear a full face and throat shield. Agitation is not required; in fact, vigorous agitation may reduce cell survival and increase the risk of explosion. The water in the water bath should be kept clean by frequently replacing the water and periodically washing the instrument. Also, getting the neck of plastic ampoules wet should be avoided to reduce contamination risk.

Open Ampoule

The ampoule is removed from the water bath and placed in a solution of 70 percent ethanol in water. Alternatively, the ampoule is sprayed with a mist of 70 percent ethanol. The ampoule is then placed in the center of a 4-in square sterile gauze sponge or a sterile towel. (The cloth provides both a sterile field and protection from sharp, jagged glass if a glass ampoule breaks unevenly.) Open the ampoule as previously directed.

Dilute Cell Suspension

In general, survival rates are better with slow dilution vs. rapid dilution (Evans et al. 1962), warm diluent (20° to 37° C) vs. cool diluent, and serum in the dilution medium vs. serum-free (Farrant and Ashwood-Smith 1980).

In most cases, dilution by adding the contents of the ampoule to a flask of culture medium will yield satisfactory results. Alternatively, the ampoule contents can be transferred to a flask and dilution medium added dropwise with gentle mixing to a final volume of 10 mL. As the cell suspension volume increases, more medium can be added at a slightly faster rate. Dawson (1992) recommends dilution over a 1–2 min period. Slow dilution reduces osmotic shock and produces improved recovery results (Strong et al. 1974).

In very difficult situations, such as repeated recovery failures and only one ampoule remaining, twofold serial dilutions can be made in a 24-well plate as follows. One mL of complete culture medium is placed in each well of a 24-well plate. One mL of cell suspension is gently added to the first well. After gentle mixing, one mL is withdrawn for addition to the next well, and so on. After up to 2 weeks of observation, outgrowth usually occurs in the 1:8, 1:16, and 1:32 wells if the recovery is successful.

Since rapid thawing is beneficial, it follows that the addition of warm diluent would be best. Dilution of the ampoule contents is merely a continuation of the dilution process that began with thawing. The osmotic stress of dilution is better tolerated by cells at room temperature, or 37° C, than at 0° C (Thorpe et al. 1976; Woolgar and Morris 1973).

Abrupt dilution is better tolerated by some cells than by others. The addition of serum or other extracellular colloid may reduce dilution trauma (Farrant 1980).

Liquid cell culture medium tends to become alkaline upon storage. The medium should be equilibrated with the appropriate amount of CO_2 to avoid excess alkalinity before mixing it with recently thawed cells.

Cryoprotectant Removal

In most cases, adequate dilution of cyroprotectant is a better choice than complete removal at culture initiation. Adverse effects of DMSO on the recovery of cells were absent at dilutions of tenfold or more when 5 percent DMSO was used in the freeze mixture (Shannon and Macy 1973). DMSO will also evaporate from the medium at 37° C (Morris 1995).

Removal of the cryoprotectant the day after thawing by changing the medium may yield poor results in some circumstances. For example, some adherent cells only slowly attach to the substrate over a period of several days. Changing the medium too soon could cause a substantial reduction in the number of viable cells needed to start the culture.

Some cell lines will recover better in the complete absence of DMSO, but this is more often the exception than the rule. Centrifugation to remove the DMSO should be avoided for the following reasons:

1. Not all cells are identical in their ability to withstand centrifugation shear forces (Farrant and Ashwood-Smith 1980). Immediately after thawing, the cell membrane is more fragile than normal, particularly if the cooling rate employed was suboptimal (Thorpe et al. 1976).

2. Centrifugation also should be avoided because it promotes cell clumping, even if the cells tolerate gentle centrifugation ($70-100 \times g$). Clumping is likely to be more of a problem for adherent dependent cells than for suspension cells. Clumping will cause poor uniformity in the monolayer formed by adherent dependent cells. In most cases dilution of the cryoprotectant is preferable.

Seeding the Cells

The cells should be added to vessels at a density somewhat higher than that attained after normal subculture. For suspension cultures, $1-2 \times 10^5$ viable cells/mL should be seeded. Flasks containing suspension cultures should be placed on their sides rather than standing upright, because gas and nutrient exchange are more efficient with more surface area. Adherent cells generally should be seeded at about $2-3 \times 10^6$ viable cells/25 sq cm flask. Seeding at a tenfold higher concentration than usual after subculture has been suggested (Dawson 1992).

Some hybridomas tolerate freeze-thaw trauma poorly but have enhanced recovery rates when seeded over a feeder layer of murine macrophages. Peritoneal macrophages condition the medium and phagocytize dead and dying cells, but have a short life span in culture. Macrophages should be utilized within 2 weeks of harvest. The peritoneal macrophages from a single Balb/c mouse will be sufficient to condition several 24-well plates at 2×10^4 cells per well (Doyle and Morris 1996). McBride (1996) outlines a procedure for the production of peritoneal macrophages.

Feeder cells can also be generated by treatment with radiation or antimitotic agents, such as mitomycin C, and cryopreserved for later use (Hay 1996b). Radiation-treated mouse and human feeder cells are available commercially (ATCC).

Postfreeze Checks

It is good practice to thaw an ampoule shortly after the freeze to confirm the short term success of the cryopreservation effort.

Viability Check

Viability can be quickly evaluated by dye exclusion testing (Hay 1992), as already discussed. However, it should only be used as an immediate and *qualitative* indication of survival. Cells that are more sensitive to freeze/thaw trauma may appear to have high viability immediately after thawing but have a greatly reduced viability the following day. Cells of lymphoblastoid origin, particularly murine T cells, and hybridomas commonly exhibit this behavior. It is likely that they have extensive damage at the time of the dye exclusion test but have not developed sufficient membrane leaks to admit the dye, and/or DMSO is inhibiting dye uptake.

For *quantitative* evaluation of cryopreservation success, dye exclusion staining has been shown to be inadequate (Harris and Griffiths 1974; Craven 1960; Farrant et al. 1972). DMSO inhibits the uptake of trypan blue. If DMSO is washed out of the cells or allowed to leach out, cells are more easily stained (Harris and Griffiths 1974). Postthaw recovery is measured most quantitatively by DNA synthesis (Farrant et al. 1972) or by using the recovery index proposed by Harris and Griffiths (1974).

Even if the cells have suffered a significant loss of viability, no attempt should be made to rid the culture of dead cell debris with a density gradient treatment. The risk of contamination from increased handling and the risk of further damage by manipulation of fragile cells are too great. In addition, this treatment could select a subpopulation of cells, altering experimental results.

Contamination Check

It is good practice to verify that the cryopreserved material is free of contaminants. Most bacterial contamination can be discovered by screening with a variety of bacterial selection media (Hay 1992). Suggested media for detecting a broad spectrum of bacterial and fungal contaminants are shown in Table 5.1. Alternatively, one ampoule can be used to inoculate a flask containing 10 mL of complete culture medium without antibiotics. After undisturbed incubation at 37° C for 2 weeks, bacterial contamination will be apparent by a milky turbid appearance of the medium. Contaminated cryopreserved cultures should be discarded. Attempts to remove contaminants are usually futile.

Species Verification

Isoenzyme analysis can confirm the species of origin and detect interspecies cell contamination. A kit is available for isoenzyme analysis (Innovative Chemistry, Inc.).

Restoration of Cell Function

After cryopreservation, cell lines should be tested for retention of the function of interest. For example, 80 percent of granulocyte cells excluded dye in viability testing following thawing, but of these cells, only 5 percent were capable of phagocytosis (Knight 1980). Cavins et al. (1965) also found that dye exclusion correlated poorly with cellular integrity as measured by phagocytic performance.

Table 5.1. Suggested Regimen for Detecting Bacterial or Fungal Contamination.*

Test Medium	Incubation Temperature (°C)	Incubator Atmosphere	Observation Time (Days)
Blood agar with fresh defibrinated rabbit blood (5 percent)	37 37	Aerobic Anaerobic	14 14
Thioglycollate broth	37 26	Aerobic Aerobic	14 14
Trypticase soy broth	37 26	Aerobic Aerobic	14 14
Brain heart infusion	37 26	Aerobic Aerobic	14 14
Sabouraud broth	37 26	Aerobic Aerobic	21 21
YM broth	37 26	Aerobic Aerobic	21 21
Nutrient broth with 2 percent yeast extract	37 26	Aerobic Aerobic	21 21

*Hay 1992.

After thawing, it is advisable to culture the cells into log phase growth before functionality testing. This may allow restoration of functions that have been affected by freezing. Surviving Chinese hamster cells that had been exposed to a freeze-thaw cycle without any cryoprotectant additive sustained sublethal damage (McGann et al. 1972), but incubation of the cells in complete culture medium for 3 hours at 37° C fully repaired the damage as determined by plating efficiency.

Most investigators have been unable to detect any gain or loss of properties after proper freezing and storage (Coriell 1979). Cells retained the same morphology, tumorigenicity, growth rate, and virus susceptibility (Craven 1960). Most biological systems exhibit no deterioration over time when they are stored at temperatures of –150° C or lower (Leibo and Mazur 1971). The stability of biological and biochemical characteristics of 42 mouse neoplasms was examined after storage at –195° C for up to 6 years, and no changes were observed in the biochemical and morphological markers and chromosomes (Wodinsky et al. 1971).

STORAGE
Temperature vs. Expected Storage Time

ATCC stores most of its 4000+ cell lines with the intent of long-term or archival storage. Mammalian cells *require* temperatures of –130° to –196° C for extended storage (Ashwood-Smith 1980; Doyle and Morris 1996; Freshney 1994; Coriell 1979; Nagington and Greaves 1962; Shannon and Macy 1973). Damaging ice crystal formation is retarded at these temperatures (Coriell 1979). ATCC has successfully recovered HeLa (ATCC CCL-2) cells that had been stored in liquid nitrogen at ATCC for more than 30 years. Viability was greater than 90 percent with good attachment and growth in culture.

Extended storage at temperatures of –80° to –70° C is inadequate (Shannon and Macy 1973; Macy and Shannon 1976; Coriell 1979) and will cause a gradual loss of viability until recovery fails. Storage at –80° C is practical for *short-term* storage only. Depending on the cell line's tolerance to cryopreservation trauma and on the freezing/thawing techniques employed, successful recovery can be expected after storage of 2 to 4 weeks. Unpredictably poor recovery results can be expected after 1 to 6 months and very poor recovery results after 6 months to 1 year. At the end of year one and year two of storage at –79° C, cell viability was only 2–3 percent compared to 80–90 percent for cells stored in liquid nitrogen (Shannon and Macy 1973). Vero cells frozen in 35-mm dishes and stored for one year at –160° C (liquid nitrogen vapor) had a survival rate of 30 percent, but storage at –80° C yielded a significantly reduced 11 percent survival rate (Ohno 1996).

Storage at temperatures substantially warmer than –80° C yields even worse results. It is not possible to store HeLa cells for more than 2 weeks at –20° C (Craven 1960). After 6 months of storage at –65° C, cells failed to recover (Shannon and Macy 1973).

Temperature fluctuation during storage has an adverse effect on cell recovery rate and should be avoided. Manipulation of the cells while in storage should be minimized to maintain both the storage temperature and the recovery viability (Doyle and Morris 1996).

Liquid Nitrogen: Vapor Phase vs. Liquid Phase

Vapor phase storage is the best choice for most situations. It offers the following benefits compared to immersion storage:

- Temperatures in the required range of –150° to –180° C
- Reduced risk of ampoule contamination with infected liquid nitrogen
- Reduced risk of ampoule explosion (see safety section)

However, one clear advantage of liquid phase immersion storage is that ampoules are held at the consistently colder temperature of –196° C. If the liquid level is properly maintained, there will be no temperature fluctuation. In contrast, vapor temperatures vary between –150° and –180° C, due to location within the freezer and liquid nitrogen levels.

As already mentioned, plastic ampoules, in general, are unsuitable for immersion storage. If plastic ampoules *must* be used with immersion storage, CryoFlex polyethylene tubing (Nunc Inc.) may reduce the risks. The ampoule is placed inside the tubing, and the tubing is heat sealed at both ends, which provides an additional barrier to liquid nitrogen entry. Virus-contaminated liquid nitrogen has been reported and has the potential to infect stored cells (Shafer et al. 1976).

Equipment

One should expect and prepare for the day that compressor-driven mechanical freezers (–80° to –135° C) will fail. Backup freezers with emergency capacity should be maintained. Irreplaceable material should be divided and stored in more than one freezer. Safe deposit of valuable research material at another site or repository should be considered. Opening and closing of freezers should be minimized to limit temperature fluctuation.

The merits of liquid nitrogen immersion storage vs. vapor storage have already been discussed. However, liquid nitrogen is a consumable, and its cost will be deducted from grant and contract accounts. In contrast, mechanical freezers operate on electricity typically supplied by institutional overhead funds. Liquid nitrogen tanks do fail but not with the frequency or certainty that mechanical freezers do. Low-liquid-level alarms (Taylor-Wharton) are available to announce freezer failure or the need for more nitrogen.

Dry Shipper nitrogen freezers (Taylor-Wharton) are designed for shipment of material at liquid nitrogen vapor temperatures. They contain a material that absorbs liquid nitrogen and holds it at the perimeter of the freezer, which permits storage of the cells at the central core of the tank with no liquid contact or spillage possible. These freezers are available in various sizes and can be a relatively inexpensive solution to liquid nitrogen vapor storage needs.

Storage Practices

The material stored in a single mechanical or liquid nitrogen freezer can be divided into 2 parts. Place the working stock cells near the top of the freezer in an easily accessible place, and maintain seed stocks or infrequently needed cells at the bottom of the freezer. This arrangement blends easy access with short exposure to temperature fluctuations. Also, if liquid nitrogen levels do fall farther than desired, seed material will be at the coldest portion of the freezer.

As protection against a storage failure in the laboratory or plant, a safe deposit at a culture collection can be made (Doyle and Morris 1996). ATCC offers state-of-the-art liquid nitrogen storage facilities in the United States for this purpose. Culture repositories in other countries may offer similar services (Hay 1996a).

SUMMARY

The primary risks for cell lines grown in continuous culture are contamination, death from use of defective growth medium, and equipment failure. Cryopreservation provides insurance that a stock of cells can be recovered when adverse or unexpected events occur. Cryopreservation should be used to maintain consistent cell stocks. Seed stock and working stock ampoules should be used to replenish continuously cultured cells every 8–12 weeks.

Freeze only one cell line at a time. This practice will *eliminate* the possibility of mixing one cell line with another and *reduce the risk* of introducing cells into ampoules that are incorrectly labeled.

Never freeze *all* available cells. Maintain some cells in culture until the frozen cells have been successfully recovered without contaminants and perform as expected.

A majority of mammalian cells can be prepared for freezing (with good recovery expectations) as follows:

- Harvest cells by centrifugation at 100 to 200 × g for 10 min.
- Resuspend the cell pellet in complete culture medium (including serum at 10–20 percent) that has been supplemented with 5 percent DMSO.
- Adjust cell density to $2-6 \times 10^6$ viable cells/mL.
- Dispense one mL aliquots into ampoules.

Although results with some cells may be improved by optimization of certain variables, the aforementioned steps have been very effective.

Mammalian cells *require* temperatures of –130° to –196° C for extended storage. Damaging ice crystal formation is retarded at these temperatures. Extended storage at temperatures of –80° to –70° C is inadequate and will cause a gradual loss of viability until recovery fails.

Vapor phase liquid nitrogen storage is the best choice for most situations. It offers the following benefits compared to immersion storage:

- Temperatures maintained in the required range of –150° to –180° C
- Reduced risk of ampoule contamination with infected liquid nitrogen
- Reduced risk of ampoule explosion

However, one clear advantage of liquid phase storage is that ampoules are held at the consistently colder temperature of –196° C.

Successful cryopreservation of mammalian cells depends on many variables, all of which are interrelated. None can be considered alone. Although the cooling rate is the single most important factor in successful cryopreservation, every step of the process affects every other step of the process. No single freezing medium and protocol will be ideal for the cryopreservation of all cells. The process of cryopreservation must be carefully validated and documented to satisfy regulatory requirements.

REFERENCES

Ashwood-Smith, M. J. 1980. Low-temperature preservation of cells, tissues, and organs. In *Low-temperature preservation in medicine and biology,* edited by M. J. Ashwood-Smith and J. Farrant, 19–44. Baltimore, MD: University Park Press.

Ashwood-Smith, M. J., C. Warby, K. W. Connor, and G. Becker. 1972. Low-temperature preservation of mammalian cells in tissue culture with polyvinylpyrrolidone (PVP), dextrans, and hydroxyethyl starch (HES). *Cryobiology* 9:441–449.

Brown, B. L., and S. C. Nagle, Jr. 1965. Preservation of mammalian cells in a chemically defined medium and dimethylsulfoxide. *Science* 149:1266–1267.

Cavins, J. A., I. Djerassi, A. J. Roy, and E. Klein. 1965. Preservation of viable human granulocytes at low temperatures in dimethyl sulfoxide. *Cryobiology* 2:129–133.

Collins, S. J., F. W. Ruscetti, R. E. Gallagher, and R. C. Gallo. 1978. Terminal differentiation of human promyelocytic leukemia cells induced by dimethyl sulfoxide and other polar compounds. *Proc. Natl. Acad. Sci. USA* 75:2458–2462.

Coriell, L. L. 1979. Preservation storage and shipment. In *Methods in enzymology, cell culture,* vol. 58, edited by W. B. Jakoby and I.H. Pastan, 29–33. San Diego: Academic Press.

Craven, C. 1960. The survival of stocks of HeLa cells maintained at –70° C. *Exper. Cell Research* 19:164–174.

Dawson, M. 1992. Cryopreservation. In *Cell culture, labfax,* edited by M. Butler and M. Dawson, 141–145. San Diego: Academic Press, and Oxford: Bios Scientific Publ. Ltd.

Doyle, A., and C. B. Morris. 1996. Cryopreservation: Basic techniques. In *Cell and tissue culture: Laboratory procedures,* edited by A. Doyle, J. B. Griffiths, and D. G. Newell. 4C:1.1–4C:1.7, New York: J. Wiley.

Evans, J. E., H. Montes de Oca, J. C. Bryant, E. L. Schilling, and J. E. Shannon. 1962. Recovery from liquid nitrogen temperature of established cell lines frozen in chemically defined medium. *J. Natl. Cancer Inst.* 29:749–757.

Farrant, J. 1980. General observations on cell preservation. In *Low-temperature preservation in medicine and biology,* edited by M. J. Ashwood-Smith and J. Farrant, 1–18. Baltimore, MD: University Park Press.

Farrant, J., and M. J. Ashwood-Smith. 1980. Practical aspects. In *Low-temperature preservation in medicine and biology,* edited by M. J. Ashwood-Smith and J. Farrant, 285–310. Baltimore, MD: University Park Press.

Farrant, J., S. C. Knight, and G. J. Morris. 1972. Use of different cooling rates during freezing to separate populations of human peripheral blood lymphocytes. *Cryobiology* 9:516–525.

Ferguson, J. 1960. Long-term storage of tissue culture cells. *Aust. J. Exp. Biol.* 38:389–394.

Freshney, R. I. 1994. *Culture of animal cells: A manual of basic technique.* 3d ed. New York: Wiley-Liss.

Glick, J. L. 1980. Cryogenic storage and recovery. In *Fundamentals of human lymphoid cell culture,* by J. L. Glick, 115–117. New York and Basel, Switzerland: Marcel Dekker, Inc.

Greene, A. E., B. H. Athreya, H. B. Lehr, and L. L. Coriell. 1970. The effect of prolonged storage of cell cultures in dimethyl sulfoxide and glycerol prior to freezing. *Cryobiology* 6:552–555.

Harris, L. W., and J. B. Griffiths. 1974. An assessment of methods used for measuring the recovery of mammalian cells from freezing and thawing. *Cryobiology* 11:80–84.

Harris, L. W., and J. B. Griffiths. 1977. Relative effects of cooling and warming rates on mammalian cells during the freeze cycle. *Cryobiology* 14:662–669.

Hay, R. J. 1992. Cell line preservation and characterization. In *Animal cell culture: A practical approach,* edited by R. I. Freshney, 95–148. Oxford: IRL Press.

Hay, R. J. 1996a. Animal cells in culture. In *Maintaining cultures for biotechnology and industry,* edited by J. C. Hunter-Cevera and A. Belt, 161–178. San Diego, New York, Boston: Academic Press.

Hay, R. J. 1996b. Feeder layers. In *Cell and tissue culture: Laboratory procedures,* edited by A. Doyle, J. B. Griffiths, and D. G. Newell, 2D:0.1. New York: J. Wiley & Sons.

Hay, R. J., J. Caputo, and M. L. Macy, eds. 1992. *ATCC quality control methods for cell lines.* 2d ed. Rockville, MD: American Type Culture Collection.

Knight, S. C. 1980. Preservation of leukocytes. In *Low-temperature preservation in medicine and biology,* edited by M. J. Ashwood-Smith and J. Farrant, 120–138. Baltimore, MD: University Park Press.

Knight, S. C., J. Farrant, and G. J. Morris. 1972. Separation of populations of human lymphocytes by freezing and thawing. *Nature New Biology* 239:88–89.

Leibo, S. P., and P. Mazur. 1971. The role of cooling rates in low-temperature preservation. *Cryobiology* 8:447–452.

Leibo, S. P., J. Farrant, P. Mazur, M. G. Hanna, Jr., and L. H. Smith. 1970. Effects of freezing on marrow stem cell suspension: Interactions of cooling and warming rates in the presence of PVP, sucrose, or glycerol. *Cryobiology* 6:315–332.

Lovelock, J. E., and M. W. H. Bishop. 1959. Prevention of freezing damage to living cells. *Nature* 183:1394–1395.

Macy, M. L., and J. E. Shannon. 1976. Freezing procedures for the preservation of animal cell cultures in liquid nitrogen. In *Biological handbooks I: Cell biology,* edited by P. L. Altman and D. D. Katz, 46. New York: Academic Press.

Martin, B. 1994. Cell preservation. In *Tissue culture techniques: An introduction,* by B. Martin, 137–142. Boston, Basel, Berlin: Birkhäuser.

McBride, B. 1996. Macrophages: Production of peritoneal macrophage feeder layers. In *Cell and tissue culture: Laboratory procedures,* edited by A. Doyle, J. B. Griffiths, and D. G. Newell, 2D:1.1. New York: J. Wiley.

McGann, L. E., and J. Farrant. 1976. Survival of tissue culture cells frozen by a two-step procedure to –196° C. II. Warming rate and concentration of dimethyl sulfoxide. *Cryobiology* 13:269–273.

McGann, L. E., J. Kruuv, and H. E. Frey. 1972. Repair of freezing damage in mammalian cells. *Cryobiology* 9:496–501.

Morris, C. B. 1995. Cryopreservation of animal and human cell lines. In *Methods in molecular biology, Cryopreservation and freeze-drying protocols,* vol. 38, edited by J. G. Day and M. R. McLellan, 179–187. Totowa, NJ: Humana Press.

Nagington, J., and R. I. N. Greaves. 1962. Preservation of tissue culture cells with liquid nitrogen. *Nature* 194:993–994.

Nelson-Rees, W. A., D. W. Daniels, and R. R. Flandermeyer. 1981. Cross-contamination of cells in culture. *Science* 212:446–452.

Ohno, T. 1996. A simple method for *in situ* freezing of anchorage dependent cells. In *Cell and tissue culture: Laboratory procedures,* edited by A. Doyle, J. B. Griffiths, and D. G. Newell, 4C:2.1–4C:2.4. New York: J. Wiley.

Ohno, T., K. Kurita, S. Abe, N. Eimori, and Y. Ikawa. 1988. A simple freezing medium for serum free cultured cells. *Cytotechnology* 1:257–260.

Ohno, T., K. Saijo-Kurita, N. Miyamoto-Eimori, T. Kurose, Y. Aoki, and S. Yosimura. 1991. A simple method for *in situ* freezing of anchorage-dependent cells including rat liver parenchymal cells. *Cytotechnology* 5:273–277.

Paul, J. 1975. *Cell and tissue culture.* 5th ed. New York: Churchill Livingstone.

Polge, C., A. U. Smith, and A. S. Parkes. 1949. Revival of spermatozoa after vitrification and dehydration at low temperatures. *Nature* 164:666.

Porterfield, J. S., and M. J. Ashwood-Smith. 1962. Preservation of cells in tissue culture by glycerol and dimethyl sulfoxide. *Nature* 193:548–550.

Shafer, T. W., J. Everett, G. H. Silver, and P. E. Came. 1976. Biohazard: Virus-contaminated liquid nitrogen. *Science* 191:24–26.

Shannon, J. E., and M. L. Macy. 1973. Freezing, storage, and recovery of cell stocks. In *Tissue culture, methods and applications,* edited by P. F. Kruse, Jr., and M. K. Patterson, Jr., 712–718. New York: Academic Press.

Simione, F. P., and E. M. Brown, eds. 1991. *ATCC preservation methods: Freezing and freeze-drying.* 2d ed. Rockville, MD: American Type Culture Collection.

Strong, D. M., A. Ahmed, K. W. Sell, and D. Greiff. 1974. Differential susceptibility of murine T and B lymphocytes to freeze-thaw and hypotonic shock. *Cryobiology* 11:127–138.

Thorpe, P. E., S. C. Knight, and J. Farrant. 1976. Optimal conditions for the preservation of mouse lymph node cells in liquid nitrogen using cooling rate techniques. *Cryobiology* 13:126–133.

Ure, J. M., S. Fiering, and A. G. Smith. 1992. A rapid and efficient method for freezing and recovering clones of embryonic stem cells. *Trends in Genetics* 8:6.

U.S. Department of Health and Human Services. 1993. *Biosafety in microbiological and biomedical laboratories.* 3d ed. Public Health Service, CDC, NIH, Superintendent of Documents, Stock No. 017-040-00523-7, Washington, DC: U.S. Government Printing Office.

Venkataraman, M. 1992. Effects of cryopreservation on immune responses: VI. An inexpensive method for freezing peripheral blood mononuclear cells. *J. Clin. Lab. Immunol.* 37:133–143.

Venkataraman, M. 1995. Effects of cryopreservation on immune responses: VIII. Enhanced secretion of interferon gamma by frozen human peripheral blood mononuclear cells. *Cryobiology* 32:528–534.

Vingerhoets, J., G. Vanham, L. Kestens, and P. A. Gigase. 1995. A convenient and economical freezing procedure for mononuclear cells. *Cryobiology* 32:105–108.

Waymouth, C., and D. S. Varnum. 1976. Simple freezing procedure for storage in serum-free media of cultured and tumor cells of mouse. *Tissue Culture Association* 2:311.

Wodinsky, I., K. F. Meaney, and C. J. Kensler. 1971. Viability of forty-two neoplasms after long-term storage in liquid nitrogen at −196° C. *Cryobiology* 8:84–90.

Woolgar, A. E., and G. J. Morris. 1973. Some combined effects of hypertonic solutions and changes in temperature of posthypertonic hemolysis of human red blood cells. *Cryobiology* 10:82–86.

SUPPLIERS

ATCC
American Type Culture Collection
Manassas, VA 20110

BioWhittaker
Walkersville, MD 21793

Chemglass, Inc.
Vineland, NJ 08360

Forma Scientific
Marietta, OH 45750

Innovative Chemistry, Inc.
Marshfield, MA 02050

Intermec Corp.
Lynnwood, WA 98046

Life Technologies, Inc.
Gibco-BRL
Gaithersburg, MD 20898

Minnesota Valley Engineering, Inc.
MVE Cryogenics
New Prague, MN 56071

Millipore Corp.
Bedford, MA 01730

Mine Safety Appliances
Pittsburgh, PA 15208

Nalgene
Rochester, NY 14602

Nunc Inc.
Naperville, IL 60563

Pharmacia LKB Biotechnology
Piscataway, NJ 08854

Shur-Bend Mfg.
St. Paul, MN 55126

Sigma Chemical Co.
St. Louis, MO 63178

Taylor-Wharton
Theodore, AL 36590

Union Carbide Industrial Gases, Inc.
Linde Division
Somerset, NJ 08873

VWR Scientific
San Francisco, CA 94120

Wheaton Scientific
Millville, NJ 08332

6

FREEZE-DRYING OF HUMAN LIVE VIRUS VACCINES

Steven S. Lee and Fangdong Yin

Merck & Co., Inc.

INTRODUCTION

Among all human vaccines available today, the live attenuated forms have historical significance - dating back as early as 1798 for Smallpox. Killed whole organisms, purified proteins or polysaccharides and genetically engineered protein complexes are also used as vaccines against various diseases (Plotkin and Plotkin, 1994). Among about eleven commercial human live virus vaccines available to public, eight are prepared in lyophilized form. Therefore, lyophilization is the desirable manufacturing process capable of providing potent commercial products with stability. Existing vaccines, such as those for Measles, Mumps, Rubella, Varicella and combination vaccines such as M-M-R ®II, share the same problems for manufacturing operations: thermal degradation in liquid and solid forms. The structure of many important human live virus vaccines, some composed of a lipid bilayer outside their protein/nucleic acid complex, is still a subject of active research. With very little known about the degradation or inactivation mechanisms, one must apply serious considerations to the design of the formulation and the corresponding lyophilization process in order to successfully preserve the potency of live viruses. This chapter presents a brief description of live virus vaccines and their unique characteristics relevant to the manufacturing process, followed by some details of current freezing/freeze-drying practice in the pharmaceutical industry.

HUMAN LIVE VIRUS VACCINES (LVVs)

Compared to killed vaccines generally considered highly safe, live vaccines are still quite attractive because active, attenuated infection mimic the real-life situation and, thereby is expected to give greater and more long-lasting resistance to possible reinfection. Many commercial vaccines are derived from live viruses, such as Polio, Hepatitis A, Measles, Mumps, Rubella, Varicella-Zoster Virus (VZV), and Influenza viruses (Ellis, 1994). Their physical and chemical characteristics are listed in Table 6.1 (Fields Virology, 1996). In general, virus particles are made of a nucleocapsid, a RNA or DNA-structural protein complex, which may or may not be enclosed by a membrane with functional proteins embedded. Their sizes are within an order of magnitude of 100 nm.

Most interestingly, they are often pleomorphic, "enveloped" viruses, i.e., in addition to the common nucleic acid-protein complex surrounded by proteins, a lipid bi-layer associated with proteins of specific enzymatic or other activities like membrane fusion is typically found (see Table 6.1). Those enveloped viruses acquire their membrane or envelope by budding through an appropriate cellular membrane such as plasma or ER membrane. The viral glycoproteins such as hemagglutinin (H) and fusion protein (F), are usually anchored in the cellular membrane by a transmembrane hydrophobic peptide that terminates with a small domain on the cytoplasmic side. Such a complicated structure, as shown in the Measles virus model in Figure 6.1 (Griffin and Bellini, 1996), presents tremendous difficulties for process development and manufacturing. Consequently, most production processes have been established based on empirical experience rather than a complete fundamental understanding. Those vaccines listed in Table 6.1 are used world-wide today. A special process is required to preserve the potency and immunological properties after storage under varying conditions.

Table 6.1. Physical and Chemical Characteristics of Some Human Live Virus Vaccines.

Virus	Morphology	Particle size (nm)	Inner structure	Outer structure	Membrane-bound proteins
Adenovirus	rounded icosahedral shape	70–100	capsid composed of 252 capsomeres: 240 hexons plus 12 pentons; polypeptide VII is the major core protein; double-stranded DNA	non-enveloped; outer shell composed of 7 known polypeptides; polypeptide II is most dominant	N.A.
Mumps	pleomorphic; spherical shape	100–600	nucleocapsid composed of single stranded, non-segmented RNA, negative polarity, and proteins NP, P, L	enveloped; lipid bilayer with membrane or matrix protein M	two glycoproteins: hemagglutinin/ neuraminidase (HN); hemolytic activity and induces fusion of lipid membrane (F)
Measles	pleomorphic; spherical shape	100–300	nucleocapsid composed of single stranded, non-segmented, RNA, negative polarity, and proteins N, P, L	enveloped; lipid bilayer	two glycoproteins: hemagglutinin (H) and fusion (F); no neuraminidase activity; causes formation of intranuclear inclusion bodies

Continued on next page.

Continued from previous page.

Virus	Morphology	Particle size (nm)	Inner structure	Outer structure	Membrane-bound proteins
Poliovirus	roughly spherical shape	24–30	single-stranded RNA (2.5x10^6 M.W.), genome surrounded by a single protein shell composed of 4 structural viron proteins (60 copies of VP1 & 3 plus ~59 copies of VP2 & 4)	non-enveloped; icosahedral protein capsid	N.A.
Rotavirus	wheel-like; spherical shape	65–75	11 segments of double-strand RNA	non-enveloped; triple-layered icosahedral protein capsid; 60 spikes composed of hemagglutinin dimers extended from the smooth outer shell	N.A.
Rubella	pleomorphic; spherical shape	60–70	nucleocapsid composed of single stranded RNA, positive polarity, and C protein only	enveloped; lipid bilayer	glycoproteins: E1 and E2
Varicella (VZV)	pleomorphic; spherical or brick-shaped	180–200	nucleocapsid assembled by 162 capsomers; linear double-stranded DNA	enveloped; lipid bilayer	glycoproteins
Yellow Fever	spherical shape	40–60	discrete nucleocapsids composed of C (capsid) protein and single-stranded RNA	enveloped; lipid bilayer	type I: E (envelope) and M (membrane) proteins

410

Figure 6.1. Structure of measles virus.

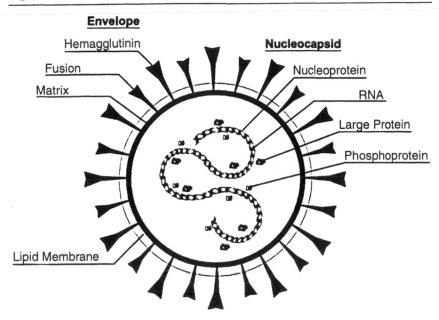

INSTABILITY OF LIVE VIRUS VACCINES

Most of the enveloped LVV are highly unstable by nature. They are characterized as pH and temperature labile and solvent sensitive by Rightsel and Greiff (1967). For example, a fragile VZV particle is subject to degradation by many physical and chemical treatments, including solvents, detergents, proteases, heating to 60° C, extreme pH's, ultrasonic disruption and even prolonged storage at temperatures of –10° C or above (Cohen and Straus, 1996). On the other hand, a non-enveloped virus, such as HAV, is stable to acid and resistant to many solvents and heat. HAV is stable at 60° C for 60 minutes and only partially inactivated after 10-12 hours at 60° C.

Cowdery et al. (1976) reported two experimental designs for assessing long-term stability of lyophilized live virus vaccines, measuring potency at either normal storage -20° C and 2-8° C or accelerated studies at elevated temperatures to simulate results by Arrhenius plots. For example, Measles virus has an average rate of 0.11 log potency loss per year at 5° C. In a different experiment, 0.17 log and 0.20 log potency loss per year were reported for

Rubella and Mumps, respectively, at 2–8° C storage. The degradation or inactivation rate is temperature dependent, apparently following the Arrhenius relationship (Bentley, 1970; Greiff and Rightsel, 1965):

$$\log_{10} \kappa = \left[\frac{-E_a}{2.303\,R} \right] \cdot \frac{1}{T+C}$$

where κ is the rate constant; E_a is the energy of activation; T is the absolute temperature (K); R is the ideal gas constant and C is a constant. A predicted 0.23 log loss for Mumps at 5° C was confirmed by experimental results of 12 months storage and subsequently, instability of live Measles, Mumps and Rubella virus vaccines were demonstrated (Mann, et al., 1983; McAleer, et al., 1980). More recently, the potency loss of lyophilized Varicella-Zoster virus was reported (Bennett, et al. 1991). Note that gelatin and/or human serum albumin were commonly used as the main ingredient for LVV (Doebbler, 1966; McAleer et al., 1980).

In addition, the stability of live virus vaccines apparently depends on the stabilizer and the strain and they lose potency as a function of time and temperature. While the exact causes are not understood, the potency loss rate increases with increasing temperature. Proper storage of lyophilized LVV under controlled refrigeration or frozen conditions is essential to ensure desirable potency up to the time of delivery to the patients, typically up to 2-3 years. Therefore, the formulation development for the freeze-drying process continues to be a major challenge today. This is briefly discussed in the following section.

FORMULATION OF LIVE VIRUS VACCINES

The formulation, stability and delivery of human live attenuated vaccines has been recently summarized in an excellent review (Burke, et al., 1999), which contains comprehensive lists of formulations for various live virus strains. Due to our limited fundamental knowledge of how and why those complex products degrade, the formulation design for various live virus vaccines were performed empirically in liquid and in lyophilized states. In general, the stabilizer compositions are complicated and extensive. For example, small organic compounds like amino acids, such as sodium glutamate, arginine, lysine and cysteine; mono-saccharides, such as glucose and mannose; disaccharides such as sucrose, maltose and lactose; sugar alcohols such as sorbitol and mannitol; are combined in varying proportions as customized "stabilizer cocktails", to be blended with one or more high molecular weight biopolymers to provide a structural matrix for the final product. Polysaccharides, such as dextran, starch, cellulose and their derivatives; proteins, such as human serum albumin and bovine serum albumin, gelatin, hydrolyzed gelatin and their derivatives; are typically used as major components which constitute the soluble, nongelling polymeric matrix known as the lyophilized cakes without pyrogenicity or immunogenicity. The stabilizing effects of any combinations of agents mentioned above are generally considered synergistic and/or additive.

One of the criteria for comparing formulations is the enhanced heat resistance or thermostability of live viruses. As one of the most challenging cases, live Varicella vaccine remains highly unstable in the presence of many known stabilizing components. Furthermore, when mixed with other live vaccines, additional potency loss in Varicella was reported (Koyama and Osame, 1992). Stability enhancement for live virus vaccines is the focus of formulation design, which may also be accomplished by the selection of more heat-resistant strains (Provost and Wadsworth, 1995). A summary of published LVV formulations is listed in Table 6.2. In general, live viruses harvested in the culture medium are blended in proportion with a specific, non-polymeric stabilizer cocktail and a major polymeric component, likely to be hydrolyzed gelatin, with some buffering capacity provided by phosphates.

LYOPHILIZATION OF LIVE VIRUS VACCINES

Equipment

Lyophilization equipment has been described and reviewed extensively elsewhere (Snowman, 1991; Trappler, 1995). A schematic diagram of a typical production scale freezer-dryer is illustrated in Figure 6.2. The single door (front), multiple shelves and horizontal condenser configuration is typical for manual loading/unloading of vials. Sterilization is achieved by 121° C steam in the chamber and cleaning (CIP) may be performed by multiple spray balls. Recent designs with full jackets, automatic shelf loading/unloading systems and a liquid nitrogen cooling system are not discussed here.

Table 6.2. Formulations of Some Human Live Virus Vaccines.

Vaccine	Special features (Examples)	Major polymeric components	Culture medium	Key non-polymeric stabilizing agents	Buffer	Reference
Cell free viruses	general lyo formula (chick embryo fibroblast material, human embryonic lung fibroblasts)	0.1–10% polyvinyl-pyrrolidone (m.w. 10,000-40,000)	modified Eagle's basal medium	2–10% sucrose, SPGE (sucrose, phosphate, [mono- or di-basic] sodium glutamate, etc.	N/A	Zygraich and Peetermans, (1975)
Live virus vaccines liquid or lyophilized	general lyo formula (measles)	partially hydrolyzed gelatin (2–5%)	199 medium (2–55%)	polyhydricalcohols: sorbitol, mannitol, or dulcital, citrate or acetate (10%)	acidic buffer phosphates	McAleer & Markus, (1979)
Live virus vaccine	general lyo formula (measles)	partially hydrolyzed gelatin (34%)	10x medium 199	sorbitol (8.6%), NaHCO$_3$, DGP solution (0.15%), a mixture of amino acid and antioxidants	phosphates	McAleer & Markus, (1981)
Adenovirus (type 5)	lyo formula	N.A.	Eagle's MEM	water	N.A.	Wigand et al. (1981)
Live virus vaccine	general lyo formula (measles)	partially hydrolyzed gelatin (34%)	10x medium 199	monosaccharide or di-saccharide, L-glutamic acid, L-Aginine	phosphates	Markus & McAleer (1982)

Continued on next page.

414

Continued from previous page.

Vaccine	Special features (Examples)	Major polymeric components	Culture medium	Key non-polymeric stabilizing agents	Buffer	Reference
Live Virus	general (lyo) formula (eg: Herpes virus, vaccine, Varicella, M-M-R-V)	modified gelatin	medium 199 medium 0	sorbitol, sodium, bicarbonate, SPGA (sucrose, phosphate glutamate, and albumin)	phosphates	McAleer, Maigetter & Markus (1991)
Attenuated Live Viruses	improved thermal stability at 45° C (measles, mumps, and rubella)	gelatin hydrolysate (m.w.3,500) 2–3% (w/v)	medium 199	lactose (2.5–5%), saccharose (2.5–5%), D-sorbitol (1.8–2%), sodium glutamate (0.1%) (w/v)	N.A.	Makino et al. (1991)
Poliovirus (Sabin Types 1, 2, 3)	oral formulation	N.A.	DMEM	1 M $MgCl_2$, or 87% D_2O at pH 7.2	phosphates	Craniac et al. (1996)
Varicella and M-M-R V	lyo vaccines free of Ca^{++} and Mg^{2+}	gelatin + hydrolyzed gelatin (free of Ca^2/Mg^{2+})	MEM free of Ca^{2+}/Mg^{2+}	ECTA (0.1g/L) sucrose (or lactose) sodium glutamate (or cysteine) NaCl/KCl D-sorbital	phosphates	Koyama and Osame (1992)
Measles, mumps, and rubella	lyo formulations with low phosphate (MMR)	hydrolyzed gelatin	medium 199	0.0075M phosphate, sorbitol amino acids, sucrose	phosphates	Provost and Nalin (1992)

Continued on next page.

415

Continued from previous page.

Vaccine	Special features (Examples)	Major polymeric components	Culture medium	Key non-polymeric stabilizing agents	Buffer	Reference
Yellow Fever (17D-204)	lyo formulation	gelatin (optional)	medium 199	sucrose or lactose, and some amino acids such as sodium glutamate, histidine or alanine	phosphates	Sood et al. (1993)
Live virus vaccines	varicella, lyo formulation	Dextran	N/A	sorbitol, mannitol saccharose, EDTA, sodium glutamate, urea, amines	N/A	Fanget & Francon (1997)

Figure 6.2. Schematic diagram of a large-scale freeze-dryer.

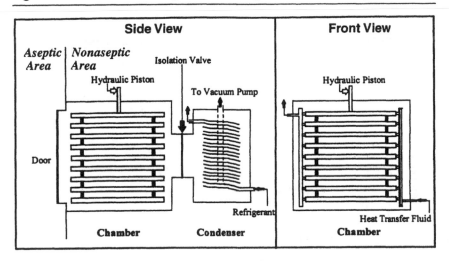

Process Overview

Lyophilization presents an important process for preservation of valuable commercial products, and is particularly suitable for live virus vaccines. There have been many comprehensive reviews on various aspects of this technology (Franks, 1990; Snowman, 1991; Nail and Gatlin, 1993) and, more specifically, on the applications to biological products such as proteins (Carpenter and Crowe, 1988; Geigert, 1989; Greiff, 1977; Hageman, 1988; Mackenzie, 1977; Pikal, 1990a, 1990b; Pikal, et al., 1991). Lyophilization of sterile biological products basically follows the same procedures for other non-biological products. In brief, vials are aseptically filled with live virus suspension (0.5-50 mL), followed by partially stoppering to allow escape of water vapor during drying. Filled, partially stoppered vials are collected in trays and transferred aseptically to a freeze dryer, which typically has multiple shelves. Multiple trays can be placed on each shelf either manually or automatically with an automatic loader. A heat transfer fluid, such as silicone oil, is circulated through the shelves to pre-chill them to typically below –40° C before vials are loaded. Then the freeze-drying process may begin. As depicted in Figure 6.3, three major stages, freezing, primary drying and secondary drying take place in sequence during

---

Figure 6.3. Temperature profile of a live virus vaccine lyophilization cycle.

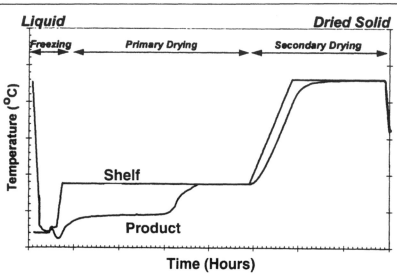

Time (Hours)

---

a lyophilization cycle for LVV. Greiff et al. (1965, 1968, 1969a, 1969b, 1972) studied the freezing and freeze-drying of various live viruses and later performed practical LVV stability evaluations (Cowdery et al., 1976; McAleer et al., 1980; Mann et al., 1983; Bennett et al., 1991).

Special Process Considerations for LVV

Freezing is a critical step in the freeze-drying process for it determines the formation of ice and solute microstructures, which in turn affect the rates of primary and secondary drying and ultimately, the quality of the product. Details of the freezing step have been well documented by Nail and Gatlin (1993). Because LVV formulations exhibit relatively low collapse temperatures, the product is usually first frozen rapidly on pre-cooled shelves at –40° C to –50° C for several hours to allow complete solidification of the contents in each vial. The actual product freezing occurs in probably less than 30 minutes, but more time is required to allow for uniform freezing. As a rule of thumb, all vials are preferably frozen

4–5° C below the phase transition or eutectic temperature. However, excessive freezing below the phase transition temperature lengthens the cycle unnecessarily. The composition/structure of the frozen matrix will impact the drying rate. In some cases, rapid freezing yields an amorphous cake with small ice crystals which allow for more intramatrix bridges of the amorphous (non-ice) ingredients, providing more integrity to the matrix, and less possibility of collapse.

Sublimation of ice begins when heat is supplied to the shelves as indicated by the rapid rise of shelf temperature from the freezing temperature (–40° to –50° C) to the primary drying temperature (for example, –15° C to –25° C). Because of their relatively low collapse temperatures, live virus vaccines require longer primary drying time. Heat transfer is achieved by conduction from the shelves through the tray, the bottom of the glass vial, and up through the frozen product, a phenomenon well characterized by Nail and Gatlin (1993). As drying proceeds, the dried solid layer grows from the top, gradually forcing the sublimation interface downward through the frozen solid. Note that as primary drying continues, the product temperature remains nearly constant until almost all presumably "unbound" or "free" water is removed. Then the product temperature begins to rise. The water content in dried product at this point is generally between 5% and 10%. For LVV, determination of the collapse temperature is critical for designing the primary drying process. The optimum primary drying is achieved by holding the product temperature as close as possible to the collapse temperature without exceeding it.

Additional drying time is required to further remove presumably "bound" or adsorbed water from the product, reducing water content to below 3%. The secondary drying phase starts with rapid ramping of shelf temperatures from the primary drying temperature (–15° C to –25° C , for example) to the secondary drying temperature (20° C to 30° C) in several hours, accompanied by almost synchronized product temperature increase. The time required to achieve sufficiently dried product is again determined empirically; usually some safety is built in to ensure the achievement of a low moisture level.

Fundamentally, the lyophilization process is a combination of dynamic, simultaneous heat and mass transfer processes, which ideally should be completely controlled by shelf temperature

Figure 6.4. Energy Balance of a Single Vial on Shelf.

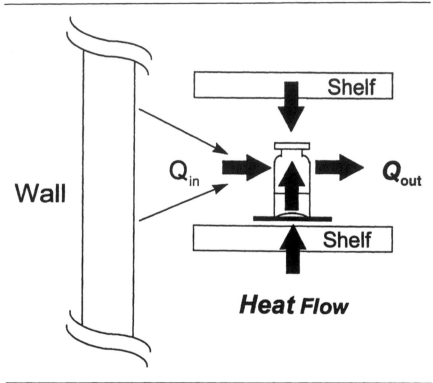

and/or chamber pressure. However, in a lyophilization cabinet heat sources other than the shelves can play a significant role. As illustrated in Figure 6.4, temperature change of the content inside each vial is the balance of total heat flux in and out. The heat balance for each vial can be expressed by the following equation

$$\frac{mC_p dT}{dt} = Q_{in} - Q_{out}$$

where m is the mass of vial contents; C_p is the heat capacity of vial and product; T is the temperature and t is time. Q_{in} and Q_{out} are heat fluxes entering and exiting the vial, respectively. Note that Q_{in} is the sum of radiation heat from the walls and the shelf above, conduction heat from the shelf below and adjacent walls. While the shelf temperatures above and below each vial are very similar, radiation heat input from non-temperature controlled cabinet

Figure 6.5. Temperature Variation During a Lyophilization Cycle (1, 5, 10 and 15 indicate the number of row from the outer perimeter of trays).

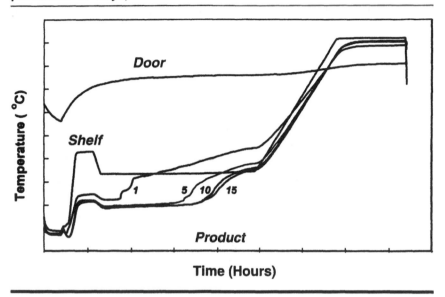

walls (about 15° C to –20° C) toward frozen vials at –40° C to –50° C can be significant. Thus, the radiation heat impact can vary with vial location. Shelf temperatures measured at the same locations of different shelves were identical (± 0.5° C), whereas temperature distribution on the same shelf was observed to vary. The location of vials dictated the extent of heat input from different sources and influenced the drying process. This location-dependent process performance can affect the homogeneity of product cake appearance and moisture content. For example, vials in row #1 on the shelf, the closest row to the cabinet wall, showed complete water sublimation much earlier than the rest of the vials. This rapid drying presumably led to higher moisture content after the completion of secondary drying due to undesirable cake structure changes such as melt back.

As shown in Figure 6.5, during an experimental run in a formulation under development, the temperatures observed in Row #1 vials rise rapidly beyond the shelf temperature level indicating a heat source other than the shelves is providing additional heat

Figure 6.6. Effect of Radiation Heat on Product Moisture.

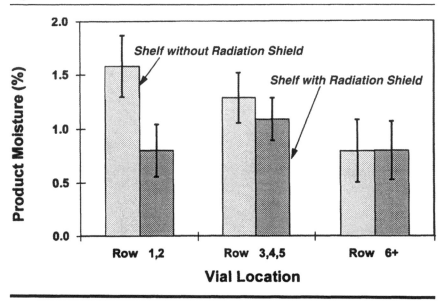

and contributes significantly to the primary drying process for these vials. Radiation from the relatively hot wall/door at 15–20° C to these very cold vials (–40° to –50° C) has been identified as the extra heat source. The sudden drop in door temperature to below zero during the freezing stage and the differences in product temperature profiles of rows #1, 5, and 10 are all evidence of radiation (see Figure 6.5).

The radiation effect on vials becomes less significant for the majority of vials farther away from the walls, such as the tenth row and beyond. This phenomenon has been commonly referred to as the "edge effect," (Nail and Gatlin, 1993). Although the edge effect was not quantitatively described before, it did impact to a varying extent the formulation development results in laboratory scale lyo cabinets and the product quality on production scale. The magnitude of such an effect may vary with the formulations and products. Moreover, the edge effect might be partly responsible for the difficulties in scaling up lyophilization processes.

To confirm the edge effect caused by radiation and evaluate its impact quantitatively, an experimental lyophilization run was performed with radiation shields. In Figure 6.6, the final moisture content of lyophilized LVV product vials are compared with and without a radiation shield. Although all the product moisture contents are significantly lower than 3%, the most obvious variation in moisture content is found in vials situated in rows #1 and 2, while such a difference diminish rapidly toward the center of the shelf. Among many possible appearance failures, cake collapse and meltback were likely observed in vials located in outer rows. As illustrated in Figure 6.7, compared to the normal cake in the vial on the right, the vial on the very left contains a collapsed cake. It is speculated that the extent of edge effect may be highly dependent upon the formulation. One way to reduce it is to control wall and door temperatures, especially during primary drying period, in lyophilization cabinets equipped with jacketed walls and/or doors.

VALIDATION OF LYOPHILIZATION PROCESS

The goal of process validation is to provide a high degree of confidence that product quality is consistently built in the final product, whatever container or dosage form it is in, given the limitation of statistical sampling and finished product testing. For lyophilization processes, successful validation of process would reduce, if not eliminate, the need to monitor each cycle with numerous thermocouples, to remove in-process samples for moisture measurements, and to control shelf temperature distribution for every run. As a cGMP requirement (21CFR211.100; 21CFR 211.110; 21CFR 211.113), compliance with proper record-keeping practice and documentation is required. The most relevant FDA guideline is the "Guide to Inspections of Lyophilization of Parenterals", (FDA, 1993) which covers formulation, filling, sterilization, transportation and lyophilization process. An excellent review on validation of lyophilization can be found in a PDA course (Trappler, 1996) and several documents (Williams and Polli, 1984; FDA, 1986; Agalloco and Akers, 1992; Avallone, 1990; Berry and Nash, 1993; Lee, 1988a and 1988b). Equipment Qualification is typically performed at the vendor's site prior to shipment followed by Installation Qualification (IQ), Operational Qualification (OQ) and Performance Qualification (PQ), at the manufacturing site. All equipment qualification is somewhat similar for all lyophilizers. Only some process validation issues more relevant to live virus vaccines are mentioned here.

Figure 6.7 Photograph of lyophilized live virus vaccine cake in 3-mL glass vials.

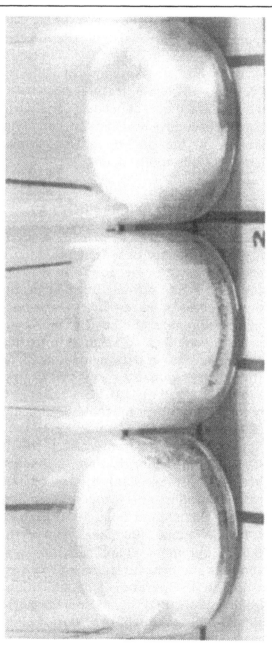

Table 6.3. Items to be considered for Lyoplilization Process Validation.

A. Product characteristics
  - Phase transition temperature
  - Collapse temperature
B. Process parameters with range
  - Temperatures: shelf, product and condenser
  - Pressures: condenser and chamber
  - Cycle control and monitoring: computer software validation
C. Product uniformity and quality
  - Moisture
  - Cake appearance
  - Release specifications
D. Stability studies
E. CIP/cleaning validation
F. Sterilization In Place

Process Validation is product-specific and should begin during development before the unit is even selected by characterization of the product. Table 6.3 lists key considerations for Process Validation. As summarized in Table 6.3, product characteristics, such as product phase transition (eutectic) temperature and collapse temperature; process parameters, such as chamber pressure, and shelf and product temperatures and their acceptable processing ranges; batch uniformity and quality measured by moisture and cake appearance; product release testing criteria and stability testing programs are all essential to the validation process.

Among many product characteristics, the physical form is most critical to product stability and appearance. The physical form of lyophilized live viruses is usually amorphous, whereas excipients such as sugar alcohols might form crystalline structures. The structure influences the performance of the freeze-drying process and the subsequent product stability thereafter. For example, compared to crystalline forms, low collapse temperature and water desorption rate of amorphous material usually result in a longer cycle time at lower shelf temperature. Formulation design may be the only way to improve or change that.

A lyophilization cycle may be controlled by several key process parameters such as temperature, pressure and timing, which should be defined and validated with acceptable limits. Particularly important is the monitoring of product temperature using calibrated thermocouples to ensure normal operating conditions, which is the critical information for identifying atypical cycles associated with power failure or equipment breakdown. Computer control based on a flow chart or logic is commonly used today to perform lyophilization runs automatically. Validation of the software follows the same criteria for other processes (FDA, 1983; PMA, 1986). To establish a reproducible end-point, according to predetermined time period and/or temperature requires extensive development efforts. For live virus vaccines, a conservative strategy employing both time and temperature may be used to ensure that a minimal secondary drying time is achieved.

Stability of an amorphous product is likely inferior to the crystalline counterparts, particularly at higher temperatures. Therefore, long term stability program up to 2 years at low storage temperatures is typically required for live virus vaccines. Samples for stability studies for the license application are usually generated during validation runs. Bioassay such as the plaque plate-count has been a standard method for potency determination.

Sterilization of a lyophilizer is generally performed through the use of moist steam (121–123° C) under pressure (15 psig), a procedure identical to autoclaving. Fully automated SIP (Sterilize-in-place) is considered a standard feature of production lyophilizers today. In addition, CIP (Clean-in-place) has become a desirable standard feature for new lyophilizers. However, the design of the CIP system varies among vendors, partly due to the complex shelf structure inside the cabinet. Both SIP and CIP are common to all products and therefore they are not discussed further.

One of the most important aspects of lyophilization process validation is to ensure the transfer of technology from R&D to production. A well-studied lyophilization cycle in the laboratory may not be readily scaled up to production scale if it is not established based on heat and mass transfer parameters. For example, the radiation effect observed in a laboratory lyophilizer is dramatically different from that observed in a production unit. In addition, homogeneity in product quality can be readily achieved in small scale lyophilization cabinets because of even impact of radiation heat on all vials. The "edge effect," hypothetically more significant in production cabinets due to the limited penetration of the wall radiation heat, tends to cause variations in product quality.

Ideally a minimum of three full scale production batches are produced using identical procedures to demonstrate process consistency. A maximum product load should be employed to adequately challenge the proposed cycle and the equipment. Subsequently, full scale batch runs are required to define critical process parameter ranges. For example, ranges (upper and lower bounds of a desired operating condition) for the shelf temperature of primary and secondary drying and the ramp-up rate between the two periods must be determined empirically based on resultant product quality. Moreover, because of concerns about temperature variationson shelves, samples chosen from proposed worst-case locations throughout the chamber should be subjected to testing.

RECOMMENDATIONS FOR CONTINUED IMPROVEMENTS

Although freeze-drying has been performed like an "art" for decades, it can be studied as a simultaneous heat and mass transfer process with a phase change. Understanding the fundamental heat and mass transfer processes taking place during the freeze-dry cycle (Nail, 1980; Pikal et al., 1984; Pikal, 1985) affords the pursuit of a detailed simulation model using a Computational Fluid Dynamic (CFD) package such as the Fluent™ software. Once product characteristics are determined (e.g. collapse temperature), many questions may be answered by simply changing process parameters and cycle development may be reduced to a great extent by running simulation models first. Fewer large-scale lyo runs need to be performed to validate model prediction; thus, streamlining the scale-up process. Different geometry, shelf/tray sizes, vacuum pump capacity, condenser capacity, etc., of various scale equipment and the radiation effect can be modeled to find optimal process parameters. For example, the laboratory scale lyophilizer used for formulation development should mimic the production scale lyophilization cycles, according to heat and mass transfer characteristics established by the simulation model. Such a scale-down, systematic approach will provide strong foundation for consistent, rapid technology transfer and process validation from Research and Development to Manufacturing. As for new lyophilizers, wall and door jackets designed for temperature control during the process may be desirable, in addition to CIP and SIP features.

Another area for improvement is process monitoring. Having more information on the process parameters of a lyophilization cycle permits extensive evaluation of formulations. For example, product moisture content in vials is one of the most important measurable targets for the lyophilization process. A mass spectrometer (MS) can be connected on-line to a lyophilizer to monitor global water removal throughout the cycle. A correlation between distribution of moisture in vials and global water partial pressure can provide direct, reliable process control with a reproducible end-point for this particular parameter.

REFERENCES

Agalloco, J., and J. Akers. 1992. Current practices in the validation of aseptic processing. *J. Parenteral Sci. and Technol.*, Technical Report no. 17.

Avallones, Henry. 1990. Current regulatory issues regarding sterile products. *J. Parenteral Sci. and Technol.*, July–August.

Bennett, P. S., R. Z. Maigetter, M. G. Olson, P. J. Provost, E. M. Scattergood, and T. L. Schofield. 1991. The effect of freeze-drying on the potency and stability of live *Varicella* virus vaccine. *Develop. Biol. Standard.* 74:215–221.

Bentley, D. L. 1970. Statistical techniques in predicting thermal stability. *J. Pharm. Science.* 59:464–467.

Berry, I., and R. Nash. 1993. *Pharmaceutical process validation.* 2d ed. New York: Marcel Dekker, Inc.

Burke, C. J., T-A. Hsu, and D. B. Volkin. 1999. Formulation, stability and delivery of live attenuated vaccines for human use. In *Crit. rev. ther. drug carrier systems* 16, edited by S. D. Bruck, 1–84. New York: Begell House Inc.

Carpenter, J. F., and J. H. Crowe. 1988. The mechanism of cryoprotection of proteins by solutes. *Cryobiology* 25:244–255.

Cohen, J. I., and S. E. Straus. 1996. *Varicella-zoster* virus and its replication. In *Fields virology*, edited by B. N. Fields et al., 2525–2585. Philadelphia, PA: Lippincott-Raven Publishers.

Cowdery, S., M. Frey, S. Orlowski, and A. Gray. 1976. Stability characteristics of freeze-dried human live virus vaccines. *Develop. Biol. Standard.* International Symposium on Freeze-Drying of Biological Products, Washington, DC 36:297–303.

Craniac, R., R. Wu, D. Otelea, M. M. Georgescu, F. Delpeyroux, S. Guillot, J. Balanant, and M. Tardy-Panit. 1996. The replacement of water with deuterium oxide significantly improves the thermal stability of the oral poliovirus vaccine. *Develop. Biol. Standard.* 87:161–166.

Doebbler, G. F. 1966. Cryoprotective compounds. *Cryobiology* 3:2–11.

Ellis, R. W. 1994. New technologies for making vaccines. In *Vaccines*, 2d ed., edited by S. A. Plotkin and E. A. Mortimer, Jr., 869. Philadelphia, PA: W. B. Saunders Company.

Fanget, B., and A. Francon. 1997. Stabilizers for live vaccines. PCT application WO97/23238.

Fields, B. N. et al., eds. 1996. *Fields virology*. 3d ed., vols. I and II. Philadelphia, PA: Lippincott-Raven Publishers.

Food and Drug Administration (FDA). 1983. *Guide to inspection of computerized systems in drug processing*. February.

Food and Drug Administration (FDA). 1993. *Guide to inspections of lyophilization of parenterals*. July.

Food and Drug Administration (FDA). 1986. Lyophilization of parenterals. *Inspection Technical Guide*, Washington, DC: U.S. Department of Health and Human Services.

Franks, F. 1990. Freeze drying: From empiricism to predictability. *Cryoletters* 11:93–100.

Geigert, J. 1989. Overview of the stability and handling of recombinant protein drugs. *J. Parenteral Sci. and Technol.* 43:220–224.

Greiff, D. 1977. Freeze-drying cycles. *Develop. Biol. Standard.* International Symposium on Freeze-Drying of Biological Products 36:106.

Greiff, D., and V. Damjanovic. 1969b. Thermodynamic analysis of the degradation of suspensions of viruses dried by sublimation *in vacuo*. In *Proceedings of the symposium on the thermodynamic aspects of freeze-drying,* Lausanne.

Greiff, D., and C. Greiff. 1972. Linear nonisothermal, single-step, stability studies of dried preparations of influenza virus. *Cryobiology* 9:34–37.

Greiff, D., and W. A. Rightsel. 1965. An accelerated storage test for predicting the stability of suspensions of measles virus dried by sublimation *in vacuo*. *J. Immunol.* 94:395–400.

Greiff, D., and W. A. Rightsel. 1968. Stability of suspensions of influenza virus dried to different contents of residual moisture by sublimation *in vacuo*. *Appl. Microbiol.* 16:835–840.

Greiff, D., and W. A. Rightsel. 1969a. Stabilities of dried suspensions of influenza virus sealed in a vacuum or under different gases. *Appl. Microbiol.* 17:830–835.

Griffin, D. E., and W. J. Bellini. 1996. Measles virus. In *Fields virology,* 3d ed., edited by B. N. Fields et al., 1269. Philadelphia, PA: Lippincott-Raven Publishers.

Hageman, M. J. 1988. The role of moisture in protein stability. *Drug. Dev. Ind. Pharm.* 14:2047–2070.

Koyama, K., and J. Osame. 1992. Stabilized live vaccine. European patent application 0 568726 A.

Lee, J. 1988a. Compliance for the lyophilization of parenterals: Part I. *Pharm. Tech.* October:54–60.

Lee, J. 1988b. Compliance for the lyophilization of parenterals: Part II. *Pharm. Tech.* November:38–42.

Mackenzie, A. P. 1977. The physical-chemical basis for the freeze-drying process. *Develop. Biol. Standard.* International Symposium on Freeze-Drying of Biological Products 36:51–67.

Makino, S., and K. Sasaki. 1991. Stabilized live attenuated vaccine and its production. U.S. patent 4,985,244.

Mann, G. F., L. M. C. Allison, and J. S. Lloyd. 1983. Stability of further-attenuated measles vaccines. *Reviews of Infectious Diseases* 5(3):482–486.

Markus, H. Z., and W. J. McAleer. 1982. Vaccine stabilizer containing L-glutamic acid and L-arginine. U.S. patent 4,337,242.

McAleer, W. J., and H. Z. Markus. 1979. Vaccine stabilizer. U.S. patent 4,147,772.

McAleer, W. J., and H. Z. Markus. 1981. Lyophilization process for live viral compositions. U.S. patent 4,273,762.

McAleer, W. J., H. Z. Markus, A. A. McLean, E. B. Byunak, and M. R. Hilleman. 1980. Stability on storage at various temperatures of live measles, mumps and rubella virus vaccines in new stabilizer. *J. Develop. Biol. Standard.* 8:281–287.

McAleer, W. J., R. Z. Maigetter, and H. Z. Markus. 1991. Stable lyophilized live *Herpes* virus vaccine. U.S. patent 5,024,836.

Nail, S. L. 1980. The effect of chamber pressure on heat transfer in the freeze drying of parenteral solutions. *J. Parenter. Drug Assoc.* 34:358.

Nail, S. L., and L. A. Gatlin. 1993. Freeze drying: Principles and practice. In *Pharmaceutical dosage forms, parenteral medications,* vol. 2, 2d ed., edited by K. E. Avis et al., 163–233. New York: Marcel Dekker, Inc.

Pikal, M. J. 1985. Use of laboratory data in freeze drying process design: Heat and mass transfer coefficients and the computer simulation of freeze drying symposium on lyophilization. *J. Parenter. Drug Assoc.* 39:115.

Pikal, M. J. 1990a. Freeze-drying of proteins. Part I: Process design. *Biopharm,* September:18–27.

———— 1990b. Freeze-drying of proteins. Part II: Formulation selection. *Biopharm,* October: 26–30.

Pikal, M. J., K. M. Dellerman, M. L. Roy, and R. M. Riggin. 1991. The effects of formulation variables on the stability of freeze-dried human growth hormone. *Pharm. Res.* 8(4): 427–436.

Pikal, M. J., M. L. Roy, and S. Shah, 1984. Mass and heat transfer in vial freeze drying of pharmaceuticals: Role of the vial. *J. Pharm. Science.* 736:1224.

PMA's Computer Systems Validation Committee. 1986. Validation concepts for computer systems used in the manufacture of drug products. *Pharm. Tech.,* October.

Plotkin, S. L., and S. A. Plotkin. 1994. A short history of vaccination. In *Vaccines,* 2d ed., S. A. Plotkin and E. A. Mortimer, Jr., 5. Philadelphia, PA: W. B. Saunders Company.

Provost, P. J., and C. Wadsworth. 1995. Thermostable *Varicella zoster* virus, W095/17503.

Provost, P. J., and D. R. Nalin. 1992. Non-stinging measles-mumps-rubella vaccine composition. European patent 0 514 208 A1.

Rightsel, W. A., and D. Grief. 1967. Freezing and freeze-drying of viruses. *Cryobiology* 3:423–431.

Snowman, J. 1991. Freeze drying of sterile products. In *Sterile pharmaceutical manufacturing—applications for the 1990s,* vol. 1, edited by M. J. Groves, W. P. Olson, and M. H. Anisfeld, 78–108. Buffalo Grove, IL: Interpharm Press, Inc.

Soods, D. K., S. B. Aggarwal, S. B. Sharma, J. Sokhey, and H. Singh. 1993. Study on the stability of 17D-204 yellow fever vaccine before and after stabilization. *Vaccine* 11:1124–1128.

Trappler, E. H. 1995. Fundamentals of lyophilization. In *Aseptic pharmaceutical manufacturing II, applications for the 1990s,* edited by M. J. Groves and R. Murty. Buffalo Grove, IL: Interpharm Press, Inc.

————. 1996. Validation of lyophilization. Course notes, Lyophilization Technology Inc., Parenteral Drug Association, Inc. June 20–21, Baltimore, MA.

Wigand, R., P. Bachmann, and G. Brandner. 1981. Stabilization of isomeric DNA viruses against thermoinactivation by lowered ionic strength. *Arch. Virol.* 69:61–69.

Williams, N. A., and G. P. Polli. 1984. The lyophilization of pharmaceuticals: A literature review. *J. Parenteral. Sci. and Technol.* 38:48–58.

Zygraich, N., and J. Peetermans. 1975. Stabilizing compositions for cell-free and cell-free virus preparations containing them. U.S. patent 3,915,794.

7

THE DESIGN AND USE OF THERMAL TRANSPORT CONTAINERS

Thomas C. Pringle

Insulated Shipping Containers, Inc.

INTRODUCTION

In the biopharmaceutical industry, a large percentage of materials that have been subjected to analytical control or that are moved through the cycle of production from raw material to end product must be preserved at some or all stages of their development. One of the most common preservation methods is freezing (cryopreservation), especially when stability is not fully understood. In these cases, as studies take place in accordance with the standard methods of establishing stability[1], materials are usually kept frozen. However, depending on what is known about the material, it may require freezing as well as storage and transportation at temperatures ranging from those of the typical commercial freezers, at –18° C, to as low as –196° C, found with preservation in liquid nitrogen or nitrogen vapor.

  In the normal research or production cycle of a biopharmaceutical, cryopreserved component materials are typically moved many times as they undergo analysis, clinical trials, and if successful, scale-up and eventual commercial production. In this chapter, the range of commonly used thermal transport containers and refrigerants will be presented along with their individual performance capabilities. Selecting a transport method to maintain the "cold-chain" in transporting temperature-sensitive materials requires a fundamental understanding not only of the material's acceptable temperature range in transit but also of the transit environment and the capabilities and capacities of the insulated transport options that are available. In addition, information will be presented on variables that must be considered when designing a packaging system and the common methods used for qualifying the selected transport system for the safe handling and transport of cryopreserved materials.

TRANSPORT SYSTEM SELECTION CRITERIA

The selection criteria for the design and development of transport systems for cryopreserved biological material encompass a wide range of factors.

Temperature Range Requirement

The first and most critical consideration is the *acceptable temperature range requirement for the material during transport*. In cases in which cell lines or cultures must be maintained at very low liquid nitrogen temperatures or in ultralow freezers within a narrow temperature range, they cannot be conveniently transported at these specific or desired temperatures. In these cases, "transport" temperature criteria may be established for the material that is not exactly the same as its "storage" temperature, since the transport period is typically of short duration. For example, certain materials have to be maintained in storage below –196° C (i.e., in liquid nitrogen storage). Although it is possible to ship in liquid nitrogen containers, it is not convenient due to the handling hazard the liquid nitrogen presents. Some freight carriers will not accept the liquid nitrogen tanks in their system, and, therefore, alternative "dry" nitrogen transport tanks are used. The dry nitrogen tanks are convenient to use, do not present a hazard, and will very easily maintain temperatures from –192° C to –140° C for 10 days or longer.[2] Therefore, in this case, the product's transport temperature is not exactly the same as the storage temperature, and an exception is made for the sake of convenience and safety as long as the quality of the material is not compromised in the process.

On the other side of the spectrum, some well-characterized, stable frozen materials that are maintained in commercial freezers (–18° C or –30° C) and transported packed in dry ice or frozen ice packs have much more liberal acceptance criteria, with the simple goal of reaching their destination still frozen (i.e., below –5° C). Most biopharmaceutical companies, however, prudently target specific temperature ranges for "families" of frozen products in their shipping criteria (e.g., < –140° C, < –70° C, and < –18° C). There are even examples of a family of products that are stored frozen but in transport are acceptable even if they reach refrigeration temperatures (e.g., < 10° C).

Refrigerant and Transport Method Options

Insulated containers and frozen transport packaging systems will be discussed in a later section of this chapter, but considerations regarding the refrigerant or temperature stabilizer factors must be addressed first. The most commonly used material for maintaining frozen temperatures of products during transport is dry ice. However, in some cases, dry ice can affect products in a negative manner. For example, if a product is stored in a bottle with a rubber stopper and seal, the dry ice can cause the rubber stopper to freeze, contract, and shrink away from the bottle opening. Then carbon dioxide (CO_2) from the sublimating dry ice will enter the bottle. Following the shipment, when the material is received and the bottle is removed from the dry ice, the rubber reverts to its original shape, sealing the bottle and also sealing the CO_2 inside the bottle. In some cases, this is not a desirable result. Examples of a CO_2-rich atmosphere affecting shelf life or altering products are known and must be considered part of the refrigerant selection process.[3]

Dry ice is considered a hazardous material and has to be used in a shipping container that meets the U.S. Department of Transportation (DOT), International Air Transport Association (IATA), and the selected freight carrier's requirements. The primary requirement is that the container system allow the CO_2 to vent during transport. The venting can create other problems in that air carriers have the right to "bump" or refuse to fly a container with dry ice for various reasons, including the safety of animals being transported in the freight area, since the CO_2 will displace O_2, thus creating a hazard to humans and animals. For shipment using dry ice, typically one-piece, molded, insulated containers made of expanded polystyrene (EPS) or rigid polyurethane are used. Both are acceptable within the handling systems of freight forwarders. If CO_2 is a problem for products, items such as barrier bags that prevent CO_2 from coming into contact with the product can be used.

Materials that cannot be shipped with dry ice due to CO_2 sensitivity or low-temperature instability can be kept frozen to a limited degree using high-quality insulated containers and commercially available frozen gel ice packs that have depressed freezing points (–10° C and –20° C). The time period that the gels can maintain the desired temperature is based on the amount of gel and the quality of the insulation of the container. Typically, these gels (water-absorbing polymers) are used for short duration shipments of 24 to 72 hours. Since the gels are composed primarily of water with chemicals added to depress the freezing point, they have a limited life for maintaining these low temperatures. When using these low-temperature phasing gels or other phase change materials with phasing temperatures below 0° C, it is important that they be conditioned (frozen completely) at temperatures well below their phasing temperature prior to use. For example, –20° C phasing gel packs must be frozen several degrees below their phasing temperature, which is colder than the common –18° C commercial freezer. In this case, the optimum conditioning freezer would be one in the range of –30° C. Additionally, due to the specific heat and large mass of the material, an efficient freezing method is required to ensure that the packs are frozen solid to have them perform (absorbing heat) at their optimum. It is not uncommon to find that a case of 12 × 48 oz gel packs takes several days at –30° C to freeze all the packs completely, but when placed in the freezer individually, the packs are easily frozen solid within 24 hours. To understand the amount of energy it takes to convert the water from liquid to solid (specific heat) and from solid to liquid (heat of fusion), it is necessary to define the specific heat and heat of fusion for water.[4] In the British system, the *specific heat* is the amount of heat that must be removed from a unit mass (1 lb) to change its temperature by 1°F; the *heat of fusion* is the amount of heat that must be supplied to change a unit amount (1 lb) at its melting point from the solid to the liquid state. Water has a specific heat of 1.0 (Btu/lb °F or kcal/kg ° C) and has a heat of fusion in its solid form (frozen ice) of 144 Btu/lb (80 kcal/kg). The lower phasing gels have less water since they have a higher percent of freezing point depressing additives by weight, which allows them to absorb more heat at a faster rate and, as a result, do not stay frozen as long as pure water for the same quantity or mass. Due to the gels' mass and the amount of latent heat they contain, the freezing time can be reduced by refrigerating the gels prior to freezing to remove a large part of that heat.

Another refrigerant used to keep products frozen during transport is liquid nitrogen, which is stored in specially made metal tanks or in "dry" nitrogen shipping containers. In the dry nitrogen container, the nitrogen is actually absorbed in the walls of the specially designed shipping vessel, so the "vapor phase" of nitrogen is used to keep the material frozen at temperatures warmer than liquid nitrogen but well below dry ice temperatures. The containers must not be kept upright to work efficiently, but are not under pressure, and there is less hazard as with CO_2 vapor. The considerations for use of these containers are the limited internal product space, and the availability and hazard of working with liquid nitrogen. Table 7.1 lists several refrigerants and commercially available insulated container and tank products for shipping cryopreserved materials.

In some cases, there is a very specific temperature required for products. It is possible, through the use of a combination of refrigeration and packaging materials, to maintain such required temperatures. For example, some products must be maintained at a temperature between –40° and –30° C—no warmer or colder. Part of the reason for such a narrow range may be that the stability of the material is unknown or under investigation. The material must be transported under these specific temperatures. With no commercially available refrigerant for use at those temperatures, the control is achieved through development of a custom packing configuration. Dry ice is placed in the bottom of an insulated container, and then an insulation barrier (e.g., 1"-thick EPS cut panel) is placed tightly over the dry ice; the product can be placed above this barrier so that the material is not actually in contact with the dry ice but adjacent to it. Figure 7.1 illustrates how the EPS barrier provides a buffer between the dry ice (–80°) and the product, allowing the material to be kept frozen but at a higher temperature.

Table 7.1. Refrigerants and Shipping Methods Used for Maintenance of Specific Frozen Temperatures in Various Transport Applications.

Refrigerant	Container	Practical Freezing Range During Transport	Practical Length of Trip
Liquid nitrogen	Custom liquid tank	−196°C	1 to 4 days
Liquid nitrogen	Dry nitrogen shipper	−192° to −140°C	1 to 10 days
Solid CO_2 dry ice (carb ice)	One-piece molded EPS or polyurethane container	−78° to −40°C	1 to 3 days (EPS) 1 to 10 days (polyurethane)
−20°C phasing gel ice	One-piece molded EPS or polyurethane container	−20° to −5°C	1 to 3 days
−10°C phasing gel ice	One-piece molded EPS or polyurethane container	−10° to −5°C	1 to 3 days

Depending on the barrier thickness, fit, and insulation quality, the payload temperature will be maintained in the −40° to −30° C range until the dry ice is completely dissipated. Therefore, by adjusting the barrier, it is possible to provide specific temperatures (+/− 10° C) for the material as necessary. It should be kept in mind that this is a cumbersome process and mistakes are frequently made because of the complicated nature of the packing, but specific freezing temperatures (warmer than dry ice temperatures) can be maintained as illustrated.

Shipping Method

After the desired temperature range of the frozen material and the refrigerant and transport container options are defined, the next important consideration is *how the material will be shipped.* This

Figure 7.1. Example of an insulated container packing configuration with a temperature performance graph demonstrating how a material can be kept frozen with dry ice in a range of –30° to –40°C.

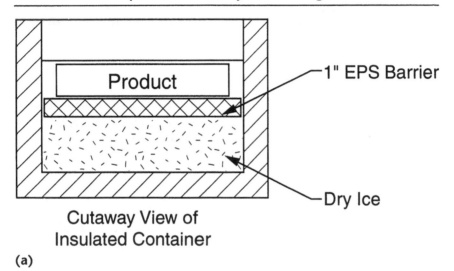

Product

1" EPS Barrier

Dry Ice

Cutaway View of
Insulated Container

(a)

(b)

includes the length of time and exposure limits of the transport cycle. Carrier limitations and expected transit times are fairly well understood for domestic and international transport. However, the variables of weather and ambient and dynamic (vibration and shock) exposure also must be taken into consideration during pre-shipment qualification (testing) of the selected method.

Internal and External Factors

The other considerations in the selection process of the proper transport packaging method are the miscellaneous *internal (company) and external (carrier and regulatory) handling issues* and the definition of the *quantity and form of the material* that will constitute a single shipment. With an understanding of the desired temperature range and time in transit, the shipping method options, and an estimate of the desired size of the frozen material payload (practical to handle and store), the next step is to consider the internal and external factors that can affect the choice of the best transport packaging method for each application.

The internal handling issues relate primarily to the known quality and safe handling requirements for a material—that is, the temperature range that must be maintained during transport, the hazard class (if any), manufacturing issues related to the product, and its replacement value. In some cases, frozen seed stock cultures or cell lines are so valuable that they are never shipped by express freight but by a special service in which a courier travels with the freight. Another example is the shipment of bulk frozen materials, such as production intermediates, that are prepared in one plant and shipped to another—maybe even to a distribution locale outside the United States—for final production, labeling, and distribution. The shipment of frozen intermediates in bulk quantities of from 10 to 50 L in carboys or sterile bags is very common and has proven to be an economical alternative to shipping finished product.

Examples of external factors are the regulatory and freight forwarder/carrier issues that affect the handling of materials, such as the freight forwarder's requirements for handling hazardous freight (flammable, toxic, or known infectious materials). The International Air Transport Association's (IATA) Dangerous Goods Regulations manual[5] describes these materials, their hazard class, and the packaging requirements for their safe shipment.

Freight carriers and forwarders should be very familiar with these regulations. Shipping methods for hazardous frozen materials are the same as for nonhazardous materials, but the insulated container that is used has to pass specific preshipment package performance tests. In addition, the package must be labeled to show that it has been tested and meets the requirements for that classification of material.

Infectious materials have to be placed in containers that can meet extremely rigorous package performance tests such as a series of 9 m (27 ft) drops and 7 kg (15.5 lb) spear punctures.[6] The primary container that actually holds the frozen infectious material has to meet pressure tests showing that the seals or stoppers will not leak. At each stage of this "distribution" package testing, the shipper is opened and the primary container examined for leaks (for frozen material, the testing uses a glycol solution that will not freeze when packed with dry ice). Frozen poison (toxic) material goes through a less rigorous series of shock, vibration, and compression package performance tests to make sure the internal package can withstand the transit environment. The IATA manual is updated annually and describes the packaging requirements for any material that would be considered hazardous. The manual is the accepted reference source on the subject for the air transport industry. It should also be noted that federal regulations require that any individual planning to ship infectious materials must complete a certified training program prior to actually packing and shipping these materials.[7]

Finally, the overall size of the material payload for a single shipment must be factored in with the various internal and external factors as part of the total dynamic for the shipping process for each frozen material.

DEFINING INSULATED CONTAINER/TRANSPORTER AND REFRIGERANT OPTIONS

Insulation Quality Considerations

It is important to understand the thermal properties (heat transfer characteristics) of the varying insulation materials used to manufacture insulated shipping containers and transport systems. The most commonly used insulation material in the construction of insulated containers is expanded polystyrene (EPS), quite often referred to by the trade name Styrofoam™. The other common insulation material used for containers and transport systems is rigid polyurethane. The latter would be used for higher performance thermal shipping applications due to its higher insulation property (resistance to heat flow), which is twice that of EPS for the same wall thickness.[8] Figure 7.2 shows an example of each type.

All insulation materials have an inherent thermal conductivity value based on the heat transfer rate for the specific material. The thermal conductivity constant is known as the material's **k factor**.[9] EPS foams can be manufactured in varying densities (e.g., from 1 lb/ft^3 to 2.0 lb/ft^3) with individual k factors for each density. Polyurethane foam k factors are based on the specific formula used in its manufacture. Note: Some polyurethane foams are known as *free-rise* foams and are not the same as the type of foam used for insulated container applications. The free-rise foams provide a cushion inside an outer container used to transport heavy machinery to keep it immobilized or protected during shipment. The polyurethane used for insulated containers is quite different and is referred to as *rigid* or closed-cell foam. Soft, open-cell, *sponge-type* polyurethane foams are also used as insulation but, due to their structure and inherent k factor, have lower thermal properties than either EPS or rigid polyurethane. **R value** is a measure of the thermal efficiency of a finished thickness of insulation material and indicates the resistance to heat flow, which is directly related to the k factor of the material. To calculate R values, the wall thickness of a material is divided by its k factor constant (Table 7.2).

The transporters used for liquid nitrogen (LN$_2$)and nitrogen vapor (dry nitrogen containers) are specially designed with dual metal walls incorporating an internal sealed vacuum. This design provides an extremely efficient barrier to the transfer of heat that is necessary for the preservation of the –196° C material. Like the liquid nitrogen transporters, the dry nitrogen containers come in a range of sizes and mouth diameters for varying product requirements (Figure 7.3). The dry nitrogen transporters utilize an absorbent material built into the inside of the containers that physically soaks and traps the liquid nitrogen in the filling process. This allows the vacuum-walled insulated transporters to protect materials in the –140° C and below range for up to 10 days. The transporters are also available with a hardshell airline transport case.

Figure 7.2. Examples of styrene and polyurethane molded containers used for the transport of frozen biological materials.

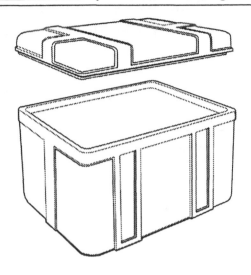

(a)

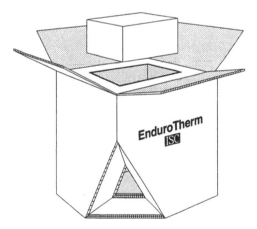

(b)

Table 7.2. R Value Comparisons for Insulation Materials.*

Insulation Density	EPS 1.0 lb/ft³	EPS 1.5 lb/ft³	EPS 2.0 lb/ft³	Polyurethane Rigid 2.0 lb/ft³	Polyurethane Open-Cell 1.5 lb/ft³
k factor	0.26	0.24	0.23	0.14	0.28
Wall thickness					
1"	R = 3.85	R = 4.17	R = 4.35	R = 7.14	R = 3.57
2"	R = 7.69	R = 8.33	R = 8.7	R =14.29	R= 7.14
3"	R = 11.54	R = 12.5	R = 13.04	R = 21.43	R = 10.71

*R values measured at 75° F

R = Thermal Resistance

$R = \dfrac{d}{k}$ d = given thickness of material
k = the material's thermal conductivity

R in units: $R = \left(\dfrac{ft^2 \cdot {}^\circ F \cdot hours}{BTU} \right)^9$

Refrigerant Considerations

When considering the insulated transport options (other than freezer trucks) for shipping frozen materials, the three practical packaging choices are molded EPS or polyurethane insulated containers or dry nitrogen transporters. When sending an item in dry ice around the corner (or perhaps within the confines of a university campus), a Styrofoam™ cooler with dry ice will work fine, even for overnight delivery. But when requirements call for specific narrow ranges of temperatures or specific performance for items that are very critical or very expensive, it is necessary to use a better insulating material such as polyurethane, the industry standard for high performance, narrow temperature range, and high protective applications. Insulated picnic coolers (e.g., Igloo or Coleman plastic or fabric body coolers[10]) are not recommended for transporting frozen materials due to their limited cold-holding ability (minimum insulation thickness) and susceptibility to damage (i.e., dry ice contact can actually cause the thin inner polyethylene walls to discolor and crack due to thermal shock). Any insulated container or transporter shipped internationally (48 hrs or longer) should use a higher grade material such as

Figure 7.3. "Dry" nitrogen biological shippers.

polyurethane due to the inconsistencies in delivery times for foreign shipments as well as potential customs delays.

With use of any refrigerant in an insulated container or transporter, proper training is mandatory to ensure that the materials are conditioned and packed according to a standard operating procedure (based on preshipment qualification testing and documentation as described in the next section). Some refrigerants present a health and safety hazard (such as dry ice and liquid nitrogen that can burn the skin upon contact). Proper labeling on the shipping container is especially important for dry ice shipments not only to identify the material but also to indicate the quantity in the container (always check with carrier regulations and airline relevant procedures and limitations). As mentioned previously, understanding the restrictions and safety requirements for using dry ice is important since CO_2 used inside the transport container will displace oxygen in a trapped environment (e.g., an airplane cargo hold or closed car or truck), creating an atmosphere that can cause death by asphyxiation for humans or animals traveling with the shipment.

Other important aspects of container and refrigerant design are whether or not the refrigerant will maintain the proper temperature and whether the physical characteristics of the refrigerant will affect the product. For example, when using low phasing gel ice packs, if the frozen material is not isolated from the gel packs inside the container, the gel packs can become dislodged internally during the normal handling process. The frozen gel packs can then physically damage the frozen material vessel by direct impact during a sudden drop or shift in the load. With dry ice shipping, the dry ice will evaporate in the process of sublimation (changing from the solid state to the vapor state) over the course of the shipment and leave a void inside the container. As the dry ice evaporates—unless the frozen material is secured inside the shipper through a partition, a divider, or an inner core pack—the material will be free to move around in the void and will be susceptible to damage if dropped or jarred during transit. The dry ice must be isolated in some way to hold the inner product contents securely in place. This is another justification for the preshipment testing of packaging—not only to prove that the container will provide the thermal protection desired but also to ensure that the steps of performing distribution or dynamic testing (vibration and shock) will indicate weakness or instability of the proposed packing configuration.

Dry ice is available in block, slab, pellet, and crushed form. The larger the surface area of the material, the faster the rate of sublimation. Therefore, for the same weight, the block form would significantly outlast the crushed form. However, the block form in some cases may not fit the available space inside the container or, due to its mass and weight, may present a threat of damage to the frozen material if the container is dropped. In these cases, the pellet option is worth considering due to the fact that it can fill in and around materials, eliminating voids and cushioning the load to some degree.

Placement of dry ice and gel packs is important to ensure that the desired temperatures are maintained during shipment. Generally, the gel packs are placed around all sides of the frozen material to make sure that the frozen packs are in adequate contact with the material. Since heat rises and will be absorbed by the refrigerant, the most logical place to locate the gels is on top of the load. However, shipping containers do not always travel in the proper orientation in the conveyances of freight carriers even when directed by the up arrows on the outside of the container. For dry ice, the same principle applies. Generally, the frozen materials are best placed in the center of the dry ice load, with a third of the dry ice in the bottom of the container and the balance on top. Examples of product loading configurations with refrigerant placements and resulting performance graphs will be illustrated later in this chapter.

Shipping Dynamics

Designing a shipping system for the safe transport of frozen materials requires the complete understanding and awareness of the various dynamics of temperature that are occurring simultaneously. The simple requirement that a material be maintained in a frozen state within a specific temperature range over a given period of time involves a very complex range of factors that must be accounted for in designing a system that will perform consistently. During packing and shipping, the dynamics that must be considered are the temperature and condition of the refrigerant, the temperature of the container, the ambient temperature in the shipping environment, and the frozen material temperature—all of which are undergoing change, and all of which have a role in temperature maintenance.

The refrigerant's primary role in the container or transporter is to absorb heat in order to create a stable temperature environment inside the container at its phasing, melting, or sublimation temperature. As the refrigerant absorbs heat, it uses energy in a dynamic process. Liquid nitrogen and dry ice evaporate, and gel ice packs melt. While the refrigerant is thawing or evaporating, it is acting to absorb the heat coming through the wall, thereby maintaining the frozen material in the range of the refrigerant's phase change temperature. The frozen material's temperature begins to rise inside the container in response to heat that has entered through the wall of the container from the ambient transport environment that is not being absorbed by the refrigerant. When the refrigerant has reached a point of complete thawing or evaporation, its heat-absorbing ability is exhausted. Use of a high-quality insulation material in the walls of the shipping container reduces the amount of heat entering the container and slows the rate of thaw or evaporation. Such insulation prolongs the range of performance for a given amount of refrigerant in the transport container. The overall size of the container (outside dimension) and resulting total surface area are also factors. The larger the surface area, the more heat is absorbed and the more refrigerant will be required to provide the desired range of temperature control. These factors are expressed in the following equation:

$$Q(\text{quantity of heat transferred}) = \frac{\text{surface area}}{\text{R value}}(\text{ambient temp.} - \text{inside temp.})$$

The conclusions to be drawn are:

- Smaller overall surface area of the transport container = Reduction in heat transfer rate.
- Higher R value of the insulation of the container = Reduction in heat transfer rate.

Insulated Container/Transporter Qualification

Any container or transporter considered for use in shipping cryopreserved materials should be preshipment tested (both thermal and distribution testing) using a simulated product load in the frozen material's actual primary container if possible. These qualification steps will provide a performance profile to make sure the transport container meets its expected or advertised characteristics for safe transport of the material. Thermal and distribution testing can be performed by a number of commercial labs[11] that follow standardized testing procedures[12] to evaluate the performance characteristics of the transport container for conformance to the desired results.

For thermal qualification, what is commonly done in the biopharmaceutical industry is to measure the thermal protection capacity of a package over the target transit period. This is done by analyzing the temperature profile of simulated product (packed in the container with the prospective type and amount of refrigerant) over time while exposed to one or more ambient conditions. Typical ambient conditions used for thermal package testing (performed in a laboratory environmental chamber) are profiles of extreme heat or cold (i.e., summer and winter) as well as a "moderate," less-extreme temperature profile. In most cases, companies evaluate both the minimum and maximum expected loads and the "worst case" environmental conditions anticipated to qualify a range of product volume for the transport container. As an example of how this testing is done, a test protocol is first developed following an industry accepted "standard" testing procedure.[13] Thermocouple probes are connected to a data logger and placed in the frozen material in the primary vial or vessel, which is then loaded into the shipper. The proposed amount of refrigerant is then added (e.g., dry ice or frozen gel packs), or in the case of the dry nitrogen shipper, liquid nitrogen is preloaded into the container prior to adding the frozen material. The transport system is then placed in the environmental chamber, which is programmed for the ambient condition to be evaluated. The evaluation period follows the expected transit time, whether a single day or a 5-day or longer shipping period. Figure 7.4 illustrates how thermal testing evaluation is performed.

The temperature data from each test provide a means to analyze the performance limitations of each shipping system. Following

---

Figure 7.4. Example of a thermal performance evaluation of a shipping container showing the elements of temperature measurement, container packing configuration with temperature probe locations, and a typical performance graph or profile of the test results.

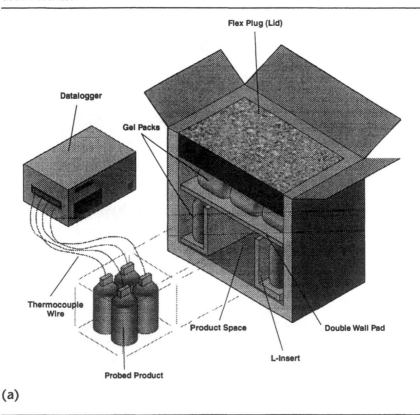

(a)

Continued on next page.

initial "screening" or "design" testing and satisfaction with the results, qualification or validation is performed using multiple test replicates (e.g., 3 to 5) of the same package to prove reproducibility of the data. The final test report with loading details serves as the basis for preparation of a standard operating procedure (SOP) for use by the internal company personnel responsible for packing and shipping the product.

In addition to thermal qualification of the selected transport container, it is recommended that the container be distribution or

Figure 7.4. Continued.

Heat test results show product and ambient temperatures.

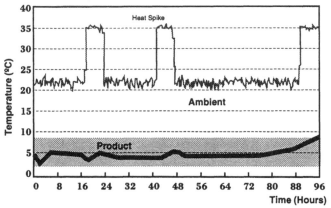

Freeze test results show product and ambient temperatures.

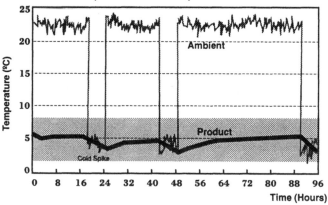

(b)

dynamically tested to make sure the frozen material is protected when exposed to the expected transit conditions. Again, standard test methods are employed simulating the nature of the expected transit conditions on simulated product in the selected container using the services of a qualified commercial laboratory. Commonly used methods consist of a series of compression, vibration, and shock (drop) tests simulating truck or air transport and handling, following precise models for a given package size and weight.

Following qualification, the final package and packing configuration should be discussed with the freight forwarder to make sure that all carrier requirements are met and that all outer markings are clearly understood by the shipper prior to the first use of the transport system container. Special outer markings on the container are necessary—indicating that the container has been tested or meets certain package performance requirements—for shipment of anything hazardous, including dry ice, infectious or toxic materials, or anything that has a UN classification as hazardous. Freight forwarders can also assist in identifying hazardous materials. The container air bill or shipping document must indicate the nature of the frozen material, and the freight forwarder will examine the package to make sure it meets the required testing; if not, the package will be returned. For transport of frozen infectious materials, the individuals responsible for the actual shipment must use appropriate containers and be certified (i.e., must have attended and passed a training program on the proper procedures for packing and shipping this class of materials) to avoid potential liability issues.

All of the steps just described are fundamental to the selection and qualification of the transport system that will best fit each application. Cost effectiveness and adequate performance engineering (not over-engineering, which will add extra cost, or under-engineering, which may put the material at risk) will be assured with the comprehensive attention to these details. Finally, as a rule in the industry, there is a very important consideration from a packaging cost standpoint: for products that have a very high value or are hard to replace, hazardous, or subject to potential damage in transit, the package development process should be as thorough as possible. The cost of a highly refined and qualified package is cheap insurance for the product's safe transit. On the other side of the coin, for products that are inexpensive, easily replaced, and don't represent a hazard, the costs of developing a qualified system may not be justified or need to be as complete as in the former case.

EXAMPLES OF THE THERMAL PERFORMANCE RANGE OF COMMERCIALLY AVAILABLE TRANSPORT SYSTEMS

The test graphs and accompanying diagrams in Figures 7.5–7.9 illustrate the performance range that can be expected with the various types of insulated containers and transporters that are

Figure 7.5. Example of a 48-hour dry ice shipment showing the packing configuration and temperature performance profile of the container.

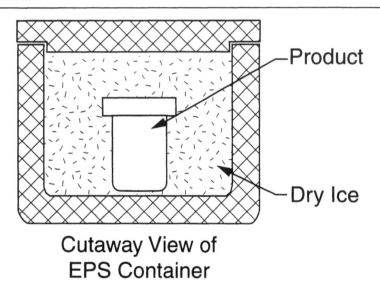

Cutaway View of EPS Container

(a)

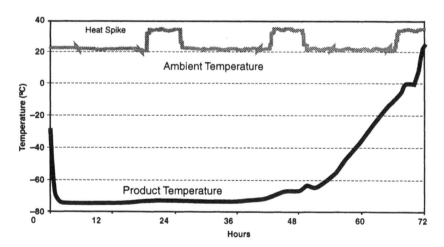

(b)

Dry Ice in an EPS Molded Container
Criteria: <–20° C for 48 Hours
(a) Packing Configuration (b) Performance Profile

Figure 7.6. Example of a 96-hour dry ice shipment showing the packing configuration and temperature performance profile of the container.

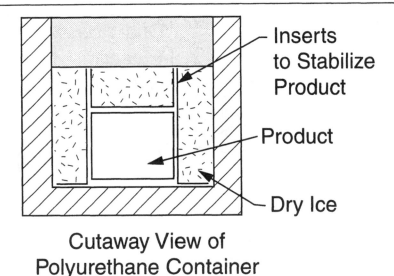

Inserts to Stabilize Product

Product

Dry Ice

Cutaway View of Polyurethane Container

(a)

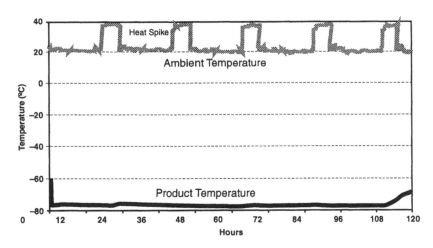

(b)

Dry Ice in a Polyurethane Molded Container
Criteria: <–60° C for 96 Hours
(a) Packing Configuration (b) Performance Profile

Figure 7.7. Example of a 48-hour shipper using a custom double phase change (–60° and –52°C phase points) eutectic solution to maintain a frozen material below –40°C without the use of dry ice.

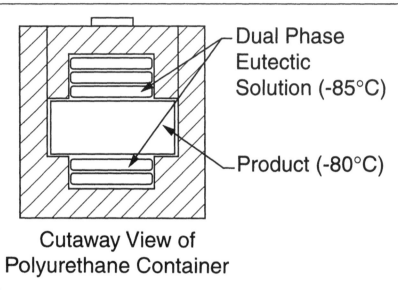

Dual Phase
Eutectic
Solution (-85°C)

Product (-80°C)

Cutaway View of
Polyurethane Container

(a)

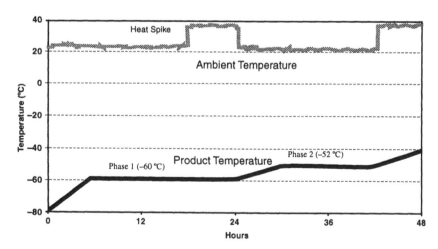

(b)

Eutectic Solution in a Polyurethane Molded Container
Criteria: <–40° C for 48 Hours without Dry Ice
(a) Packing Configuration (b) Performance Profile

Figure 7.8. Example of a 48-hour low phasing gel ice shipment showing the packing configuration and temperature performance profile of the container.

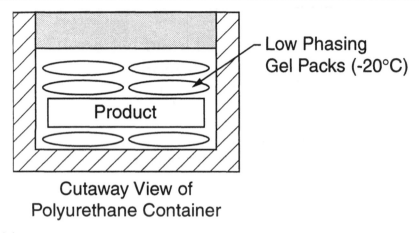

Cutaway View of
Polyurethane Container

(a)

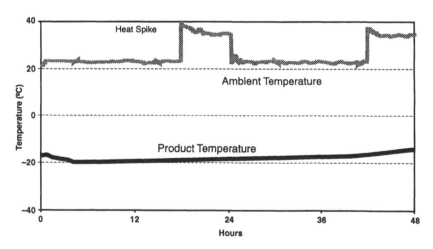

(b)

Low Phasing Gel Ice in a Polyurethane Molded Container
Criteria: <–15° C for 48 Hours without Dry Ice
(a) Packing Configuration (b) Performance Profile

Figure 7.9. Example of a dry nitrogen shipping container system showing the packing method and temperature performance profile of the shipping container.

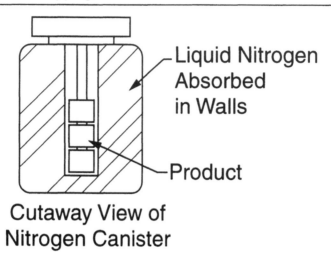

Cutaway View of
Nitrogen Canister

(a)

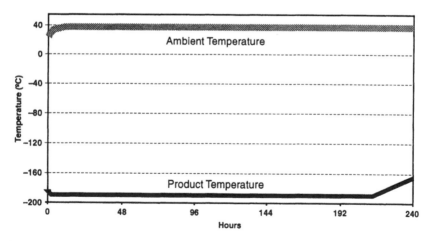

(b)

Liquid Nitrogen in a "Dry" Nitrogen Shipping Container
Criteria: <–140° C for 10 Days
(a) Packing Configuration (b) Performance Profile

used to ship cyropreserved materials. The thermal testing was conducted according to standardized test methods,[14] and in all cases the temperature indicated on the performance graphs for the "product" is the actual reading of the thermocouple inserted in the frozen material, as opposed to being the temperature of the air adjacent to the material. It should be noted that "air" temperatures inside the transport container are usually quite different from the frozen material temperatures, especially when the frozen material has a significant mass, which decreases its rate of temperature increase when exposed to heat.

Note: The following test data are provided as examples of the performance profiles that can be expected with some of the most commonly used methods for transporting frozen material and represent specific sizes of primary containers, product, and refrigerant amounts, as well as specific conditioning and ambient exposure temperatures for the packing configuration being evaluated. It must be emphasized that each material, transport container, and packing method must be evaluated independently to assess the performance capability of the entire package for each application.

Any thermal evaluations performed on transporters and insulated containers must be done by qualified individuals using standardized procedures and equipment that will ensure reproducible results. Only in this way can reliable conclusions be drawn regarding the performance capability of each container and packing configuration under consideration.

TEST METHODS FOR EVALUATING PACKAGE PERFORMANCE

Thermal Performance Testing—Design to Qualification

The initial stage of evaluating a transport container's thermal performance capability is the "design" phase. In the design phase, testing is performed as single tests to give a rough idea of the performance profile. For validation or qualification, testing is performed using multiple replicates of the same package to establish reproducibility of the system. In the design phase, as in the qualification phase, it is recommended that testing be conducted by a qualified laboratory according to established procedures based on the current standard test methods already referenced.

Thermal analysis of transport containers is carried out by exposing the complete insulated container, refrigerant, and frozen material to the range of ambient conditions that are expected to be encountered in transit. The ambient exposure profile can be developed by monitoring current shipments (mounting an electronic temperature monitor on the outside of the shipper) in the exact transit environment or by working with industry associations that have performed studies in this area.[15] The most important concern for frozen material shipments is the monitoring of summer heat conditions since this will represent the most extreme case. As in any transit environment, weather and the variables in the carrier's systems are the unknowns from a temperature exposure standpoint. But if the sampling is extensive enough (e.g., daily shipments to the warmest destinations over the 3 warmest summer months), it is likely to develop a reasonable database that will include the majority of the variables in the profile. Additionally, freight carriers may be able to provide information on a very general scale on the exposure ranges the shipment is likely to encounter in their systems. Ambient monitoring studies have been conducted in car trunks and in 2-day airfreight in the systems of UPS and FedEx[16] that help define the range of temperature exposure to be encountered in transit.

In the actual thermal testing, the transport package being evaluated is exposed to the temperature profile (developed and supported by the shipper) that best fits the conditions expected to be encountered during shipping. Thermocouples are placed in real or simulated material, and the package is loaded with the material and the chosen amount of dry ice or gel ice packs. The package is then placed in an environmental chamber that has been programmed with the desired ambient profile. The material and ambient temperatures are logged at 15- to 30-min intervals over the selected test period and the results evaluated for conformance to the desired goal. In the design phase of testing, the amount of refrigerant or product load can be adjusted as necessary to reach the intended goal. Once the goal is attained in the design phase, the package can then be tested in the qualification process. Qualification usually encompasses a more complicated testing scheme involving a signed off testing protocol (i.e., both client and laboratory agree on the test program) and the use of calibrated thermocouples and datalogger. In addition, the environmental chamber should be thermally mapped to define the areas of the chamber (for location of the package undergoing evaluation) that will provide the most consistent temperatures correlating to the programmed ambient profile.

Some biopharmaceutical companies refer to this package qualification process as "validation." However, validation requires the use of the actual product material, where as the typical thermal test normally uses "simulated" materials instead of the real product.

To establish reproducibility, the existing thermal test standard for evaluating the insulation quality of packages—ASTM Standard D3103—describes a testing program that incorporates 5 replicates to establish reproducibility of the results. More commonly, the accepted criterion is testing in triplicate for reproducibility. Once the thermal testing is complete, a comprehenisve test report is produced that describes the full details of the testing and the exact way the container or transporter was packed for the qualification. This report and description of the packing configuration should be used to establish training, quality assurance evaluations, and standard operating procedures (SOPs) for use by the logistics department charged with the actual shipment of the material, to make sure the packing procedure conforms to the way the container was tested.

Following selection of a transport packaging method, the next step is to do the distribution testing. This will provide the information required about the stability of the inner packing configuration to determine if the frozen material will be susceptible to damage in the transit environment.

Distribution Test Methods

Distribution testing utilizes selected profiles of compression, vibration, and shock based on models that were developed in the transportation industry for this purpose. There are several accepted methods, including **ASTM Standard 4169-96**, the **International Safe Transit Association (ISTA) Procedures 1, 1A, 2, and 2A**, and the **National Motor Freight Classification Item 180** test procedure (also known as Procedure 5 under ISTA classification).[17] The selected testing methodology for each of these procedures is very similar in scope and is based on the preferences of the shipper (e.g., based on the size of the load, material value, fragility, or on the specific area of the world to which the shipment will be made). The selected testing should be performed by a qualified package test laboratory using calibrated equipment to assure that the testing standards are met. To understand why this testing is valuable, consider the following example. Dry ice and the frozen material being transported usually have different resonating frequencies. When the package containing these materials is exposed to a standard vibration profile in the test lab, the two materials will move inside the container in accordance with their respective frequencies. Depending on the packing configuration, a frozen material buried in dry ice pellets can actually move above the pellets, which will not have been the way the package was tested thermally and may not be the most desirable location for the material in general. From these distribution studies, improvements can be made to secure the material by adding inner packaging components to stabilize the configuration. Additionally, once the package meets the distribution testing criteria, the results of the testing can be submitted to the ISTA for certification. The ISTA can then provide legal support to the shipper for cases in which damage occurred in transit but the package testing had demonstrated that it met the standard accepted transit extremes.

USER'S RESPONSIBILITIES

The thermal transport system user has the responsibility of performing quality audits on its suppliers at regular intervals to verify that the container and refrigerant (dry ice, gel ice, and LN_2) manufacturers provide consistent products. Insulated containers utilize foam insulation (expanded polystyrene or polyurethane) that can have various properties depending on its density, chemical composition (e.g., polyurethane component formula), or dimensions (inner dimension and thickness of insulation walls). Refrigerants such as dry ice can have varying properties such as unit surface area (e.g., outside dimensions of pellets or blocks), age, and amount of water or other chemicals (e.g., glycol) present. Low phasing gel ice packs can have variations in total weight per unit bag or bottle (e.g., ± 2 oz), chemical makeup (e.g., amount of water and other additives per unit volume), or seal integrity. These variables may result in shorter or longer phasing (thawing) rates, cause them to phase at a higher than desired temperature, or cause them to leak during transit due to broken seals or punctures.

Once a transport system has been qualified, the shipper should demand certificates of compliance from suppliers of the components to make sure their products are of consistent characteristics and quality. If any changes to their products are made by the suppliers, the users must be notified and equivalency testing must be performed to make sure there is no difference in the performance based on the changes made—no matter how insignificant they may seem. Equivalency testing is normally not as extensive as qualification and involves repeating one or more of the thermal or distribution tests that would be most indicative of any negative change in the performance of the system.

The responsibility of companies supplying the transport system components is fundamentally to provide a quality product with consistent properties. Adoption of good manufacturing practices (GMP)[18] by the supplier is highly desirable in order to maintain this goal. Once a specific refrigerant material (i.e., brand, size, physical characteristics) has been qualified, it should be used routinely rather than changing from one type of refrigerant pack or dry ice type to another assuming it will perform the same. Liquid nitrogen (LN_2) of a specific grade is very consistent and will not be subject to the same issues of quality as dry ice or gel packs.

Another key responsibility of the user is to ensure that the loading procedure for the transport system is being performed consistently. This can be enhanced through the development of training programs and the SOPs already described. Generally, the preshipment qualification procedures that have been performed on a package are translated into an SOP for whatever department is using the transporter. Insulated containers come in various sizes, and since most companies use several sizes, each type of container should be marked with an individual company item code or part number to differentiate it from other similar containers to prevent mix-up in the shipping department.

USE OF TEMPERATURE MONITORS IN FROZEN SHIPMENTS

The monitors most commonly used to track frozen shipments are microprocessor-based electronic monitors with probe extensions that separate the electronics from the low temperatures of the frozen payload (due to the electronics requirement of operating in a range above –40° C). Chemical indicators on card tags are also used to monitor frozen shipments. They contain chemicals that melt and migrate on blotter paper into a visible window on the tag when the chemical is warmer than a specific response temperature (e.g., –18° C). Whenever the temperature falls below the response temperature, the migration of the chemical stops. These types of monitors have limitations, such as not providing a trip record, and can be subjective in their visual interpretation (wide range of temperature sensitivity). Some examples of monitors are shown in Figure 7.10.

The electronic monitors (using temperature probe extensions) most commonly used to track shipments of frozen materials are available from several sources, are reusable, and come with similar features, such as a comprehensive software package, precalibrated within 1° C, and temperatures readable down to the –90° C range. They come in various sizes and shapes—from matchbox to credit card to hockey puck size.

For use, the electronic component is placed on the outside of the insulation, farthest away from the refrigerant source: as the frozen product is placed in the transporter or insulated container, the thermocouple probe is placed adjacent to or as close as possible to the frozen product. The refrigerant is then added around the material and the probe; the probe transmits the temperature reading of its location next to the frozen material (which may not be completely representative of the frozen material's temperature since it is sampling the air temperature next to the material). The readings are transmitted to the electronic monitor at specific intervals (minutes to hours) that are programmed into the monitor prior to use. This information can then be downloaded upon receipt or returned to the shipper for downloading. It is common to download this information into a database that is established to track the shipping history as well as to provide the general acceptance or rejection criteria for each shipment based on temperatures staying within a prescribed range. Monitors are used most frequently when shipping valuable materials or products that have to be maintained within a prescribed range (e.g., < –60° C). For shipments that are just to be kept "frozen" in shipment (e.g., < –5° C), the simplest, most reliable method is not to monitor them but, as they are received, to note whether there is still dry ice present in the container and whether the product appears frozen.

COMMON PROBLEMS IN THE COLD-CHAIN

Maintenance of the desired temperature of a cryopreserved material for a specific amount of time is the primary goal of any transport

Figure 7.10. Examples of commercially available electronic and chemical temperature monitors used for frozen shipments.

(a)

Continued on next page.

system. As described in earlier sections, some of the uncontrollable events that can affect temperature maintenance must be understood and anticipated. Occurrences such as a container with dry ice being bumped (or delayed) in transport due to airline procedures are well known. Any shipment that is labeled hazardous due to dry ice or because of its nature has the potential of being delayed by the forwarder or carrier for a wide variety of reasons. Delays due to customs clearance for international shipments, weather, or unforeseen air transport delays (e.g., strikes, etc.) can affect a product's movement through the cold chain. In some cases, the recipient has not been notified or is not available when the package arrives, so

Figure 7.10. Continued.

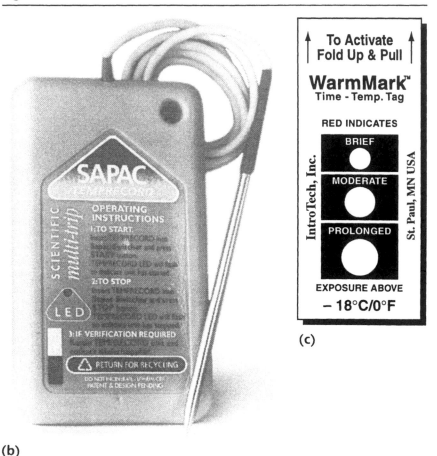

(c)

(b)

the package cannot be delivered. Other examples are poor container labeling or miscommunication from the shipper (or freight forwarder), which can cause a package to sit unopened for a day or more, allowing the material to thaw and become spoiled. It is critical that the container be identified properly in the transit system of the carrier and that the recipient be alerted to expect the shipment. In addition, the recipient must clearly understand the exact storage conditions that are necessary once the material is received.

It is possible to identify some of the problems in the shipping process through electronic monitoring using the tools described previously. To gain the maximum amount of information, some monitors are available with a "dual" monitoring function. They use a thermocouple probe inserted in the frozen material area (as previously described) to record the product temperature during shipment, and, they also have a thermister in the body of the monitor (which is located on the exterior of the container) that records the ambient transit temperatures at each step in the transport cycle. Understanding the internal product temperature in relation to the external ambient conditions can be helpful in the complete analysis of a transport container. A database can be created with the time and temperature information that will be an invaluable asset for understanding and monitoring the cold chain forwarding system from shipment through receipt and will be useful for future design of additional transport packaging for frozen materials.

SOURCES OF PRODUCTS AND TESTING SERVICES FOR SHIPMENT OF FROZEN MATERIALS

Insulated Shipping Containers

EPS

Polyfoam Packers Corp.
2320 S. Foster Ave.
Wheeling, IL 60090

Tuscarora Inc.
800 Fifth Ave.
New Brighton, PA 15066

Polyurethane

Insulated Shipping Containers, Inc.
5240 W. Buckeye Road
Phoenix, AZ 85043

Charles Engineering
P.O. Box 428
Belcamp, MD 21017

Low Phasing Gel Ice Packs (–10° and –20° C)

Polyfoam Packers Corp.
2320 S. Foster Ave.
Wheeling, IL 60090

Pelton•Shepherd Industries
2721 Transworld Dr. Suite A
Stockton, CA 95213

Dry Ice/Liquid Nitrogen

Consult local yellow pages for local dealers.

Dry Nitrogen and Liquid Nitrogen Shippers (Canisters and Tanks)

Taylor-Wharton Div.
Harsco Corp.
4075 Hamilton Blvd.
Theodore, AL 36590

MVE Inc.
8011 34th Ave. S.
Bloomington, MN 55425

Package Testing Services

Distribution/UN

GH Package and Product Testing
325 Commercial Dr.
Fairfield, OH 45014

Distribution/UN/Thermal

Ten-E Packaging Services
1666 County Road 74
Newport, MN 55055

Thermal/Design

Insulated Shipping Containers, Inc.
5240 W. Buckeye Road
Phoenix, AZ 85043

Temperature Monitors

Chemical

IntroTech, Inc.
702 Birchwood Ave.
St. Paul, MN 55110

Electronic (microprocessor based)

Sensitech, Inc.
800 Cummings Center
Suite 258X
Beverly, MA 01915

Monitor Co.
P.O. Box 4958
Modesto, CA 95352

Airline Package Courier Services

World Courier Inc.
1313 Fourth Ave.
New Hyde Park, NY 10040

IATA Approved Dangerous Goods Training Schools

Emory Worldwide Airlines
Environmental Affairs
One Emory Plaza
Vandalia, OH 45377

REFERENCES

1. *ICH Harmonized Tripartite Guideline for Stability Testing of New Drug Substances and Products* (ICH Q1A). September 23, 1994.
   The Center for Drugs and Biologics Guideline for Submitting Documentation for Stability of Human Drugs and Biologics. 1987.

2. Test data on file at ISC, Inc., 5240 W. Buckeye Rd., Phoenix, AZ 85043.

3. Pringle, Thomas C. Clinical chemistry human plasma-based controls, with preassigned values for laboratory instruments, may have values altered by pH change caused by CO_2-rich atmosphere in vials due to shipment in dry ice.

4. Beiser A. 1973. *Physics* 16(4):308: Cummings Publishing Co.

5. *IATA Dangerous Goods Regulations,* 39th Ed. 1998. Geneva, Switzerland: Labelmaster Div. An American Labelmark Co. 800 621–5808 (773 478–0900).

6. *IATA Dangerous Goods Regulations,* 39th Ed. 1998. UN Packing Performance Tests 6:481–492.

7. *IATA Dangerous Goods Regulations,* 39th Ed. 1998. Packing Instruction 602, 5:366–367; Package Performance Test 6.5:489–491.

8. Data on file at ISC, Inc., 5240 W. Buckeye Rd., Phoenix, AZ 85043.

9. Holman, J.P. 1997. *Heat Transfer,* 8th Ed.: McGraw-Hill.

10. Coleman Outdoor Products, Inc., Wichita, KS; Igloo Products Corp., Houston, TX.

11. IATA Dangerous Goods Regulations, 39th Ed. 1998. Packaging Testing Facilities, F:650–654.

12. American Society of Testing and Materials, ASTM Committee on Standards, 1916 Race St., Philadelphia, PA 19103; International Safe Transit Assoc., 1400 Abbott Road, Suite 310, East Lansing, MI 48823; National Motor Freight Classification, Item 180, 1995, American Trucking Assoc., 2200 Mill Road, Alexandria, VA 22314.

13. ASTM Test Standard D 3103–92; Annual Book of ASTM Standards, vol. 15.09.

14. Standard test procedures on file at ISC, Inc., 5240 W. Buckeye Rd., Phoenix, AZ 85043.

15. IoPP Subcommittee on Packaging for Controlled Temperature Shipping, Transport Packaging Committee, Institute of Packaging Professionals, 481 Carlisle Dr., Herndon, VA 20170; ASTM D 10.23 Subcommittee on Natural Environmental Test Methods.

16. *Summer Ambient Study.* 1997. Data on file at ISC, Inc., 5240 W. Buckeye Rd., Phoenix, AZ 85043.

17. ISTA, ASTM, and NMFC test methods: see note 12.

18. Code of Federal Regulations, Title 21; 21 CFR 820, 1998. General Provisions for Good Manufacturing Practices for Medical Devices.

INDEX